Information Theory Applied To Space-Time Physics

Information Theory Applied To Space-Time Physics

Dr. Henning F. Harmuth

The Catholic University of America
School of Engineering and Architecture
Department of Electrical Engineering
Washington, DC

World Scientific
Singapore • New Jersey • London • Hong Kong

Published by

World Scientific Publishing Co. Pte. Ltd.

P O Box 128, Farrer Road, Singapore 9128

USA office: Suite 1B, 1060 Main Street, River Edge, NJ 07661

UK office: 73 Lynton Mead, Totteridge, London N20 8DH

INFORMATION THEORY APPLIED TO SPACE-TIME PHYSICS

A Russian edition of this book was produced by the Publishing House MIR, Moscow, in 1989.

ISBN 981-02-1278-X

Printed in Singapore.

To the late Ladislaus L. Marton (1901–1979)
Editor of Advances in Electronics and Electron Physics
for publishing my books.

Contents

Equations are numbered consecutively within each of Sections 1.1 to 12.11. Reference to an equation in a different section is made by writing the number of the section in front of the number of the equation, e.g., Eq.(1.2-3) for Eq.(3) in Section 1.2.

Illustrations and tables are numbered consecutively within each section, with the number of the section given first, e.g., Fig.1.3-2, Table 6.3-4.

References are characterized by the name of the author(s), the year of publication, and a lower case Latin letter if more than one reference by the same author(s) is listed for the year.

Preface

The concept of a space-time continuum has dominated physics ever since Aristotle argued it so convincingly in his book *Physica* some 2300 years ago. Newton's mechanic, Maxwell's electrodynamic, Einstein's theories of relativity, and quantum mechanics are based on it. The success of these theories is a strong argument for the space-time continuum.

Nevertheless, doubts have been expressed about the use of a continuum in a science squarely based on observation and measurement. No one has ever been able to show by direct observation that there is a space-time continuum, and it is exceedingly difficult to think how such an observation could possibly be made. However, this obvious argument against a space-time continuum is generally not accepted as sufficiently convincing in view of the success of continuum physics.

An exact science requires that qualitative arguments must be reduced to quantitative statements. The observability of a continuum can be reduced from qualitative arguments to quantitative statements by means of information theory. Information is as measurable and as invariant a quantity as, e.g., the number of apples in a basket. The observation of a continuum would provide us with nondenumerably infinite information. Both the observation of the infinitesimal and of the infinite require infinite information. These concepts are thus outside the realm of the observable, and they can be used as an approximation only in a science based on observation.

Information theory was developed during the last decades within electrical communications, but it is almost unknown in physics. The closest approach to information theory in physics is the *calculus of propositions*, which has been used in books on the frontier of quantum mechanics and the general theory of relativity*. A discussion of the principles of information theory required in this book is given in Chapter 2. However, one should keep in mind that the ability to think readily in terms of a finite number of discrete samples is developed over many years of using information theory and digital computers, just as the ability to think readily in terms of a continuum is developed by long use of differential calculus.

*Jauch (1968); Misner, Thorne, and Wheeler (1973).

The author wants to thank the many fellow scientists who have helped him over the years. The following ten are named personally: Chang Tong, Tsinghua University, Beijing, China; M. Heller, Pontifical Faculty of Cracow, Poland; R. Meister, The Catholic University of America, Washington DC; D. Olson, St.Olaf College, Northfield, Minnesota, USA; A. Z. Pataschinskiî, Siberian Section of the Russian Academy of Sciences, Akademgorodok, Russia; J. R. Seberry, University of Sydney, Australia; L. M. Soroko, Joint Institute of Nuclear Research, Dubna, Russia; A. P. Street, University of Queensland, Brisbane, Australia; H. Überall and C. W. Werntz, both of The Catholic University of America, Washington, DC.

1 Historical Review

1.1 Roots in the Greek World

The three-dimensional space around us is considered to be one of the most fundamental experiences. To turn this personal experience into a scientific concept, we have to perform a number of tasks. First, we have to describe in more detail the three-dimensional space experienced in everyday life. Second, we have to extend the concept to distances much larger than those of our immediate experience, that is, to astronomical distances. Third, we have to extend the concept to very small distances, of the order of nuclear and smaller distances. We have been working on these tasks since the beginning of science, and there is not yet any end in sight.

Our concept of space has several roots in the Greek world. The best known one is the geometry of Euclid (c. 330–c. 275 B.C.). This geometry and its critical analysis led in the nineteenth century to the concept of metric, which in turn led to the particular concept of space-time on which the general theory of relativity is based. The transition from Euclidean to non-Euclidean geometry will be discussed in more detail in the following section.

A second root was the concept of a flat or a spherical Earth[1] as center of the world; both concepts provided a distinguished system of reference for the motion of all objects. This distinguished system disappeared when Nicholas Copernicus (1473–1543) introduced his heliocentric system, particularly since the center of his system was not the Sun but a point in the vicinity of the Sun (Ptolemy, 1952, "On the revolutions ..."). Isaac Newton (1642–1727) reintroduced the distinguished system of reference by means of the axiom of the absolute space (Newton, 1971, vol. 1, p. 6, II). The absolute space and thus the distinguished system of reference was eventually

[1] Pythagoras of Samos (c. 582–c. 507 B.C.) is usually accepted to have been first to claim a spherical Earth, but this hypothesis is also attributed to Parmenides who was born about 514 B.C. An excellent review of the development of the ideas about the shape of the Earth in the Greek and Roman world is given by Brown (1949). Koestler (1968) traces the development from the Greek times to the end of the Middle Ages. Both books contain extensive references.

discredited by Ernst Mach (1838–1916), but we cannot say that the concept has finally and definitely disappeared from our thinking (Mach, 1907).

The third root from the Greek world is the concept of the *continuum.* The beginning of this concept dates back at least to the time of Zeno of Elea (c. 490–c. 430 B.C.), who advanced paradoxes that were supposed to show that infinite divisibility of space and time were not possible (e.g., the race between Achilles and the turtle, or the arrow that cannot fly). Aristotle (384–322 B.C.) makes it clear that the concept of the continuum was discussed by others before he wrote his *Physica*:

> Now a motion is thought to be one of those things which are continuous, and it is in the continuous that the infinite first appears; and for this reason, it often happens that those who define the continuous use the formula of the infinite, that is, they say that the continuous is that which is infinitely divisible [Apostle, 1969, Book III (Γ), 1, § 2].

Aristotle argued the concept of the continuum for space, time, and motion so successfully, that it has been challenged very rarely since[2]. We will see later on that Newton apparently took this concept so much for granted that he did not even mention it, even though he was very meticulous in listing and elaborating his assumptions.

Today it is usual to define the continuum with the help of the real numbers on the numbers axis. One says that a one-dimensional continuum has the same topology as the real numbers. It is further usual to map any real number x by means of the exponential function e^{iyx} onto the unit circle in the complex plane. The resulting map $\{e^{iyx}\}$ is called the *character group of the topologic group of real numbers.* Why do we map the real numbers on the circle and not on an ellipse or a polygon? The definition of a character group does not require the circle, it only permits it[3]. A shift along the numbers axis becomes a rotation along the unit circle. Only the circle remains unchanged under differential as well as finite rotations. If we used only the integers rather than the real numbers along the numbers axis, only finite shifts would have a meaning, and the circle could be replaced by a polygon that is reproduced by finite rotations. Hence, the concept of the continuum and the real numbers call for the mapping onto the circle.

The argument is frequently presented in the reversed order. Since the circle is reproduced by differential rotations, and since it is THE—rather than a—character group of the topologic group of real numbers, one must use the real numbers along the numbers axis, which in turn implies the topology of the continuum for the time axis and "for time", as well as for

[2] Book V(E) 3, § 5 defines continuous; Book VI(Z) 1, 2 elaborates the concept of the continuum further; Zeno is refuted in Book VI(Z) 2 and 9 (Apostle 1969, Aristotle 1930).

[3] van der Waerden (1966), Part 1, § 54.

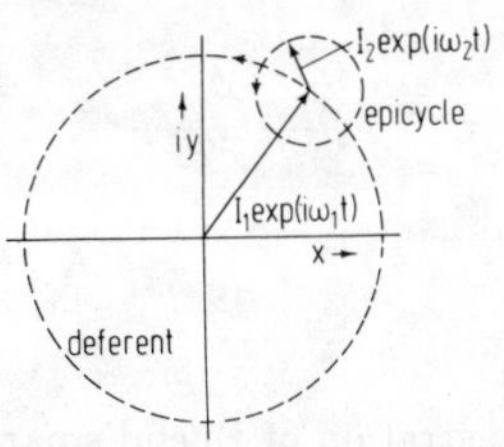

FIG.1.1-1. Superposition of circles in astronomy and in the Fourier series using complex notation.

the axes of a Cartesian coordinate system and "for space". It is difficult to see why the topology of time and space should be defined by the circle, but the reason is probably the *dogma of the circle*[4] which has influenced western thinking at least since Plato (c. 428–c. 348 B.C.). It is usually connected with astronomy, but its influence goes much further. Claudius Ptolemy (c. 90–c. 168 A.D.) stated it as follows for astronomy:

> ... we believe it is the necessary purpose and aim of the mathematician to show forth all the appearances of the heavens as products of regular and circular motions (Ptolemy, 1952, Almagest, Book III, 1; p. 83, second paragraph).

In order to apply this dogma to the orbits of the Sun, the five known planets, and the Moon around the Earth, Ptolemy assumed that they moved on primary circles called *deferents*, as shown in Fig.1.1-1. A second circle, called *epicycle*, was superimposed on each deferent. A second epicycle was then superimposed on the first epicycle, and so on. In today's language, we would say the orbits were represented by a superposition of circles.

The deferents and epicycles survived the transition to the heliocentric system of Copernicus. It was not until Johannes Kepler (1571–1630) published his book *Astronomia Nova* in 1609 that the superpositions of circles were replaced by ellipses. The dogma of the circle in astronomy came to an end, but this does not mean it disappeared from science.

In much of physics and electronics the circle lives on under the new name exponential function $e^{i\omega t}$. By writing $I_1\exp(i\omega_1 t)$ for the deferent and $I_2\exp(i\omega_2 t)$ for the epicycle in Fig.1.1-1 we obtain a typical illustration of the superposition of two sinusoidal oscillations. Indeed, Fig.1.1-1 is a standard illustration for single sideband modulation of a sinusoidal carrier by a sinusoidal voltage in electrical communications. Speaking more generally, the superposition of circles became the Fourier series expansion in complex notation. The representation of a function by its Fourier expansion is as pervasive today as the description of the orbits of celestial objects by

[4]See Koestler (1968), Part 1, IV, 2.

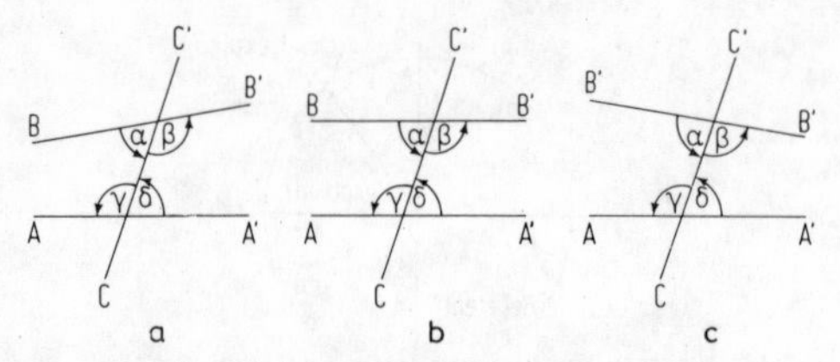

FIG.1.2-1. Illustration of Euclid's parallel postulate.

a superposition of circles before Kepler.

The exponential function $e^{i\omega t}$ and the character group $\{e^{iyx}\}$ of the topologic group of real numbers differ in notation and name only. Hence, a space-time with the topology of the continuum will distinguish the exponential function or the sine–cosine functions. Vice-versa, if the exponential function or the sine–cosine functions are sufficiently distinguished, we will have a strong argument in favor of the continuum. There is no need to elaborate that the exponential function, Fourier series, and Fourier transform are ubiquitous both in physics and in electrical communications. The question is, whether this is due to more than the long shadow of Plato's dogma of the circle. One should keep in mind that Copernicus could not free himself from this dogma, and Kepler only succeeded after much labor and failed attempts to preserve the dogma of the circle. At this time one may strongly suspect that the axiom of the continuum is proved by the dogma of the circle, while the dogma of the circle is upheld by the axiom of the continuum.

1.2 From Euclidean to Non-Euclidean Geometry

The *Elements* of Euclid dominated geometry for more than 2000 years (Euclid, 1956). From the very beginning its Postulate V attracted attention as being not sufficiently evident to be accepted without proof[1]. Its English version is usually stated as follows: *If a straight line falling on two straight lines makes the interior angles on the same side less than two right angles, the two straight lines, if produced indefinitely, meet on that side on which are the angles less than two right angles.*

Figure 1.2-1 illustrates this postulate. For $\alpha+\gamma<\pi$ the lines BB' and AA' meet on the left of the line CC' (Fig.1.2-1a), while for $\beta+\delta<\pi$ they meet on the right (Fig.1.2-1c). For $\alpha+\gamma=\beta+\delta=\pi$ the two lines meet neither on the left nor on the right, and they are thus parallel (Fig.1.2-1b).

[1]The number V appears in the critical edition of the *Elements* by Heiberg and Menge (Euclid, 1916) and in its translation by Heath (Euclid, 1956), but Bolyai (1832) calls it Axiom XI, while others give it the numbers 12 and 13, or call it the parallel axiom.

Figure 1.2-1 is quite evident as long as one agrees what a straight line is, what a plane is, and that the three lines are to be drawn in a plane. The modern reader has no difficulty visualizing the line CC' as the equator of the Earth, and AA' as well as BB' as meridians. The lines AA' and BB' can then be "produced indefinitely", the sums $\alpha + \gamma = \beta + \delta = \pi$ indicate that they are parallel, but they meet nevertheless on both sides of the equator at the North and the South Pole.

Postulate V was considered not to be sufficiently evident even by the first commentators of Euclid's work. Posidonius of Apamea (c. 135–c. 51 B.C), Ptolemy, Proclus (419–485) and many others tried to either derive it from other postulates or to replace it by a more evident one, but success proved to be elusive[2]. Lagrange (1736–1813) found it still worthwhile to affirm the independence of spherical trigonometry from Euclid's postulate (Bonola, 1955, p. 52), even though mariners had to use spherical trigonometry for efficient navigation at least since Magellan's circumnavigation of the Earth in 1519–1522. It is very difficult to understand today how spherical trigonometry could be widely used while at the same time some of the greatest mathematicians devoted their time to proving Euclid's parallel postulate.

Karl Friedrich Gauss (1777–1855) started out like all the others trying to find a proof for the parallel postulate (Bonola, 1955, p. 64–75). Over a period of some thirty years he became eventually convinced that a logically consistent geometry without the parallel postulate could be developed. He called this new geometry first *anti-Euclidean geometry*, then *astral geometry*, and finally gave it the modern name *non-Euclidean geometry*. In 1831 he gave in a letter[3] the circumference of a circle with radius r as:

$$L = \pi k(e^{r/k} - e^{-r/k}) = 2\pi k \sinh(r/k) \tag{1}$$

Using the series expansion for $\sinh(r/k)$ one obtains from this expression the circumference of the circle in a plane for $k \to \infty$:

$$\lim_{k\to\infty} 2\pi k \sinh(r/k) = 2\pi r \tag{2}$$

Consider a sphere with radius R. A circle with radius r on this sphere has the circumference:

$$L = 2\pi R \sin(r/R) = i\pi R \left(e^{-ir/R} - e^{+ir/R}\right) \tag{3}$$

[2] The definitive account of the development of non-Euclidean geometry is a book by Bonola (1955). See also Stäckel (1895).

[3] Gauss did not publish his investigations on non-Euclidean geometry. They are contained in correspondence with Schumacher, Taurinus, and others (Gauss, 1919, 1969).

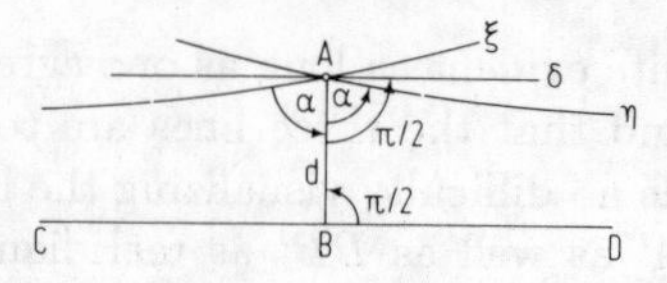

FIG.1.2-2 Illustration of Lobachevskii's parallel postulate.

The replacement of the real radius R by an imaginary radius $k = iR$ transforms Eq.(3) into Eq.(1). Franz Taurinus (1794–1874) used this substitution to make the transition from Eq.(3) to Eq.(1), and generally from spherical geometry to what he called *logarithmic-spherical geometry*, while the modern name is pseudo-spherical geometry (Bonola, 1955, p. 77). In this geometry, the sum of the angles of a triangle is always less than π, while it is larger than π for spherical geometry. The sum of the angles approaches π in both geometries when the length of the sides of the triangles approach zero. Euclidean geometry was thus shown to be an intermediate case when the radius R of a sphere increased to infinity, passed there from the real value R to the imaginary value iR, and decreased again to finite values of iR. Taurinus used the substitution $k = iR$ to derive a number of other formulas from spherical geometry, such as the volume and the surface of a sphere (Stäckel, 1899; Bonola, 1955, p. 81).

Nicolai Ivanovitsch Lobachevskii (1793–1856) published in 1830 a paper "On the Principles of Geometry", in which he developed a geometry that permitted two parallels to a given straight line through a point (Engel, 1899). This was followed by more papers and two summaries, "Geometrische Untersuchungen zur Theorie der Parallellinien" and "Pangéométrie" (Lobachevskii, 1840, 1856). Refer to Fig.1.2-2 for an explanation of his approach. The line CD and the point A are given. The vertical AB to the line CD through the point A is constructed; it has the length d. The line δ, vertical to AB is then constructed, which would be the only parallel in Euclidean geometry. However, Lobachevskii now constructs two more line, ξ and η, which have the angle $\alpha \neq \pi/2$ with the vertical AB, and calls them parallels to CD. The angle α is called angle of parallelism, and it is a function of the distance d. Any line between the lines ξ and η will never intersect the line CD. Lobachevskii succeeded in deriving trigonometric formulas from this definition of parallel lines. His geometry turns out to be identical with the logarithmic-spherical geometry of Taurinus, if one makes the substitution

$$k = 1/\ln d \tag{4}$$

Taurinus had known that his results were logically consistent, but he considered his geometry to be an exercise in abstract mathematics. Lobachevskii, on the other hand, was convinced that Euclidean geometry had no a *priori* claim to be the only useful one in the physical world. He writes:

> The fruitlessness of the attempts made since Euclid's time, for the space of 2000 years, aroused in me the suspicion that the truth, which it was desired to prove, was not contained in the data themselves; that to establish it the aid of experiment would be needed, for example, of astronomical observations, as in the other laws of nature. When I had finally convinced myself of the justice of my conjecture and believed that I had completely solved this difficult question, I wrote, in 1826, a memoir on this subject (Engel, 1899, p. 67; English version, Bonola, 1955, p. 92).

In today's language we would say that the metric of space-time must be obtained from observation, not from mathematical definitions, since mathematics can provide us only with models, and we must determine by observation which models correspond best to reality. We will see later that this obvious statement becomes controversial in our time, if we replace the concept of metric by that of topology. Before Lobachevskii, the Euclidean geometry of physical space was questioned about as much or as little as the topology of the continuum of the physical space-time is questioned today.

We turn to Lobachevskii's contemporary Johann Bolyai (1802–1860). He shares with Lobachevskii the distinction of not only realizing that the parallel postulate was not necessary and deriving the fundamental formulas of pseudo-spherical geometry, but of comprehending the importance of his results for the physical space. "I have created a new universe from nothing," he writes in 1823 (Stäckel and Engel, 1897; Stäckel, 1901; Bonola, 1955, p. 98), and the title of his publication is no less clear, "The Science of Absolute Space" (Bolyai, 1832)[4]. After 2000 years of attempting to prove or disprove the parallel postulate, the solution was found independently by Bolyai and Lobachevskii at essentially the same time.

1.3 Metric and Differential Geometry

The work of Bolyai and Lobachevskii was difficult to understand, since a lucid representation was lacking[1]. This difficulty was overcome by the development of geometries on curved surfaces by Gauss and Bernhard Riemann (1826–1866)[2]. To understand Riemann's geometries we must first

---

[4]The complete title is "Scientiam spatii absolute exhibens: a veritate aut falsitate Axiomatis XI Euclidei (a priory haud unquam decidenda) independentem: adjecta ad casum falsitatis, quadrature circuli geometrica". It was published as an appendix to a mathematical book by his father Wolfgang Bolyai.

[1]It is puzzling that Taurinus, Bolyai, and Lobachevskii elaborated the pseudospherical geometry rather than the spherical geometry, but they did not know that their geometry could be explained so easily on the surface of a pseudosphere.

[2]Gauss (1827), Riemann (1854, 1861).

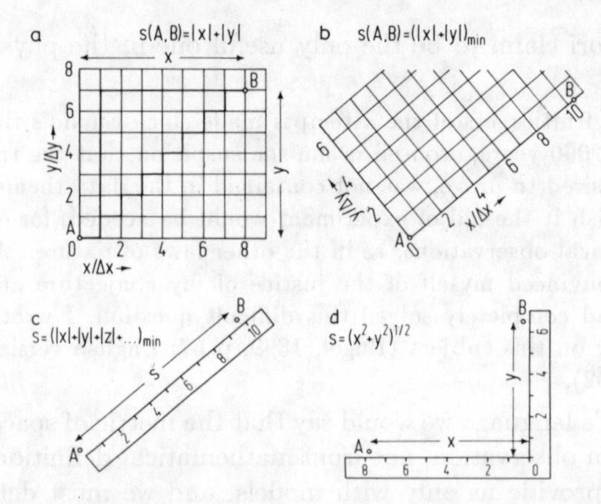

FIG.1.3-1. Various definitions of a distance $s(A,B)$. (a) The distance between two street crossings A and B of a city with checkerboard pattern is $|x|+|y|$. (b) The definition $s(A,B) = (|x|+|y|)_{\min}$ calls for a rotation of the coordinate system so that B lies either on the x-axis or the y-axis. (c) The rotation of the coordinate system in (b) corresponds to the way we measure distances with a ruler or by means of a radar. (d) The Pythagorean definition of distance is much used for theoretical work but not often for practical measurments.

elaborate the concept of metric. A metric defines the distance $s(A,B)$ between two points A and B. The following conditions must be satisfied by this distance $s(A,B)$:

$$
\begin{aligned}
&1.\ s(A,B) = 0 \ \text{ for } A = B \\
&2.\ s(A,B) \neq 0 \ \text{ for } A \neq B \\
&3.\ s(A,B) = s(B,A) \\
&4.\ s(A,B) + s(B,C) \geq s(A,C)
\end{aligned}
\qquad (1)
$$

Refer to Fig.1.3-1 for a simple example of a metric. The two points A and B are shown in a Cartesian coordinate system; for simplicity, point A is located at the origin of the coordinate system. Point B has the x-coordinate x and the y-coordinate y. We may define a distance $s(A,B) = |x|+|y|$. Although this distance does not have the usual significance of our concept of distance, it does satisfy the four conditions of Eq.(1).

We can give the distance $s(A,B) = |x|+|y|$ a practical meaning by interpreting Fig.1.3-1a as a street map of a city with streets according to the pattern of a checkerboard. The points A and B are then two street crossings and one must travel at least the distance $|x|+|y|$ to get from A

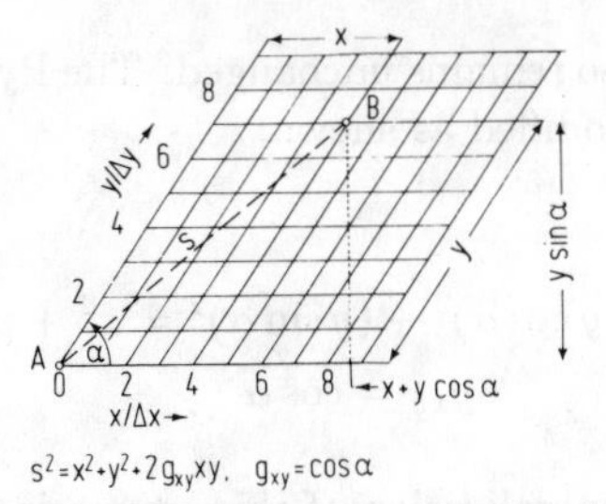

FIG.1.3-2. The Pythagorean distance $s(A, B)$ of two points A and B in a slanted Cartesian coordinate system.

to B or from B to A. One may readily see that there are many shortest routes—all having the same distance $|x| + |y|$—between A and B. These shortest routes are called *geodesics*; two of them are shown by heavier lines in Fig.1.3-1a.

Another example of a metric is provided by the definition $s(A, B) = (|x|+|y|)_{\min}$. This definition requires that the coordinate system in Fig.1.3-1a is rotated so that the sum $|x| + |y|$ becomes a minimum. One may readily see from Fig.1.3-1b that B then lies either on the x- or the y-axis. Indeed, only one axis is needed. The same applies to two points A, B in a three-dimensional space, where the distance is defined as $s(A, B) = (|x|+|y|+|z|)_{\min}$. The concept of *dimension* is not required for the definition of a distance in this form.

Figure 1.3-1c shows how the distance between A and B is measured with a ruler[3]. The ruler is rotated around either point A or point B until the other point lines up with the ruler. Hence, the practical measurement of Fig.1.3-1c corresponds to the rotation of the coordinate system according to Fig.1.3-1b and the definition $s(A, B) = (|x| + |y| + |z|)_{\min}$ for the distance.

The same value for the distance $s(A, B)$ is also obtained by the Pythagorean definition of distance, $s = (x^2 + y^2)^{1/2}$, shown in Fig.1.3-1d. For a practical measurement, this definition calls for two rulers intersecting at a right angle. Such a ruler would be unnecessarily complicated compared with the one of Fig.1.3-1c, but the Pythagorean definition is very practical if a coordinate system is given and the distance s has to be calculated from the known coordinates of A and B.

If the x-axis in Fig.1.3-1a is not perpendicular to the y-axis, one obtains the slanted coordinate system of Fig.1.3-2. The definition $s(A, B) = |x|+|y|$ is applicable to this case, but cities with streets according to a slanted checkerboard pattern are rare. The definition $s(A, B) = (|x| + |y|)_{\min}$ of Fig.1.3-1b remains applicable without any change, and its interpretation

[3]Distance measurement by means of a radar is also done in this way.

according to Fig.1.3-1c also remains unchanged. The Pythagorean definition of Fig.1.3-1d has to be modified as shown:

$$s^2(A,B) = (x + y\cos\alpha)^2 + (y\sin\alpha)^2 = x^2 + y^2 + 2g_{xy}xy$$
$$g_{xy} = \cos\alpha \tag{2}$$

Riemann used the generalization of this expression from two variables x, y to n variables x^0, x^1, ... , x^{n-1}, and also postulated that the points A and B should only have the infinitesimal distance ds rather than the finite distance $s = s(A,B)$. The square ds^2 of the infinitesimal distance becomes:

$$ds^2 = \sum_{i=0}^{n-1}\sum_{k=0}^{n-1} g_{ik}dx^i dx^k \tag{3}$$

Riemann geometries found their great application in Einstein's general theory of relativity. The variable x^0 is there interpreted as the time t, while x^1, x^2, and x^3 stand for the space variables x, y, z. The summation signs are generally left out for simplification; the double occurrence of an index i or k in Eq.(3) serves as a sufficient instruction to sum over $i = 0, 1, 2, 3$ and $k = 0, 1, 2, 3$.

In matrix notation we may write Eq.(3) as follows:

$$ds^2 = (dx^0\ dx^1\ dx^2\ dx^3)\begin{pmatrix} g_{00} & g_{01} & g_{02} & g_{03} \\ g_{10} & g_{11} & g_{12} & g_{13} \\ g_{20} & g_{21} & g_{22} & g_{23} \\ g_{30} & g_{31} & g_{32} & g_{33} \end{pmatrix}\begin{pmatrix} dx^0 \\ dx^1 \\ dx^2 \\ dx^3 \end{pmatrix}$$
$$g_{ik} = g_{ki} \tag{4}$$

One may either multiply first the line matrix $(dx^0\ dx^1\ dx^2\ dx^3)$ with the square matrix $\mathbf{g}$ and the resulting line matrix then with the row matrix $(dx^0 \ldots dx^3)$, or do the multiplications in reversed order. Only the ordering of the terms in the sum is affected by the sequence of multiplications.

The coefficients g_{ik} of the matrix $\mathbf{g}$ are called the components of the metric tensor $\mathbf{g}$, while the coefficients dx^i of the line and row matrices are called components of the four-vector $d\mathbf{x}$. Greek letters are often used instead of the indices i and k if they assume the values 0, 1, 2, 3 only (Misner *et al.*, 1973).

Riemann's differential geometries describe the concept of space that is generally accepted today by physicists. The next step in the development of our ever changing concept of space and space-time is best introduced by way of information theory, which taught us that the information obtained through measurements is always finite. This finiteness of information is not reconcilable with a differential geometry, which is based on the infinitesimal distance ds between two points. Riemann was aware that his geometries might fail at short distances, and he wrote[4]:

> Hence, it is quite possible to think that the metric of space at infinitesimal distances does not satisfy the assumptions of geometry, and one would indeed have to assume this if the observations could then be explained in a simpler manner (Riemann, 1854, Section 3, end of second paragraph).

Before entering upon information theory and the *geometries of finite differences*, we will briefly review the philosophical development of the concepts of space and time[5].

1.4 Physical Space-Time

The various geometries discussed so far are all mathematical constructions. Their development was guided by the desire to obtain usable mathematical models of an observable physical space, but one needs some concept of this physical space before one can judge whether a mathematical model is usable.

We have already pointed out in Section 1.1 that Aristotle argued very convincingly the concept of continuum for space, time, and motion. A modern commentator sums up his contributions as follows:

> What Aristotle says about the continuum still belongs to the foundations of today's physics, even where atomistic hypotheses are used. However, before Planck, their implications were never investigated in a way that would have permitted a challenge of the concept of the continuum that was fundamental for the basic assumptions of Galilee and Newton (Wieland[1], 1961, p. 278).

Let us turn to Newton and see what he has to say about space, time, and motion:

> I. Absolute, true and mathematical time, of itself, and from its nature, flows equably without relation to anything external, and by another name is called duration; relative, apparent, and common time,

[4]This is a free translation by the author of the following statement: "... es ist also sehr wohl denkbar, daß die Maßverhältnisse des Raumes im Unendlichkleinem den Voraussetzungen der Geometrie nicht gemäß sind, und das würde man in der That annehmen müssen, sobald sich dadurch die Erscheinungen auf einfachere Weise erklären ließen".

[5]See also the book by Raine and Heller (1981).

[1]Translation by the author.

is some sensible and external (whether accurate or unequable) measure of duration by means of motion, which is commonly used instead of true time; such as an hour, a day, a month, a year.

II. Absolute space, in its own nature, without anything external, remains always similar and immovable. Relative space is some movable dimension or measure of the absolute space, which our senses determine by its position to bodies, and which is commonly taken for immovable space; such is the dimension of a subterraneous, an aerial, or celestial space, determined by its position in respect to the Earth.

III. Place is a part of space which a body takes up, and is according to the space, either absolute or relative.

IV. Absolute motion is the translation of a body from an absolute place into another; and relative motion, the translation from one relative place into another (Newton, 1971, p.6).

Although Newton was very careful in the definition of his terms, and explained in some detail what he meant with time, space, place, and motion, he could not recognize that he assumed the Euclidean metric $s^2 = x^2 + y^2 + z^2$ for space, since Euclidean geometry was the only one known. Furthermore, he assumed without saying so that space and time had the topology of the continuum. This is more surprising. Newton invented, together with Gottfried Wilhelm Leibnitz (1646–1716), differential calculus. The application of this mathematical method to the description of events in physical space-time requires that one assumes the topology of the continuum for space and time. Apparently, the authority of Aristotle was questioned on the space-time continuum no more than the authority of Euclid on geometry. David Hume (1711-1776) writes at about this time:

'This certain then, that time, as it exists, must be composed of indivisible moments. For if in time we could never arrive at an end of a division, and if each moment as it succeeds another, were not perfectly single and indivisible, there would be an infinite number of coexistent moments, or parts of time; which I believe will be allow'd to be an arrant contradiction (Hume, 1888, book I, part II, § II).

Let us turn to Immanuel Kant (1724–1804). In his discussion of space we find the following statement:

2. Space is a necessary representation a *priori*, forming the very foundation of all external intuition. It is impossible to imagine that there should be no space, though one might very well imagine that there should be space without objects to fill it.

3. On this necessity of an a *priori* representation of space rests the apodictic certainty of all geometric principles, and the possibility of their construction a *priori*. For if the intuition of space were a concept gained a *posteriori*, borrowed from general external experience, the first principles of mathematical definition would be nothing but perceptions.

> They would be exposed to all the accidents of perception, and there being but one straight line between two points would not be a necessity, but only something taught in each case be experience (Kant[2], 1922, p. 19).
>
> Because the statements of geometry are all apodictic, which means we are aware of their necessity, e.g., space has only three dimensions, such statements cannot be empiric judgements or judgements based on experience, neither can they be deduced from them (Kant[3], 1956, p. 6, line 27).

These statements by Kant, made just before the development of the non-Euclidean geometry by Bolyai and Lobachevskii, illustrate the domination of science by Euclidean geometry and the enormous change of our thinking brought about by its generalization.

Kant's statement on time are very similar to those about space, but they sound much more acceptable to us. The reason seems to be that nothing as drastic as the development of non-Euclidean geometry happened to time[4]:

> I. Time is not an empirical concept deduced from any experience, for neither coexistence nor succession would enter into our perception, if the representation of time were not given a *priori*. Only when this representation a *priori* is given, can we imagine that certain things happen at the same time or at different times.
>
> II. Time is a necessary representation on which all intuitions depend. We cannot take away time from phenomena in general, though we can well take away phenomena out of time. Time is therefore given a *priori*. In time alone is reality of phenomena possible. All phenomena may vanish, but time cannot be done away with.
>
> III. . . Time has one dimension only; different times are not simultaneous, but successive, while different spaces are never successive. but simultaneous. Such principles cannot be derived from experience, because experience could not impact to them absolute universality nor apodictic certainty. We could only say that common experience teaches us that it is so, but not that it must be so (Kant, 1922, pp. 24, 25).

Kant lived about 80 years later than Newton[5]. The concept of space

---

[2]This third paragraph is contained in the first edition of *Kritik der reinen Vernunft* of 1781, but not in the second edition of 1787. We can only speculate whether Kant developed doubts about Euclidean geometry between those years; the time was ripe.

[3]This sentence appears in the edition of 1787 but not in the one of 1781. For this reason it is not included in the English edition (Kant, 1922). The translation is by the author.

[4]The greater stability of the concept of time is reflected by the fact that the chapter "Von der Zeit" differs in a few words only in the editions of 1781 and 1787 of *Kritik der reinen Vernunft*, whereas the chapter "Von dem Raum" contains considerable changes.

[5]Kant was born 82 years later and died 77 years later than Newton.

and time of both were very similar. About the same span of time separated the lifetimes of Riemann and Kant[6], but the changes are now drastic[7]:

> I have given myself the task, to construct the concept of a multidimensional quantity from general concepts of quantities. It will follow from this, that a multidimensional quantity may have various measures and the space is thus only a special case of a three-dimensional quantity. But it follows necessarily from this, that the statements of geometry cannot be derived from general concepts of quantities, but that those features which distinguish space from all other thinkable three-dimensional quantities, can come from experience only (Riemann, 1854, pp. 272, 273).
>
> That space be an unbounded three-dimensional manifold is an assumption which is used in any perception of the exterior world, according to which at any moment the range of observations is augmented and the possible locations of a searched-for object is constructed, and which is constantly confirmed by these applications. Hence, the unboundedness of space possesses more empirical certainty than any other exterior experience. But infinity does not follow from this by any means (Riemann, 1854, p. 284).
>
> The question about the applicability of the assumptions of geometry at infinitesimal distances is connected with the question about the inner reasons of the measure relationships of space. With this question, which one may perhaps still include in the study of space, the previous remark is used, that a discrete manifold contains the principle of the measure relationships already in the concept of this manifold, but it must come from somewhere else for a continuous manifold. Hence, the reality on which space is based must either be a discrete manifold, or the reasons for the measure relationships must be searched outside in attractive forces acting on them (Riemann, 1854, p. 285, last paragraph).

For our purposes there are three important differences in the thinking of Riemann compared with Newton and Kant. (a) Space and its features are no longer an "a *priori* matter" but a matter of observation and experience. (b) Space is no longer infinite, but merely unbounded; it has "measure relationships" that we now call metric, and describe in the form shown in the preceding section. (c) Whether space is continuous or discrete has become a reasonable question; this is the beginning of what we call now the question of the topology of space-time.

Metaphysical arguments and the term *a priori* disappeared from the discussion of physical space-time after Riemann. Ernst Mach writes in 1883:

---

[6] Riemann was born 102 years later and died 62 years later than Kant.

[7] This is a free translation by the author which uses technical terms not usual 125 years ago, when Riemann wrote the original.

> As the outcome of the labors of Lobachevskii, Bolyai, Gauss, and Riemann, the view has gradually obtained currency in the mathematical world, that that which we call *space* is a *particular, actual* case of a more *general*, conceivable case of multiple quantitative manifoldness. The space of sight and touch is a threefold manifoldness; it possesses three dimensions; and every point in it can be defined by three distinct and independent data. But it is possible to conceive of a quadruple or even multiple spacelike manifoldness. And the character of the manifoldness may also be differently *conceived* from the manifoldness of actual space. We regard this discovery, which is chiefly due to the labors of Riemann, as a very important one. The properties of actual space are here directly exhibited as objects of *experience*, and the pseudotheories of geometry that seek to excogitate these properties by metaphysical arguments are overthrown (Mach, 1907, p. 493).

While the concept of space underwent such a spectacular development since Kant, the concept of time does not seem to have changed until Albert Einstein (1879–1955) combined space and time to space-time in 1905. Two quotations may characterize this stage of our concept of space and time:

> The psychological subjective feeling of time enables us to order our impressions, to state that one event precedes another. But to connect every instant of time with a number, by the use of a clock, to regard time as an one-dimensional continuum, is already an invention. So also are the concepts of Euclidean and non-Euclidean geometry, and our space understood as a three-dimensional continuum (Einstein and Infeld, 1938, p. 311).

This view that space and time are "inventions" is even more clearly stated by Schlick:

> The assertion that *all* motions and accelerations are relative is equivalent to the assertion that space and time have no physical objectivity. One statement comprehends the other. Space and time are not measurable in themselves: they only form a framework into which we arrange physical events. As a matter of principle, we can choose this framework at pleasure; but actually we do so in such a way that it conforms most closely to observed events; we thus arrive at the simplest formulation of physical laws (Schlick, 1920, pp. 65, 66).

Over a span of some 200 years Newton's absolute space and time were modified so much that they lost all physical meaning and became convenient words for the measurement of distances by rulers, radar, clocks, triangulation, and so on. The implicit assumption of the metric of Euclidean space was mathematically overcome by Bolyai, Lobachevskii, and Riemann; Einstein's general theory of relativity turned the mathematical results into a physical theory. Newton's second implicit assumption, the topology of the continuum for space and time, was mentioned in two quotations of Riemann (pp. 11, 14), Einstein calls it an invention, while Schlick denies physical objectivity of space and time, and by implication of any features ascribed to them. Nevertheless, a physical space and time with the topology of the continuum is assumed in most publications on the general theory of relativity and quantum theory. This is made obvious by the use of differential calculus in both theories and emphasized in the general theory of relativity by its foundation on Riemann's differential geometries. Most books assume this topology without mentioning the fact, mention it as an established fact beyond the need of discussion, or state it as an hypothesis:

> In Einstein's theory of gravitation matter and its dynamical interaction are based on the notion of an intrinsic geometric structure of the space-time continuum. The ideal aspiration, the ultimate aim, of the theory is not more and not less than this: A four-dimensional continuum endowed with a certain intrinsic geometric structure, a structure that is subject to certain inherent purely geometrical laws, is to be an adequate model or picture of the 'real world around us in space and time' with all that it contains and including its total behavior, the display of all events going on in it (Schrödinger, 1950, p. 1).

> We are not yet ready to discuss whether the concepts of rectangular Cartesian coordinates (x, y, z) and time t are acceptable in relativity. Actually, we shall accept them later with important reservations. For the present let us think about them, but accept as a fundamental hypothesis of relativity the statement that *the totality of all possible events form a four-dimensional continuum.* This continuum we call *space-time* (Synge, 1965, p. 6).

> In conclusion, when one deals with space-time in the context of classical physics, one accepts (1) the notion of "infinitesimal test particles" and (2) the idealization that the totality of identifiable events forms a four-dimensional continuous manifold (Misner *et al.*, 1973, p. 13).

The following quotes show that the viewpoint of Einstein and Schlick is not universally accepted, and space-time is considered to be something "real" with the topology of the continuum:

> From the essence of space remains in the hands of the mathematician, using such abstraction, only one truth: that *it is a three-dimensional continuum*[8] (Weyl, 1921; 1968, vol. II, p. 213).

> Our considerations have shown that the determination of the topological properties of space are closely related to the problem of causality; we assume a topology of space that leads to normal causal laws. Only in this way does the question about the topology of space constitute a well-determined question. It must be called an empirical fact that there

[8] Vom Wesen des Raumes bleibt dem Mathematiker bei solcher Abstraktion nur die eine Wahrheit in Händen: daß *er ein dreidimensionales Kontinuum ist.*

> is one kind of topology that leads to normal causality; and it is of course an empirical fact which topology yields this result (Reichenbach, 1957, pp. 80, 81).
>
> So let us conclude that space has a definite real intrinsic structure in its metric, affinity, and topology. This means it has a shape and size in a way I have tried at length to make clear. It shows just how much space is a particular thing (Nehrlich, 1976, p. 211).
>
> It is now generally taken for granted that public time is both infinitely divisible, or "dense" as the mathematicians term it, and continuous; that is, not only can we always consider any interval as made up of smaller ones, but we are entitled to apply even irrational numbers to the measurement of time ... Our concepts are not immune to revision; and in the case of time, we are already prepared, in some locations, to speak of it as though it were discrete. But to do so consistently would require a fairly radical revision of the concept. We should have to unthink as far back as Aristotle (Lucas, 1973, pp. 29, 32, 33).

We will now turn to the question of topology, even though this means that we "have to unthink as far back as Aristotle." A considerable number of papers have been written by mathematicians and physicists on this topic[9], but they had little effect on the mainstream of physics. There seem to be two reasons for it. First, no convincing reason is given why the continuous space-time should be abandoned after centuries of good service. Second, a decisive experimentally verifiable result that differs from continuum theory has been lacking; the ideal would be a result comparable to the explanation of the perihelion movement of Mercury, which was the first verifiable result in favor of the general theory of relativity (Einstein, 1915, 1955). The first problem can be overcome quite convincingly by the introduction of results of information theory—which was developed within electrical communications—to physics. We will discuss in the following chapter the limitation on measurements of distances in space-time if the results of information theory are observed.

[9] Snyder (1974a, b), Flint (1948), Schild (1949), Hellund and Tanaka (1954), Hill (1955), Das (1960; 1966a, b, c), Yukawa (1966), Atkinson and Halpern (1967), Cole (1970; 1971; 1972a, b; 1973a, b), Hasebe (1972), Welch (1976), Harmuth (1977, 1980), Kadishevsky (1978).

2 Information Theory Applied to Measurements

2.1 The Concept of Information

We all have some notion what information is, just as we have some notion what time and temperature is before we learn to read the clock or the thermometer. To turn the notion of information into a scientific concept one must make information measurable. This was first done by Hartley (1928). Refer to Fig.2.1-1 for an explanation of his approach. It shows certain characters of the binary teletype alphabet consisting of 5 pulses with amplitudes +1 or −1. Instead of +1 and −1 one usually writes 1 and 0. The digit 1 then represents the answer YES to the question "Is a positive current flowing?", while the digit 0 represents the answer NO. With 5 digits one can represent $2^5 = 32$ characters. According to Hartley, the information H of one teletype character is measured by the logarithm of the number of possible characters,

$$H = \log_2 2^5 = 5 \tag{1}$$

and one says that the information of one character equals 5 bit or bits. Alternately, one may say the information 5 bits is the number of answers YES or NO to the question "Is a positive current flowing?" at the times $t = 0, \Delta T, \ldots, 4\Delta T$ in Fig.2.1-1. This second interpretation fits the requirements of the *calculus of propositions*, under which the concepts of information theory enter quantum physics (Jauch, 1968) and the general theory of relativity (Misner *et al.*, 1973, § 44.5 and Box 44.5).

Instead of using the binary digits with the two possible amplitudes +1 and −1 in Fig.2.1-1 one may use ternary digits with three possible amplitudes +1, 0, and −1, or generally digits with n possible amplitudes. The logarithm with base 2 in Eq.(1) is replaced by the logarithm with base n as measure for the information in this case, and the unit bit must be replaced by a new unit that refers to the base n rather than 2. This generalization creates difficulties if applied to the calculus of propositions, since our logic assumes that something either "is" (=YES) or "is not" (=NO). This means our logic is two-valued. One can develop an abstract n-valued logic and

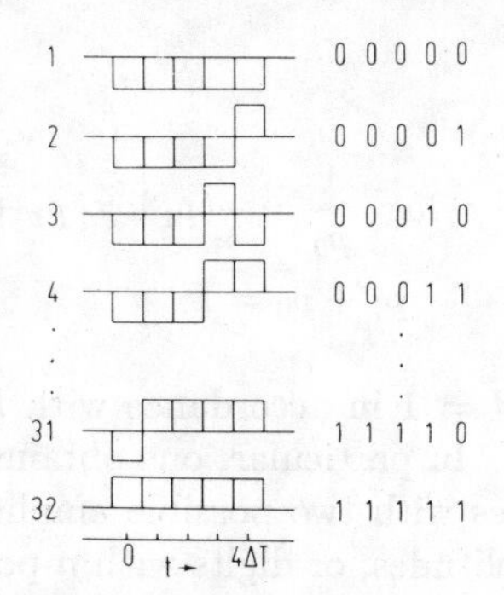

FIG.2.1-1. The 32 characters of the binary teletype alphabet with 5 pulses having the amplitude +1 or −1. Instead of +1 and −1 one usually writes 1 and 0. The digit 1 may be interpreted as the answer YES to the question "Is a positive current flowing?", while the digit 0 stands for the answer NO.

an n-valued calculus of propositions, but this is very hard on our thinking. Information theory is currently general enough for such a generalization of logic, and we have no trouble replacing the pulses or digits with two possible values in Fig.2.1-1 by pulses or digits with n possible values.

The complete set of the 32 teletype characters according to Fig.2.1-1 contains $5 \times 32 = 160$ pulses. Half of them have the amplitude +1 and half the amplitude −1. If we receive a message by teletype we will, however, not receive equally many positive and negative pulses. The reason is that the character representing the letter e (=10100) is more frequent than other characters, e.g., z (=01011). Hence, the probability of receiving a positive pulse (digit 1) is not necessarily the same as receiving a negative pulse (digit 0). The definition of Eq.(1) applies only if the probabilities p_1 for a digit 1 and $p_0 = 1 - p_1$ for a digit 0 are equal, $p_1 = p_0 = 1/2$. A generalization for $p_1 \neq p_0$ was introduced by Shannon[1] (1948). Let us first write Eq.(1) for one rather than five digits:

$$H = \log_2 2^1 = 1, \; p_1 = p_0 = 1/2 \tag{2}$$

The generalization of this formula according to Shannon equals:

[1]This generalization is based on three assumptions that may or may not appear to be plausible. The main reason for using this particular generalization seems to be that it yields formulas equal to those for entropy in statistical mechanics. The connection with statistical mechanics becomes readily recognizable if one replaces the pulses with amplitudes +1 or −1 in Fig.2.1-1 by the black and white billiard balls commonly used to derive basic concepts in statistical mechanics.

$$H = p_1 \log_2 \frac{1}{p_1} + p_0 \log_2 \frac{1}{p_0} = -(p_1 \log_2 p_1 + p_0 \log_2 p_0)$$
$$p_1 + p_0 = 1 \tag{3}$$

For $p_1 = p_0$ one obtains $H = 1$ in accordance with Eq.(2), while $p_1 \neq p_0$ yields smaller values for H. In particular, one obtains $H = 0$ for $p_1 = 0$ or $p_1 = 1$. If the binary pulses with two possible amplitudes are replaced by pulses with n possible amplitudes, or digits with n possible values $i = 1, 2, \ldots, n$ one obtains the generalization

$$H = -\sum_{i=1}^{n} p_i \log_n p_i, \quad \sum_{i=1}^{n} p_i = 1 \tag{4}$$

of Eq.(3), where p_i is the probability of a digit with value i. If one has m digits rather than one digit, one must sum the information transmitted by each one of the m digits. If the probabilities p_i do not change from one digit to the next, the total information equals m-times the values given by Eqs.(3) or (4). For the binary case we write this explicitly:

$$\begin{aligned} H(m) &= -\sum_{k=1}^{m} \left(p_{1k} \log_2 p_{1k} + p_{0k} \log_2 p_{0k}\right) \\ &= -m\left(p_1 \log_2 p_1 + p_0 \log_2 p_0\right) \quad \text{for } p_{1k} = p_1,\ p_{0k} = p_0 \\ &= m \quad \text{for } p_1 = p_0 \end{aligned} \tag{5}$$

So far we have assumed that the binary numbers in Fig.2.1-1 define characters of the teletype alphabet. However, it is evident that these numbers may represent anything else, such as voltages, distances, or times shown by a clock. Any quantity that can be represented by a binary number carries information whose value is determined by Eq.(5). According to this equation, the information of a binary number with m digits can be at most m bits. Since we can record, process, or transmit binary numbers with a finite number of digits only, the information we can record, process, or transmit must always be finite. Infinite information has no more place in physics than infinite energy[2].

[2]For a more detailed discussion of information and information theory the reader is referred to the following books: Ash (1965), Brillouin (1956), Feinstein (1958), Fey (1963), Reisbeck (1964), Reza (1961), Singh (1966).

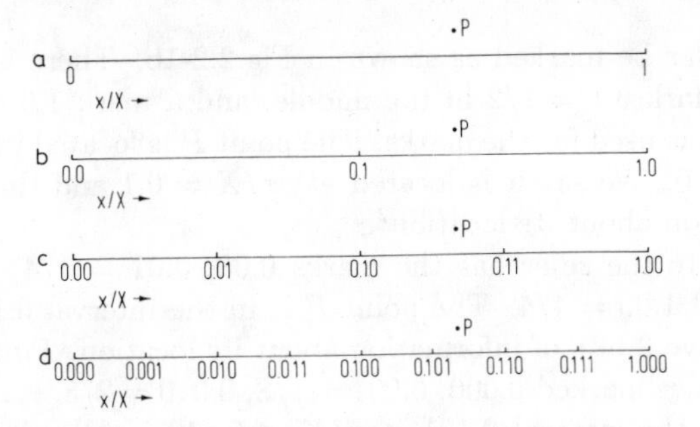

FIG.2.2-1. Information about the location of the point P for $x < X$. (a) 1 bit, (b) 2 bits, (c) 3 bits, (d) 4 bits.

The unit of information, the bit, is a pure number without dimension. Hence, information must be independent of coordinate transformations. For illustration consider 5 apples. The number 5 is independent of any coordinate transformations, while the weight, the mass, or the diameter of the apples is not. Pure numbers are thus of great interest in physics. Nevertheless, information theory has had very little influence on physics. There seem to be two tangible reasons for it. First, great expectations were once based on the mathematical equality of the equations defining information and entropy, but these expectations have so far been disappointed. An excellent book on this topic is due to Brillouin (1956). Second, a widely acclaimed result, the transmission capacity of communication channels, turned out to be correct from the standpoint of abstract mathematics but wrong when applied to an experimental science, which reduced the confidence in results of information theory generally[3]. We will neither need any connection between information and entropy nor the concept of transmission capacity, but only investigate what finite information and finite information flow[4] imply for physical measurements.

2.2 Finite Information and Finite Resolution

Let us investigate how much information we receive by measuring the location of a point—or the distance between two points—by means of a ruler. Refer to Fig.2.2-1 for an explanation of how the location x of a point relative to a ruler can be expressed in bits. Figure 2.2-1a shows a ruler with arbitrary length X. The ruler is marked at its beginning (0) and end (1). We assume—for the time being—that P is between these two points.

[3] A discussion of this point is given in Section 12.1.

[4] Information flow is the information transmitted per unit time. It is usually measured in bits per second.

Let the ruler be marked as shown in Fig.2.2-1b. There is a mark 0.0 at the left, a mark $0.1 = 1/2$ in the middle, and a mark 1.0 on the right; binary notation is used for the marks. The point P is located in the interval $0.1 < x/X < 1.0$. We say it is located at[1] $x/X = 0.1$ and that we have 1 bit of information about its location[2].

In Fig.2.2-1c the ruler has the marks 0.00, $0.01 = 1/4$, $0.10 = 2/4$, $0.11 = 3/4$, and $1.00 = 4/4$. The point P is in the interval $0.10 < x/X < 0.11$, and we have 2 bits of information about its location. Finally, Fig.2.2-1d shows the ruler marked 0.000, $0.001 = 1/8$, $0.010 = 2/8$, ... The point P is located in the interval $0.101 < x/X < 0.110$ and we have 3 bits of information. The information in bit is measured by the number of binary digits to the right of the binary point required to denote the equally spaced marks on the ruler[3].

Let us turn to the measurement of information by yes-no decisions as is appropriate for the calculus of propositions. The following questions have to be asked:

1. The interval in which P is known to be is divided into two intervals by the mark 0.1 of the ruler. Is P to the right of this mark? Yes, 1.
2. The interval in which P is known to be is divided into two intervals by the mark 0.11 of the ruler. Is P to the right of this mark? No, 0.
3. The interval in which P is known to be is divided into two intervals by the mark 0.101 of the ruler. Is P to the right of this mark? Yes, 1.

Let the number of intervals of the ruler in Fig.2.2-1 increase from 2^3 to 2^4, 2^5, ... The information increases to 4, 5, ... bits. It is immediately apparent that the number of bits can be very large, but it must always be finite. Otherwise we would have to write—or generally to transmit, process, or store—an infinitely long string of binary[4] digits 0 and 1.

Generalizing from the point P with coordinate x to an event with coordinates $x^k = x^0, x^1, x^2, \ldots$ we obtain the fundamental axiom I:

I. *The information about the position of an event relative to a coordinate system is always finite.*

---

[1]We could denote the interval $0.1 < x/X < 1.0$ rather than the mark on its left side by $x/X = 0.1$. This notation provides a number of advantages, but denoting the marks rather than the intervals is more usual and will be seen to be advantageous for multidimensional coordinate systems.

[2]This is true only if the probability of P lying in the interval $0.1 < x/X < 1$ is equal to the probability of P lying in the interval $0 < x/X < 0.1$.

[3]The subdivision of the ruler into 2^m equal intervals yields the information m bits about the location of P. Generally, one may divide the ruler into s^m equal intervals and obtain the information m units of the base s.

[4]The transition from 2^n to s^m intervals would not change this result. The number m of *digits of base s* would still have to be finite for a finite value of s; an infinite value of s would require infinitely many symbols instead of the two symbols 0 and 1 for $s = 2$.

This axiom appears rather obvious, but it is generally disregarded. For an explanation of this statement let us denote the distance between two marks in Fig.2.2-1d by $\Delta x/X$. Let Δx decrease to zero in such a way that the number of marks becomes denumerable infinite. This is already a violation of our axiom, but let us go one step further and let Δx decrease so that the number of marks becomes *non*denumerable infinite. We write now dx instead of Δx, and the information is nondenumerable infinite[5]. This is an even worse violation of our axiom, but it brings us to differential calculus and the one-dimensional continuum[6]. The use of differential calculus and the concept of a continuum implies that we can obtain nondenumerable infinite information about events.

Let the ruler in Fig.2.2-1 have 2^m marks. The information about the location of the point P then equals m bits. Since the point P is located between two marks x_P and $x_P + \Delta x$ we say its uncertainty equals $\Delta x = X/2^m$, or $\Delta x/X = 1/2^m$ in normalized notation.

Consider a second point R. The information about its location equals again m bits. The information about the distance between the points P and R also equals m bits, since the difference between two binary numbers with m digits is a binary number with m digits[7]. The uncertainty of the distance equals $2\Delta x = 2X/2^m$, since the one point lies between x_P and $x_P + \Delta x$, and the other between x_R and $x_R + \Delta x$, which implies a minimum[8] distance $x_R - (x_P + \Delta x) = x_R - x_P - \Delta x$ and a maximum distance $(x_R + \Delta x) - x_P = x_R - x_P + \Delta x$ for $x_R > x_P$.

Consider once more Fig.2.2-1. The arbitrary length X of the ruler permits us to measure distances $x < X$ from $x = 0$, but not distances $x > X$. This limitation is remedied in Fig.2.2-2. The distance of the point Q from $x = 0$ is much larger than X in Fig.2.2-2a. To locate Q, the length of the ruler is doubled in Fig.2.2-2b; the marks 0 and 1 are relabeled as 00 and 01, and the mark $x/X = 10 = 2$ is added. The point Q is still out of

---

[5] "Nondenumerable infinite information" is a short expression for "nondenumerably many bits of information" or generally "nondenumerably many units of the base s of information".

[6] *Continuum* is here a short term for *ruler with marks having the topology of the continuum* or the *usual topology of the real numbers*. Note that it is the ruler—or by implication a clock—which has the marks; space and time are not even mentioned.

[7] Zeroes at the beginning or the end of the numbers must be retained: 100.101-011.011=001.010.

[8] If the points P and R may be located anywhere with equal probability, the minimum distance $x_R - x_P - \Delta x$ and the maximum distance $x_R - x_P + \Delta x$ are less probable than an intermediate distance such as $x_R - x_P$. Hence, one can derive a probability distribution for the distance between P and R. This results in an apparent more accurate knowledge of the distance between P and R, but no additional information is gained since the probability distribution follows strictly from the features of rulers and is not the result of any measurement.

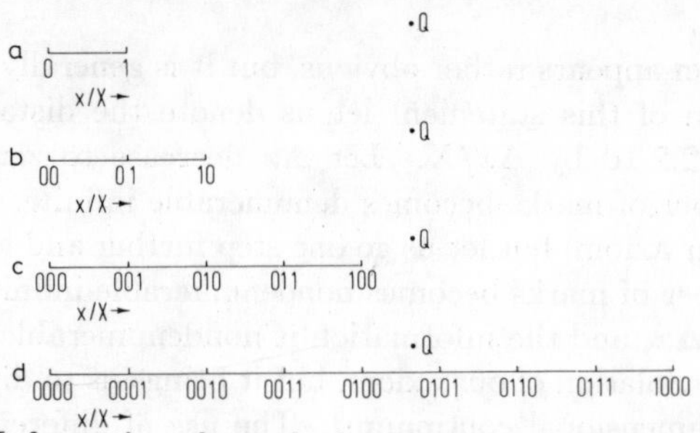

FIG.2.2-2. Information about the location of a point Q for $x > X$.

range. Again, the length of the ruler is doubled in Fig.2.2-2c; the marks are relabeled 000, 001, ... , 100. The point Q is still out of range. The third doubling of the length of the ruler in Fig.2.2-2d brings Q into range, and its distance from $x/X = 0$ equals $x/X = 0100$, meaning x/X is larger than 0100 but smaller than 0101.

The number 0100 tells us the distance of Q from $x/X = 0$ in multiples of X (0100=4). On the other hand, the number 0.101 in Fig.2.2-1d told us the distance of P from $x/X = 0$ in fractions of X (0.101=5/8). Generally, a distance is stated in the form $x/X = 0100.101$. The digits to the left of the binary point give the multiples of X, the ones to the right the fractions of X. One must include all the digits 0 in the number; for instance, 0100 and 0.010 must not be shortened to 100 and 0.01, since the shortened numbers contain less information.

The length X was chosen arbitrarily, and the number $x/X = 0100.101$ is thus arbitrary too. Let us substitute the new unit $2X$ for X. We obtain the number $x/2X = 010.0101$ instead of $x/X = 0100.101$. The change of the unit length from X to $2X$ or $X/2$ only shifts the binary point, the number of binary digits is not affected. Hence, the information—which equals the number of binary digits—does not depend on the choice of the arbitrary unit X. This is quite evident for a change from X to $2^n X$ or $X/2^n$, but it applies to other changes too. The binary number is changed in the general case by more than a shift of the binary point.

The independence of the information from the choice of a unit length X or a unit time T is a must to make it a quantity significant for physics, since information is a pure number without dimension.

A conclusion that one must draw from Fig.2.2-2 is that there can be no distance x/X that is infinite, as long as the unit length X is finite, since one would need infinite information and thus infinitely many binary digits to record such a distance. Hence, the requirement for finite information eliminates from physics the infinitely long as well as the infinitely short.

Laccess to , but arbitrarily large

one end ≠ ∞?

2.3 Finite Information Flow

Let us turn from information to information flow or transmission rate of information, which is usually measured in bits per second. The fine structure constant[1]:

$$\alpha = 72981 \times 10^{-7} = 11\,101\,111\,001\,001\,010 \times 2^{-11000}$$

does not change with time, and we are thus interested in the information $17 + 1 + 5 = 23$ bits contained in it[2], but not in an information flow. The distance between two objects, on the other hand, may change with time[3]. In this case we are interested in receiving the information about the distance before the change of the distance makes the information obsolete. The following axiom applies to the transmission rate of information:

II. *A finite time is required for the transmission of information.*

We will see later on that this requirement of a finite time would be imposed by the availability of a finite energy for the transmission of information, even if there were a means to generate and receive information instantly[4]. For the time being it is only important to accept that the time is finite regardless of the means of transmission of information[5]. We may transmit information by acoustic waves, by photons, by gravitons, by muons, by neutrinos, by mail, and so on, but it always requires a finite time. Refer to Fig.2.3-1 for the discussion of some of the consequences.

---

[1] The number 2^{-11000} should be written 10^{-11000} in binary notation, but 10 = two is too easily confused with 10 = ten in decimal notation.

[2] The negative sign of the exponent contains the information 1 bit since it is obtained by a yes-no decision between the two possible numbers -1100 and $+1100$.

[3] Time means strictly what a clock measures. We avoid the use of the concept space-time if it means more than the measurement of a distance with a ruler according to Figs.2.2-1 and 2.2-2, or the equivalent measurement of time differences with a clock. We will discuss the concept of space-time in detail later on.

[4] One could claim that axiom I already states that information is neither generated nor received in an infinitesimal time. Axiom II goes further by stating that a finite time is required that may well be longer then the uncertainty due to axiom I.

[5] Many scientists have investigated quantum mechanical restrictions on information and information rate of flow (Mundici 1981; Bekenstein 1981a, b). Investigations on limits of the accuracy of measurements are often made in connection with receivers for gravitational waves (Braginsky *et al.* 1980). It is widely accepted at this time that there is no bound on the rate at which information can be processed. This is in line with our axiom that a finite time ΔT is required for the transmission of information, without any assumption that ΔT must be larger than some minimum time. No matter how small one makes ΔT it will never become zero or a differential dt. The distinction between "arbitrarily small but finite" and a differential is often not made, but finite differences lead to difference equations while differentials lead to differential equations, and the solution of difference equations do not necessarily converge to those of a differential equation for decreasing values of the differences. We will discuss this later on in great detail.

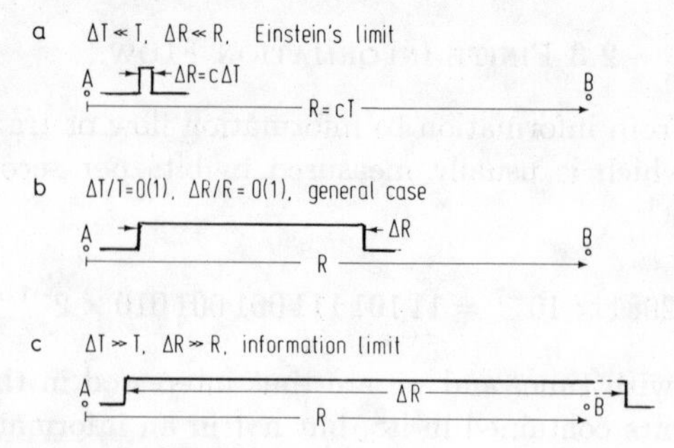

FIG.2.3-1. Duration ΔT of information carrying signals and propagation time T of the signals from point A to point B.

A space ship A transmits a "light flash" to a space ship B at a distance R in Fig.2.3-1a. The propagation time $T = R/c$ is finite. This is a typical illustration used in textbooks on the special theory of relativity discussing the concept of time and the synchronization of clocks in systems of reference with relative velocity. An important part of these discussions is that the duration ΔT of the light flash is short compared with the propagation time of the light flash or signal. Alternately, we can say that the signal has a length $\Delta R = c\Delta T$ which is short compared with the distance R between transmitter A and receiver B.

The axiom "a finite time is required for the transmission of information" implies that the duration ΔT of a signal is never zero. Furthermore, a "light flash" is either there or it is not there; hence, it carries one bit of information[6]. A light flash of duration ΔT represents an information flow of 1 bit per time interval ΔT.

In general, much more than 1 bit of information has to be transmitted. For instance, in order for the Earth to follow its orbit, it must constantly receive information about the ratio mass/distance $= m/\mathbf{r}$ of all the other gravitating masses[7]. If this information is not sufficiently accurate—too few bits of information—or if the information flow is too small, the orbit of the Earth would show statistical fluctuations around its mean value, and might actually become unstable. We have not observed such instabilities in the macro world, presumably because the energy required for a sufficient information flow is small compared with the total energy radiated.

The need for information flow for proper orbiting or other forms of

[6]This assumes that the probability for a light flash being there equals 1/2. Otherwise the information is lower according to Eq.(2.1-3).

[7]This is in terms of Newton's theory. The general theory of relativity requires that the ten independent components g_{ik} of the metric tensor of Eq.(1.3-4) caused by all gravitating masses are received.

interaction does not decrease when we go from the macro to the micro world of atoms and nuclei, but the available energy decreases drastically. Hence, we have here a better prospect of finding effects due to the limitation of the information flow.

Typical propagation times T for atomic distances[8] of 100 pm and nuclear distances of 10 fm are $T = 10^{-10}/c = 3 \times 10^{-19}$ s and $T = 10^{-14}/c = 3 \times 10^{-23}$ s. Hence, if the duration of the signal is not small compared with these times, we will not have the case $\Delta T \ll T$ of Fig.2.3-1a, but the cases of Fig.2.3-1b and c where ΔT is of the order of T or large compared with T.

Let us first observe that at the nuclear level information is typically transmitted one bit at a time. A photon, an electron, a muon, a neutrino, and so on is either received or not received, which requires one yes-no decision or one bit[9]. A first estimate of the signal duration ΔT and thus the information flow 1 bit/ΔT may then be made via the uncertainty relation in the energy-time form:

$$\Delta E \Delta T \geq \hbar \tag{1}$$

If the energy of the particle used for information is E, one must make $\Delta E < E/2$ for a decision between "the energy E was received" and "the energy 0 was received". In reality, ΔE must be much smaller than $E/2$ so that the probability of having made the correct decision between "energy E received" and "energy 0 received" is almost 1, otherwise the information will be significantly less than 1 bit. Hence, we use the following estimate for the time ΔT for the transmission of 1 bit of information:

$$\Delta T \gg \hbar/E \text{ or } \Delta T \gg h/E \tag{2}$$

For a photon,

$$E = h\nu \tag{3}$$

we obtain the condition

$$\nu \gg 1/\Delta T \tag{4}$$

[8] pm = picometer = 10^{-12} m, fm = femtometer = 10^{-15} m, am = attometer = 10^{-18} m.

[9] The situation is more complicated if one can distinguish between different kinds of the same particle, for example electrons with spin +1/2 or −1/2. Two yes-no decisions have to be made in this case: (1) Was an electron received? (2) Did it have spin +1/2? The second question makes no sense unless the first was answered with YES.

for the lowest frequency that permits the transmission of 1 bit during the time ΔT. Using the value $\Delta T = T = 3 \times 10^{-19}$ s we obtain

$$\nu \gg 3 \times 10^{20}\,\text{Hz} \quad \text{or} \quad \lambda \ll 10^{-10}\,\text{cm} = 1\,\text{pm} \tag{5}$$

for atomic distances. For nuclear distances one obtains with $\Delta T = T = 3 \times 10^{-23}$ s:

$$\nu \gg 3 \times 10^{24}\,\text{Hz} \quad \text{or} \quad \lambda \ll 10^{-16}\,\text{cm} = 100\,\text{am} \tag{6}$$

Photons with frequencies that satisfy these conditions will transmit information according to Fig.2.3-1a, where the signal duration ΔT is small compared with the propagation time T, while for photons with lower frequency the transmission time $T + \Delta T$ of information will be significantly longer than the propagation time T. Putting it differently, only if information is transmitted with very high energy photons or other particles, can one ignore the duration ΔT of the signal at atomic and nuclear distances.

Consider further the possible information flow from an electron with rest mass m_0 and energy m_0c^2. In order to transmit k bits of information during a time interval ΔT, the interval ΔT must have at least the following duration according to Eq.(1):

$$k/\Delta T \leq \Delta E/\hbar = m_0c^2/\hbar$$

$$\Delta T \geq k\hbar/m_0c^2 = k(10^{-34}/9) \times 10^{31} \times 9 \times 10^{16} = 1.2k \times 10^{-21}\,\text{s} \tag{7}$$

Light will travel the distance

$$c\Delta T \geq 3.6k \times 10^{-13}\,\text{m} \tag{8}$$

during this time. Hence, an electron can transmit at most one bit of information during the time 1.2×10^{-21} s, which is the time required for light to travel 3.6×10^{-13} m. For $k = 1000$ bits of information, we obtain from Eq.(8) the distance

$$c\Delta T \geq 3.6 \times 10^{-10}\,\text{m} = 3.6 \times 10^{-8}\,\text{cm},$$

which is a typical atomic distance. This is an enormous amount of information; for instance, for distance measurements one would need a ruler with $2^{1000} \approx 10^{301}$ marks to obtain that much information. Such an accuracy is far beyond the most accurate measurements ever made, and far beyond the machine accuracy of any digital computer.

Consider, on the other hand, the time required by light to travel a typical nuclear distance of 10^{-11} cm $= 10^{-13}$ m or less. According to Eq.(8)

not even $k = 1$ bit of information can be transmitted during this time. The electron could not announce its existence in even the crudest form and we would say that it does not exist.

For protons and neutrons we have to multiply Eq.(8) by about 1800:

$$c\Delta T \geq 2 \times 10^{-16} \text{ m} \tag{9}$$

Hence, protons and neutrons can inform us about their existence at nuclear distances. Summing up, we may conclude that information flow is not a restriction at atomic distances, but it becomes significant at nuclear distances. We will see this result confirmed in Section 10.4 by a rigorous calculation of the eigenfunctions of a boson in a Coulomb field, based on a finite resolution of location and time.

The title of this chapter, "Information Theory Applied to Measurements", seems to apply to measurements that are made by a human observer; Figs.2.2-1 and 2.2-2 were discussed in this way. However, when we advanced from information to information flow, it became clear that the need for information was more general. This need for the transfer of information is connected to the causality principle. It may be stated in the following modern form: *Every effect requires a sufficient cause.* Part or all of this causality principle may be restated as the information principle: *Every effect requires sufficient information.* In addition to information one may or may not require something else, like energy, to produce an effect, but we certainly need the information. The Earth could not follow its orbit if insufficient information about the other masses were transferred to determine the components g_{ik} of the metric tensor; an electron could not follow its course through a cathode ray tube without the transfer of information about the deflecting voltages and currents; and so on. We use the metric field in the one case and the electromagnetic field in the other case to explain this transfer of information, but what is common is the transfer of information and the propagation velocity c of the signals that transfer the information.

2.4 Finite Resolution Δx and Δt

Figures 2.2-1 and 2.2-2 explained how the location of points P and Q relative to the ruler with discrete marks is determined. The results can be applied to time measurements by replacing the notation x/X of the marks by t/T and using a point P that moves along the ruler, like the pointer of a clock moves along the face of the clock. The implied assumption is that P moves with constant velocity, but it is obviously difficult to say what a constant velocity is before one has a clock. The concept of constant

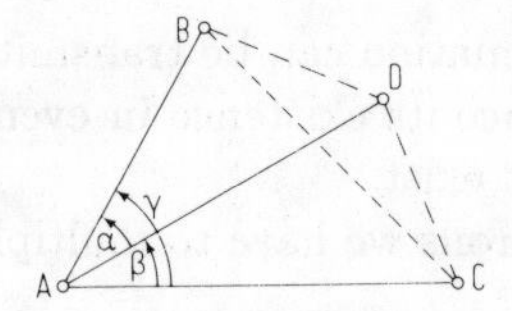

FIG.2.4-1. Description of the relative location of 4 points by means of $3+2+1=6$ distances by the 3 distances and the $2+1$ angles observable at one point.

velocity has to come from somewhere else. It was usual to assume that the rotation of the Earth around its axis or around the Sun provides us with constant periods of time, from which one could derive a definition of constant velocity, but we are no longer satisfied with astronomical constancy, and we use currently atomic clocks for the definition of constant time intervals. Time is—so far—a concept that we do not understand very well[1], and we end the discussion for the moment with a quote from a book by Misner *et al.* (1973, p. 23): "Time is defined so that motion looks simple."

We will use the term time for the finite intervals *produced*[2] by the bests clocks we can build. Which type of clock is best can be determined by building several clocks of the same type, and comparing them with several clocks of another type. The type of clock that produces the smallest differences between the several clocks is the best type[3].

With a ruler and a clock we can describe the distances between points—which may be elementary particles, planets, or stars—and the change of these distances with the time produced by the clock. Consider Fig.2.4-1. The location of the four points A, B, C, and D relative to each other is determined fully by the six distances between them. One may think that five distances would be enough, but this would introduce the assumption that the four points are located in a plane. By only talking about the distances one avoids the concept of plane, and later the concepts of three- and more-dimensional spaces.

Let now A be the Earth, while B, C, and D are celestial bodies. The ruler can no longer be some mechanical device, but we can measure the distances from the Earth A to B, C, and D by radar[4]. We cannot measure

[1] There exists an International Society for the Study of Time (Fraser *et al.*, 1972, 1975, 1978).

[2] Note that the finite time intervals and thus the time used for measurements are *produced* by the clocks; one could not very well claim that the clocks *measure* the finite time intervals.

[3] This is the actual way in which the very best clocks are checked against each other (Lübke, 1958; Kartaschoff, 1978).

[4] This is actually possible for the Moon and several planets as well as man-made satellites. Note that distance measurements by radar can be done from one point, whereas triangulation requires observation at two points. Radar requires an assumption about

the distances BC, BD, and CD. However, we can measure the angles α, β, and γ, using—if we want—the same radar that measured the distances[5]. Instead of the three non-measurable distances we have three measurable angles, and we can again describe the location of B, C, and D relative to A without ambiguity[6].

The measurement of the angles α, β, and γ is *not* equivalent to the measurement of the distances BC, BD, and CD. We cannot calculate these three distances from the three angles and the measured distances AB, AC, and AD without making assumptions. The usual assumption is, of course, that the axioms of Euclidean geometry apply. If observers at B, C, and D measure distances and angles observable at their locations and transmit this information to A, one can decide there which assumptions about geometry will reconcile the observed distances and angles. One is tempted to say that a metric is defined in this way, but one has to make the assumption that the conditions of Eq.(1.3-1) are satisfied before one can say a metric is defined. Let us emphasize the need to transmit information from the points B, C, and D to the center at point A; our previous statement about information having to be finite to be transmittable clearly applies here.

Let the points A, B, C, and D in Fig.2.4-1 move relative to each other. At point A we shall have a radar that measures distances as well as the angles between the *rays* of these distances. Furthermore, there shall be a clock at A, which permits us to record distances and angles as function of the time defined or produced by this clock. We can thus—at least in principle—describe the motion of the celestial bodies relative to the radar on Earth A by means of the concepts distance, angle, and time produced by the clock. There is no reference to a space or a time *measured* by a clock, and no mention of orbits. If we had described the movement of celestial bodies in this way from the beginning, there would never have been a need for absolute space, absolute time, a metric, or the general theory of relativity. We could predict from the measured distances and angles what distances and angles we would observe when the clock showed a later time.

This is of course what is actually done. We predict angles and distances at a future time from observations at a previous time. The concepts

---

the (constant) velocity of a propagating wave, while triangulation requires an assumption about the (straight) path of a propagating wave.

[5]The number of distances between n points equals $(n-1)+(n-2)+\ldots+1$. The number of distances measurable by a radar from one of the n points equals $n-1$, and the number of angles between the rays of these $n-1$ distances equals $(n-2)+(n-3)+\ldots+1$. Hence, the number of all distances is always equal to the number of distances and angles observable at one of the n points.

[6]This description of objects in relation to each other is often referred to as *relational space*, even though the concept of space is not required. The relational space is usually traced back to Leibnitz.

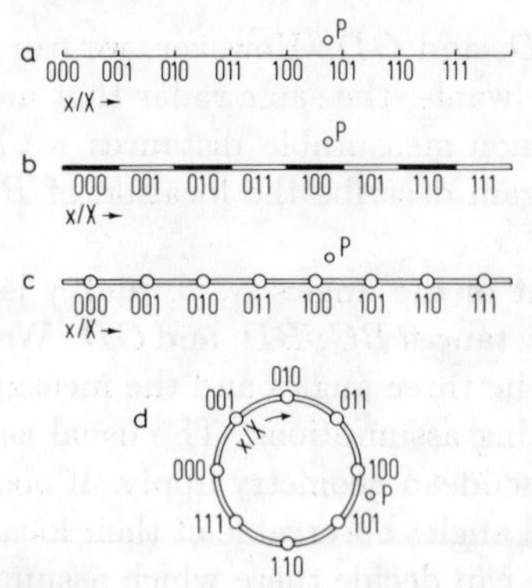

FIG.2.4-2. Various representations of a ruler with discrete marks. (a) Basic representation, (b) interval representation, (c) representation for multidimensional coordinate systems, (d) finite but unbounded ruler.

of space and planetary orbits only make such predictions more understandable to us. However, making something more understandable is of enormous importance. We celebrate the replacement of epicycles by Kepler's ellipses because of the resulting simplification, and not because one must use the concept of elliptical orbits to predict future observations of angles and distances from past ones.

Let us then introduce a concept that is not necessary but has proved to be exceedingly fruitful, the concept of a coordinate system. We will develop coordinate systems from the ruler with discrete marks in Fig.2.2-2.

Figure 2.4-2a shows again such a ruler, with the marks denoted $x/X =$ 000, 001, ... In Fig.2.4-2b the marks are replaced by intervals denoted 000, 001, ..., while Fig.2.4-2c shows a rod with spheres as marks. The representation of Fig.2.4-2b suggests the statement "P is located in the interval 100", whereas Fig.2.4-2c suggests to say instead "P is closest to the mark 100". The second statement will turn out to be more lucid for coordinate systems with more than three dimensions. The representation of Fig.2.4-2c will also be easier for plotting and comprehending than the one of Fig.2.4-2a for coordinate systems with more than two dimensions.

Figure 2.4-2d shows the ruler of Fig.2.4-2c made unbounded but finite. We will use such *rings* to construct unbounded but finite coordinate systems in the following chapter.

Figure 2.4-3a–c shows the creation of two-dimensional Cartesian coordinate systems from the rulers of Fig.2.4-2a–c. No particular difficulty is encountered. However, this is not so for the finite but unbounded ruler of Fig.2.4-2d. A possible generalization to two dimensions is shown in Fig.2.4-4a. The ends of the rods of Fig.2.4-3c are bent back to their be-

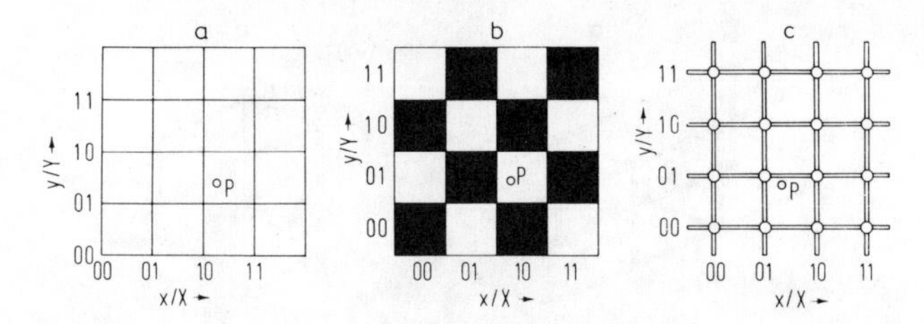

FIG.2.4-3. The extension of the rulers or one-dimensional coordinate systems of Fig.2.4-2a–c from one to two dimensions.

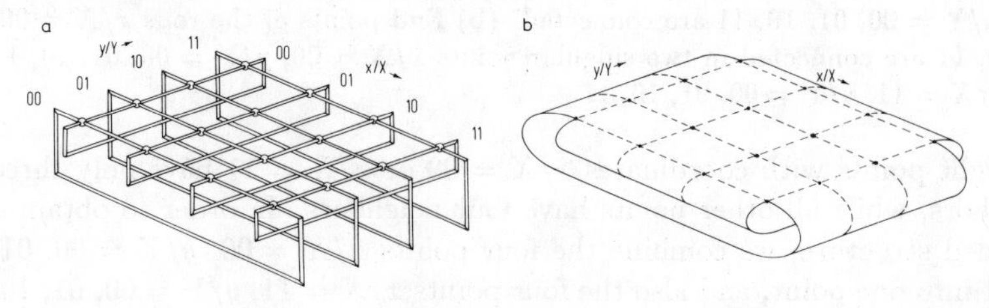

FIG.2.4-4. (a) A possible extension of the finite but unbounded one-dimensional coordinate system of Fig.2.4-2d to two dimensions and (b) the surface obtained by replacing the 16 discrete points by nondenumerably many points.

ginnings, and a two-dimensional, finite and unbounded coordinate system is obtained[7]. This coordinate system defines a *multiple connected, discrete surface*. The term multiple connected surface becomes understandable by a look at Fig.2.4-4b, which shows the 16 discrete points of Fig.2.4-4a replaced by nondenumerably many points[8].

In the one-dimensional case of Fig.2.4-2 there was only one way to produce an unbounded but finite coordinate system: the two ends of the ruler had to be connected. In two dimensions, the number of possible ways to convert a bounded coordinate system into an unbounded one is enormous. Figure 2.4-4a shows one possibility. Another is shown in Fig.2.4-5a. The end points of the rods denoted $y/Y = 00$, 01, 10, 11 in Fig.2.4-3c have been connected, but the end points of the rods denoted $x/X = 00$, 01, 10, 11 have not been connected. This is not yet a bounded coordinate system, since

[7]One may object that the distance between any two adjacent points in Fig.2.4-4a is not always the same, while in Fig.2.4-2d the distance between the points 111 and 000 is the same as between the points 110 and 111. This is strictly due to the difficulty of drafting such structures with equal distances. The difficulty will be overcome in Chapter 3.

[8]Note that the discrete coordinate system of Fig.2.4-4a is unbounded since each one of the 16 points x/X, y/Y has four neighbors. The same is not true for the continuous surface of Fig.2.4-4b, which has either "four edges and four corners" or "one edge with four nondifferentiable points"; this surface is bounded. We have here a good example of the significant (topological) difference between discrete and continuous structures.

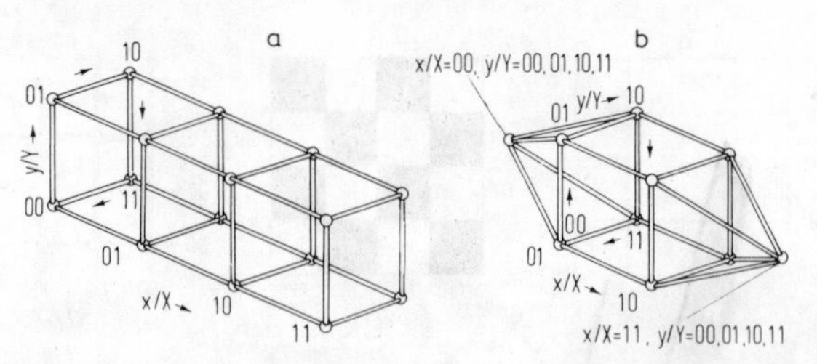

FIG.2.4-5. Changing the bounded coordinate system of Fig.2.4-3c into an unbounded one with the topological features of the sphere. (a) End points of the rods $y/Y = 00$, 01, 10, 11 are connected. (b) End points of the rods $x/X = 00$, 01, 10, 11 are connected in two singular points $x/X = 00$, $y/Y = 00$, 01, 10, 11 and $x/X = 11$, $y/Y = 00$, 01, 10, 11.

the eight points with coordinates $x/X = 00$ or $x/X = 11$ have only three neighbors, while all other points have four neighbors. In order to obtain a bounded structure, we combine the four points $x/X = 00$, $y/Y = 00$, 01, 10, 11 into one point, and also the four points $x/X = 11$, $y/Y = 00$, 01, 10, 11 into one point. The result is the structure of Fig.2.4-5b. Each point has now four neighbors.

It is not difficult to imagine the rods of Fig.2.4-5b bent so that they become sections of circles. The four rods $y/Y = 00$, 01, 10, 11—which are running in the direction x/X—may then be called meridians, while the two rods for $x/X = 01$, 10 may be called circles of latitude. The singular points $x/X = 00$, $y/Y = 00$, 01, 10, 11 and $x/X = 11$, $y/Y = 00$, 01, 10, 11 become then the north and the south pole (Fig.2.4-6). No such singular points occurred in Fig.2.4-4a. However, the discrete unbounded structure of Fig.2.4-4a leads to the continuous, *bounded* surface of Fig.2.4-4b, whereas the discrete unbounded structure of Fig.2.4-5b leads to the continuous *unbounded* surface of the sphere in Fig.2.4-6, if the number of points is made nondenumerably infinite. We see from this example that an unbounded, discrete coordinate system can avoid singular points, even though the continuous coordinate system cannot[9].

A more complicated unbounded, two-dimensional coordinate system derived from the bounded one in Fig.2.4-3c is shown in Fig.2.4-7. The coordinate system of Fig.2.4-3c with $m = 4$ points for the variable x/X and also $m = 4$ points for the variable y/Y has

$$\binom{2}{2} 2^2 (m-2)^0 = 4$$

corner points and

[9]Compare page 10 of Misner *et al.* (1973).

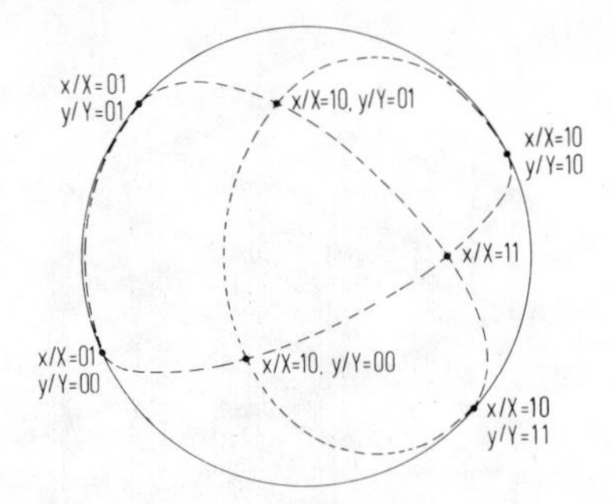

FIG.2.4-6. Spherical surface obtained by bending and stretching the rods in Fig.2.4-5b, and replacement of the 10 points by nondenumerably many.

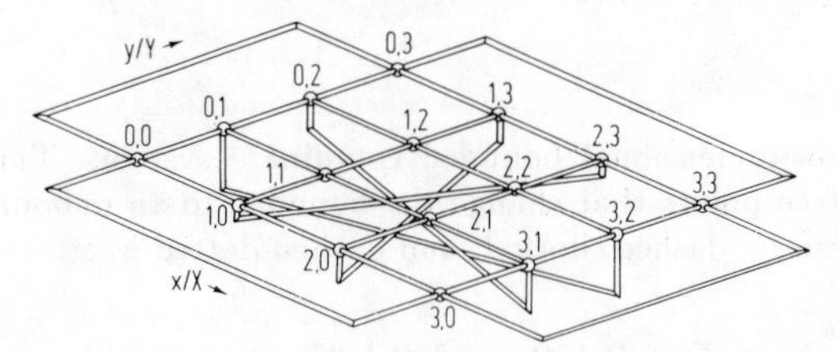

FIG.2.4-7. A third example for the change of the bounded coordinate system of Fig.2.4-3c into an unbounded coordinate system.

$$\binom{2}{1}2^1(m-2)^1 = 8$$

edge points. Each corner point must be connected to two other points and each edge point to one other point to give each point the same number of neighbors and thus produce an unbounded structure. "The same number of neighbors" means that each point in Fig.2.4-7 must be connected by the same number of rods to other points. Table 2.4-1 lists the four corner and eight edge points of Fig.2.4-3c, their neighbors in Fig.2.4-3c, and their additional neighbors in Fig.2.4-7.

The continuous surface corresponding to Fig.2.4-7 is similar to Fig.2.4-4b, but the two "return loops" are twisted 180° like the well known Möbius strip. In addition, the four points 0,0; 0,3; 3,0 and 3,3 are connected differently from the others. Without this different treatment the point 0,3 would be connected by two rods with the point 3,0; a corresponding remark would apply to the point pair 0,0 and 3,3. We do not give examples of unbounded coordinate systems with multiple connected points here, but the principle is quite legitimate, and will be encountered when dyadic coordinate systems are discussed.

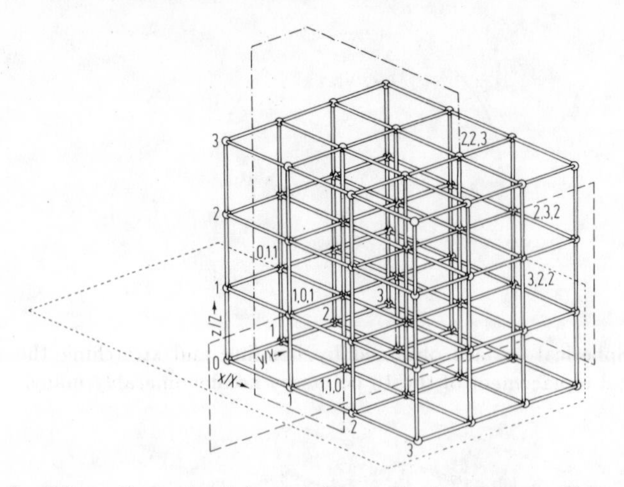

FIG.2.4-8. Three-dimensional bounded coordinate system. Three typical connections of surface points that change the bounded to an unbounded coordinate system are shown by dashed, dotted, and dashed-dotted lines.

Let us look at Figs.2.4-4b and 2.4-6. One would usually call these structures a bounded or open surface (Fig.2.4-4b) and an unbounded or closed surface (Fig.2.4-6) with a coordinate system painted on the surface. On the other hand, the corresponding discrete structures of Figs.2.4-4a and 2.4-5b are clearly coordinate systems only with no reference to a concept of "surface". Hence, the mathematical abstraction from rulers with finite distances Δx between the marks to differential distances dx and nondenumerably many marks introduces the concept of surface. By implication, we will expect that the concepts of three-dimensional space and hyperspace are introduced in the same way, but we will discuss this in more detail later on. A "change of coordinates" in Fig.2.4-4a would include a transition to Figs.2.4-5b or 2.4-7, whereas a change of coordinates in Fig.2.4-4b would usually be understood to mean a change of the painted coordinate lines but not a change of the surface, for example, to the sphere of Fig.2.4-6. Hence, the concept of surface takes on a meaning of its own, even though it is no more than one obtains by drawing so many rods in Fig.2.4-4a that they touch each other, and making the rods vanishing thin at the same time.

Before we extend the results obtained for two-dimensional, discrete coordinate systems to systems with three and more dimensions, we want to introduce a definition of dimension that can be verified by using no more than the coordinate system itself. In Fig.2.4-3c one can progress from the *corner point* $x/X = 00$, $y/Y = 00$ to the two points $x/X = 01$, $y/Y = 00$

TABLE 2.4-1

CONNECTION OF CORNER, EDGE, AND INTERIOR POINTS THAT TRANSFORM THE BOUNDED COORDINATE SYSTEM OF FIG.2.4-3c INTO THE UNBOUNDED SYSTEM OF FIG.2.4-7. $p = 1, 2$; $q = 1, 2$; $\overline{p} = 3 - p$; $\xi = x/X$; $\eta = y/Y$.

number and name of points	point		connected in bounded coordinate system to two, three, or four points with coordinates ξ, η								additional connections in unbounded coordinate system			
			1		2		3		4		5		6	
	ξ	η	ξ	η	ξ	η	ξ	η	ξ	η	ξ	η	ξ	η
$\binom{2}{2}2^2(4-2)^0 = 4$ corner points	0	0	1	0	0	1					3	0	0	3
	3	0	2	0	3	1					0	0	3	3
	0	3	1	3	0	2					3	3	0	0
	3	3	2	3	3	2					0	3	3	0
$\binom{2}{1}2^1(4-2)^1 = 8$ edge points	0	p	1	p	0	$p-1$	0	$p+1$			3	$\overline{p}$		
	p	0	p	1	$p+1$	0	$p-1$	0			$\overline{p}$	3		
	3	p	2	p	3	$p-1$	3	$p+1$			0	$\overline{p}$		
	p	3	p	2	$p-1$	3	$p+1$	3			$\overline{p}$	0		
$\binom{2}{0}2^0(4-2)^2 = 4$ interior points	p	q	$p-1$	q	$p+1$	q	p	$q-1$	p	$q+1$				

and $x/X = 00$, $y/Y = 01$; hence, the corner point has two neighbors. The *edge point* $x/X = 01$, $y/Y = 00$ has three neighbors $x/X = 00, y/Y = 00$; $x/X = 10$, $y/Y = 00$ and $x/X = 01$, $y/Y = 01$. Finally, the *surface point* $x/X = 01$, $y/Y = 01$ has four neighbors. No point exists that has more than four neighbors. A coordinate system in which the maximum number of neighbor points equals $2n$ will be called a system with n dimensions; the points with $2n$ neighbors will be called *interior points.* If all of the points of the coordinate system have the same number of neighbors they are all interior points, and the system will be called unbounded or closed. One may readily verify that all points in Figs.2.4-4a, 2.4-5b, and 2.4-7 have four neighbors, which makes these structures two-dimensional, unbounded coordinate systems.

Figure 2.4-8 shows a three-dimensional discrete coordinate system with 4^3 marks. The coordinate of point P relative to this system is x/X, y/Y, z/Z if P is closer to the mark with these coordinates than to any other. A coordinate system of this type with m^3 rather than 4^3 marks has the following four kinds of points:

$$\binom{3}{3}2^3(m-2)^0 = 8 \quad \text{corner points}$$

$$\binom{3}{2}2^2(m-2)^1 = 12(m-2) \quad \text{edge points}$$

$$\binom{3}{1}2^1(m-2)^2 = 6(m-2)^2 \quad \text{surface points}$$

$$\binom{3}{0}2^0(m-2)^3 = (m-2)^3 \quad \text{body points}$$

$$m^3 = [2+(m-2)]^3 = \binom{3}{3}2^3(m-2)^0 + \binom{3}{2}2^2(m-2)^1 + \binom{3}{1}2^1(m-2)^2 + \binom{3}{0}2^0(m-2)^3$$

The body points have $2 \times 3 = 6$ neighbors, and are thus interior points. The surface points have five neighbors, the edge points four, and the corner points three. In order to change the bounded coordinate system into an unbounded one, the surface points must get one additional neighbor, the edge points two, and the corner points three. The dashed line between the points 1, 0, 1 and 2, 3, 2 in Fig.2.4-8 shows how the additional neighbor can be provided for surface points on the surface $y/Y = 0$ and $y/Y = 3$, while the dotted and dashed-dotted lines show the same for the surfaces $x/X = 0$, $x/X = 3$ and $z/Z = 0$, $z/Z = 3$.

Contemporary scientific publications on space-time are generally based on the concept of the *manifold* with several dimensions, and it is understood that manifold means a continuous manifold which permits use of differential calculus. The coordinate systems used in this book can just as well be called discrete manifolds, if one wants to stress the connection with the more conventional terminology. The term coordinate system, on the other hand, emphasizes that we are dealing with something man-made rather than provided by nature[10]

Let us look at our results from a very general point of view. We have shown that information theory is at odds with the concept of the space-time continuum and—by implication—with any other kind of continuum. Such a difference must be reconciled. We take here the position that the space-time continuum is a mathematical abstraction that cannot be observed, and that our concept of space-time must be made to conform with information theory. This is a typical point of view for an experimental scientist. An abstract scientist may claim that information theory has to be generalized instead. Which point of view—there may actually be more than those two—is right can only be decided by experience. This book tries to make the experimental point of view plausible, and beyond that it shows in Section 10.5 that one is lead to results that differ from the current theory. This provides the chance to decide by observation whether the theory has merit.

[10]Let us emphasize that the elimination of the space-time continuum implies the elimination of the concept of continuum from everything that is measurable, including probability distributions. The measured value of a function $V_0 f(t)$—where V_0 may represent a voltage—can only assume a finite number of discrete values, as becomes obvious when one uses a digital voltmeter. The information obtained by such a measurement equals m bits if the voltmeter resolves 2^m voltage intervals and the measured voltage may have any value with equal probability. It is all in complete analogy to Fig.2.2-1 except that we would draw the scales there vertically instead of horizontally, and write V instead of X, where V is the largest voltage measurable by the voltmeter. A continuum is never encountered when we make measurements. The primary advantage of the continuum is that it makes the use of differential calculus possible, but this is a mathematical argument not a physical one.

3 Coordinate Systems

3.1 Coordinate Systems Based on Rings

It is general practice to use coordinate systems that have at least one variable that is unbounded. A Cartesian system lets all variables x, y, z, ... go to infinity; a polar or spherical systems permits the radius go to infinity; etc.

We stated at the end of Section 2.2 that the requirement for finite information eliminates from physics the infinitely long as well as the infinitely short. The infinitely short can be avoided by using arbitrarily short but finite distances Δx instead of infinitesimal distances dx between the marks of rulers or along coordinate axes as in Fig.2.4-3 or 2.4-8. To eliminate the infinitely long we introduce finite but unbounded coordinate systems.

Cartesian coordinate systems are based on the numbers axis. The usual ones, permitting infinitesimal distances dx, are based on the numbers axis with all real numbers, while the discrete ones with finite distances Δx are based on the numbers axis of the integers. The numbers axis may be considered to be the limiting case of *numbers rings*. Since numbers rings[1] are finite but unbounded, one can derive finite but unbounded Cartesian-type coordinate systems from them.

Let us first point out that the most common numbers ring is the clock face with numbers 1 to 12. This ring is obviously finite since it does not go beyond 12, but it is also unbounded since the clock does not stop at 12 but goes on to 1. Another example of a ring in everyday use is the magnetic compass with numbers from 1 to 360.

Figure 3.1-1 shows several rings with integer numbers. The smallest one has only the $2^N = 2$ numbers 0 and 1; the next larger has the 3 numbers 0, 1, 2; then comes the ring with the $2^N = 4$ numbers 0, 1, 2, 3. The rings with 5, 6, or 7 numbers are not shown, since we will be primarily interested in rings with 2^N numbers, where N is an integer, since they permit most readily to express the information about the location of a point in bits, in

[1] Books on algebra use instead the expression "rings Z_n of integers modulo n for some integer n".

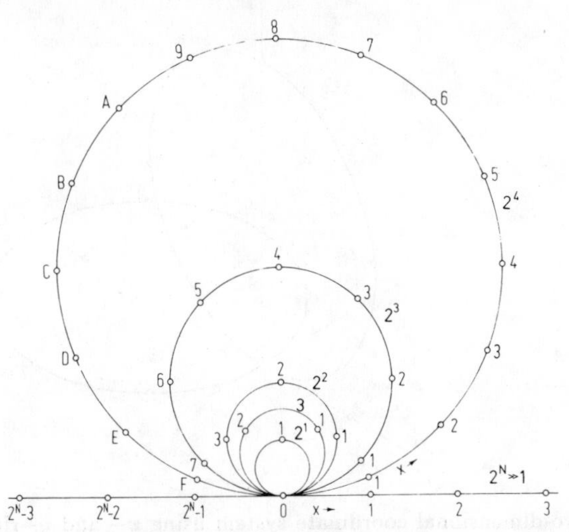

FIG.3.1-1. The numbers rings with $2^N = 2, 3, 4, 8, 16$ numbers, and the limiting case $2^N \gg 1$ that approaches the integer numbers axis.

analogy to Figs.2.2-1, 2.2-2, and 2.2-4.

The ring for $2^N \gg 1$ in Fig.3.1-1 is almost undistinguishable from the integer numbers axis. The only difference are the numbers $2^N - 1$, $2^N - 2$, ... which would be written as $-1, -2, \ldots$ on the integer numbers axis. It is quite evident that negative numbers are a way to avoid the notation $\infty - 1$, $\infty - 2, \ldots$.

The ring with $2^N = 2$ numbers in Fig.3.1-1 is called the binary or dyadic ring. The others are called ternary, quaternary, octonary, and hexadecimal. The hexadecimal ring is widely used for computers—in particular microprocessors— and the notation $0, 1, \ldots, 9, A, B, C, D, E, F$ for the 16 numbers of this ring is generally used.

A two-dimensional Cartesian coordinate system has the $x-$ and the $y-$axis that intersect perpendicularly. The equivalent for rings is shown in Fig.3.1-2. An $x-$ring and a $y-$ring, with $2^N = 8$ numbers or marks each, intersect at a right angle. Also shown are short sections of $x-$ and $y-$rings with many marks, $2^N \gg 1$. Clearly, the usual Cartesian coordinate system can be viewed as the limit of a coordinate system based on rings and a finite number of marks.

A two-dimensional Cartesian coordinate system does not only consist of the $x-$ and the $y-$axis with marks on them, but of a grid obtained by drawing straight lines parallel to the $y-$axis through the marks of the $x-$axis, and straight lines parallel to the $x-$axis through the marks of the

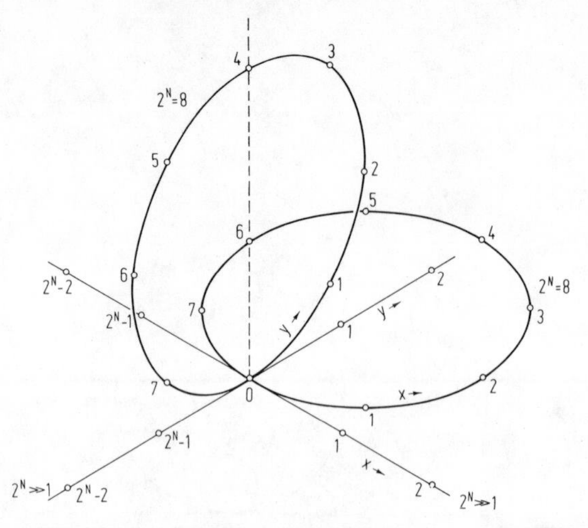

FIG.3.1-2. Two-dimensional coordinate system using $x-$ and $y-$rings instead of $x-$ and $y-$axes.

$y-$axis, as shown in Fig.2.4-3a. This is shown in more detail for the simplest case, only one line each parallel to the $x-$ and the $y-$axis, in Fig.3.1-3. The equivalent for the rings with 8 marks of Fig.3.1-2 is shown in Fig.3.1-4. A ring parallel to the $x-$ring of Fig.3.1-2 is drawn through the mark 1 of the $y-$ring of Fig.3.1-2. Similarly, a ring parallel to the $y-$ring of Fig.3.1-2 is drawn through the mark 1 of the $x-$ring of Fig.3.1-2. The four points[2] $y = 0$, $x = 0$ (0,0=00), $y = 1$, $x = 0$ (1,0=10), $y = 1$, $x = 1$ (1,1=11), and $y = 0$, $x = 1$ (0,1=01) correspond to the four points with the same marking in Fig.3.1-3.

The term *parallel* used here requires some explanation. Let us return to Fig.2.4-3c. A long rod with marks for the $x-$axis ($y/Y = 0$) and a similar one for the $y-$axis ($x/X = 0$) shall be given. These rods define the concept *straight line* for the $x-$ and the $y-$direction. In addition, we shall have many short rods of length $\Delta x = \Delta y$. We can then build the grid of Fig.2.4-3c. If the $x-$ and the $y-$rods are straight in the Euclidean sense, one obtains a grid that defines a plane in the Euclidean sense, and the grid lines $y/Y = 01, 10, 11, \ldots$ are parallel to the $x-$axis $y/Y = 0$ in the Euclidean sense.

Let now the $y-$rod be straight in the Euclidean sense, but let the $x-$rod be a circle. The short rods shall still have the same length $\Delta x = \Delta y$, but the ones used in the direction of the $y-$axis shall be straight like the $y-$axis

[2]We use the notation x, y instead of the more rigorous notation x/X, y/Y used in previous sections in order to simplify the illustrations.

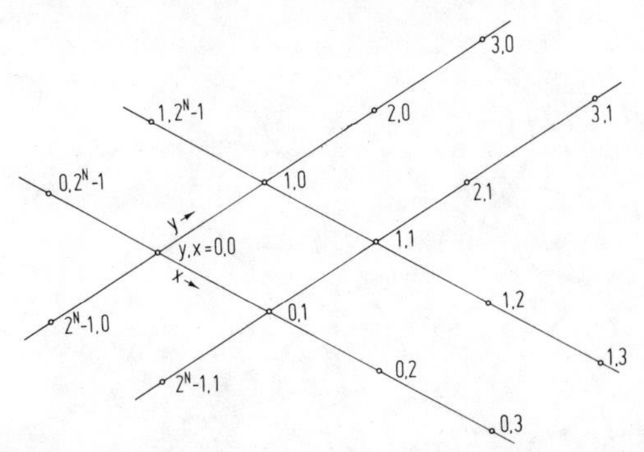

FIG.3.1-3. The $x-$ and the $y-$axes for $2^N \gg 1$ of Fig.3.1-2 supplemented by a grid line $y = 1$ parallel to the $x-$axis, and a grid line $x = 1$ parallel to the $y-$axis.

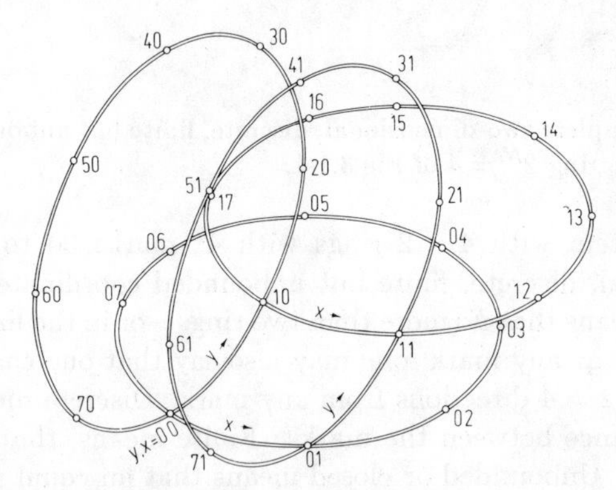

FIG.3.1-4 The analogue of Fig.3.1-3, with the straight axes and grid lines replaced by rings.

while the ones used in the direction of the $x-$axis shall be curved like the $x-$axis. A grid constructed in this way will define the surface of a cylinder rather than a plane. The grid lines $y/Y = 01, 10, 11, \ldots$ are circles parallel to the $x-$axis or $x-$circle $y/Y = 00$, while the grid lines $x/X = 01, 10, 11, \ldots$ are straight lines parallel to the straight $y-$axis $x/X = 0$.

The term parallel is thus used here only to help explain Figs.3.1-3 and 3.1-4. This term is not needed if we complete the coordinate system of Fig.3.1-4. For the completion we would have to draw eight $y-$rings through the marks $x = 0, 1, \ldots, 7$ of the $x-$rings, and eight $x-$rings through the marks $y = 0, 1, 2, \ldots, 7$ of the $y-$rings. A total of 8×2 rings with 8^2 marks is thus required. Such a construction is too difficult to draw. In order to simplify drafting, we do not use the ring $2^N = 8$ in Fig.3.1-1 but the simpler ring $2^N = 4$. Figure 3.1-5 shows the complete two-dimensional

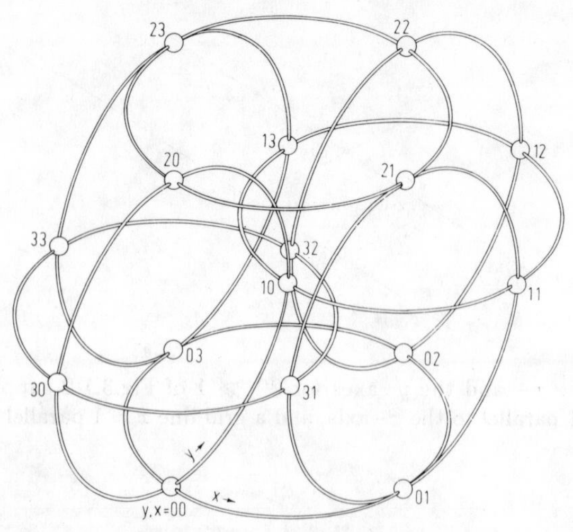

FIG.3.1-5. A complete two-dimensional, discrete, finite but unbounded coordinate system using the ring $2^N = 4$ of Fig.3.1-1.

coordinate system with 4×2 rings with 4^2 marks 00 to 33. This is a two-dimensional, discrete, finite but unbounded coordinate system. Two-dimensional means that no more than two rings—or in the limit two straight lines—intersect at any mark; one may also say that one can proceed in no more than $2 \times 2 = 4$ directions from any mark. Discrete means, that there is a finite distance between the marks. Finite means, that the number of marks is finite. Unbounded or closed means that an equal number of rings intersects at any mark. Every mark is like all the other marks except for its numbering or *address*. There are no boundary points or marks, and there are no singular points or marks.

The coordinate system of meridians and parallel circles used to give a location on the surface of the Earth has singularities at the north and the south pole. Figure 3.1-5 is very different. It does not define any surface in the usual sense. For instance, a rigid frame of meridians and parallel circles yields a spherical surface if a rubber membrane is stretched over it. The 4^2 points 00, ..., 33 in Fig.3.1-5 cannot all be touched by a rubber membrane that one may try to stretch over them. The points 10 and 32 are not touched, all the others are. This statement follows readily from Fig.3.1-6, which shows the same coordinate system as Fig.3.1-5, except that the axonometric view is replaced by a front and top view. Only the points 00, ..., 33 but not the circles that connect them are assumed to hold the rubber membrane. If the circles are rigid too, the points 01, 02, 03, and 20, 21, 23 would not be touched by the rubber membrane either.

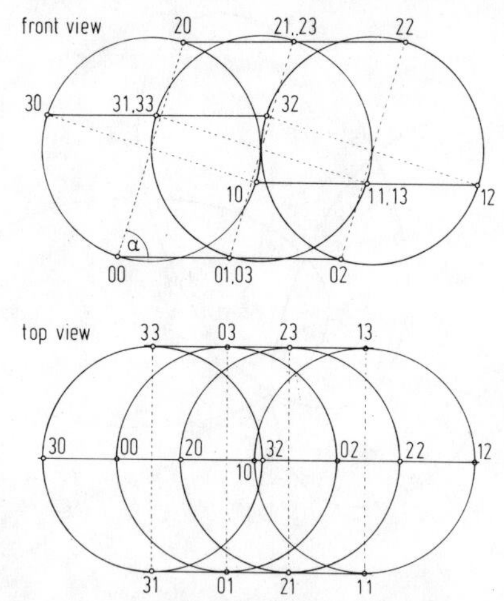

FIG.3.1-6. Front and top view of the coordinate system of Fig.3.1-5. Note that the grid points 10 and 32 would coincide if the angle α were chosen equal to 90°.

The angle $\alpha \neq 90°$ is required in Fig.3.1-6 to avoid that the points 10 and 32 coincide. This brings up the question in which ways this coordinate system can be modified without changing the rings and the marks on them. The Cartesian coordinate systems of Fig.2.4-3 can be changed by choosing the angle between $x-$ and $y-$axis unequal 90°. The result are slanted coordinate systems, like the one shown in Fig.1.3-2. The equivalent for the angle between $x-$ and $y-$axis in Fig.3.1-5 is the angle between the tangents of the $x-$ and $y-$rings in the point[3] $y, x = 00$. In addition, a second angle has to be specified in Fig.3.1-5 but not in Fig.1.3-2. One can choose the angle between the tangent of the $y-$ring in the point $y, x = 00$ and a ray vertical to the tangent of the $x-$ring in the same point, lying in the plane defined by the $x-$ring.

The generalization of coordinate systems based on rings from two to three dimensions is shown in Fig.3.1-7. Three coordinate rings x, y, and z intersect at the point 0. Each ring has $2^N = 8$ marks 0, 1, ..., 7. For $2^N \gg 1$ the three rings approach the usual three coordinate axes x, y, and z.

The drawing of a three-dimensional grid based on rings poses great practical difficulties. If we want to construct a grid with r dimensions we need r rings as in Fig.3.1-7 for $r = 3$ or in Fig.3.1-2 for $r = 2$. If each

[3]The concept of tangent is used here for simplicity. If one wants to avoid a concept based on differential calculus, one could use the length of a circle centered in the point $y, x = 00$ and extended from the point 10 to the point 01 to define an angle.

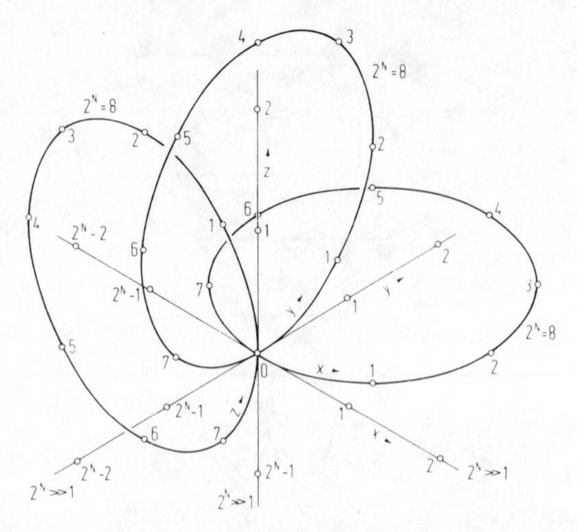

FIG.3.1-7. Three-dimensional coordinate system using $x-$, $y-$, and $z-$rings instead of $x-$, $y-$, and $z-$axes.

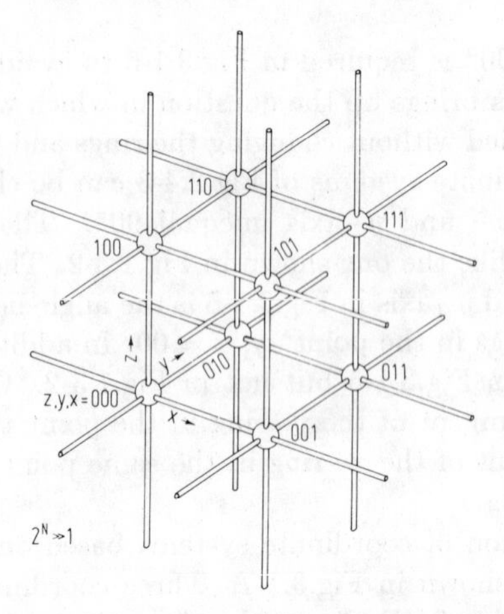

FIG.3.1-8. The $x-$, $y-$, and $z-$axes of Fig.3.1-7 supplemented by 3 grid lines ($x = 0$, $y = 1$; $x = 1$, $y = 0$; $x = 1$, $y = 1$) parallel to the $z-$axis, 3 grid lines ($y = 0$, $z = 1$; $y = 1$, $z = 0$; $y = 1$, $z = 1$) parallel to the $x-$axis, and 3 grid lines ($x = 0$, $z = 1$; $x = 1$, $z = 0$; $x = 1$, $z = 1$) parallel to the $y-$axis.

ring has n marks one needs rn^{r-1} rings for a complete grid. Table 3.1-1 shows how rapidly this number increases with r and n. Figure 3.1-8 shows the smallest possible section of a three-dimensional grid for $2^N \gg 1$ in analogy to Fig.3.1-3. One needs 12 rods or rings to construct this

TABLE 3.1-1
THE NUMBER rn^{r-1} OF RINGS REQUIRED FOR A COMPLETE COORDINATE SYSTEM WITH r DIMENSIONS AND n MARKS PER RING.

r	rn^{r-1}	$n=2$	$n=3$	$n=4$	$n=8$
1	$1n^0$	1	1	1	1
2	$2n^1$	4	6	8	16
3	$3n^2$	12	27	48	192
4	$4n^3$	32	108	256	2048
5	$5n^4$	80	405	1280	20480

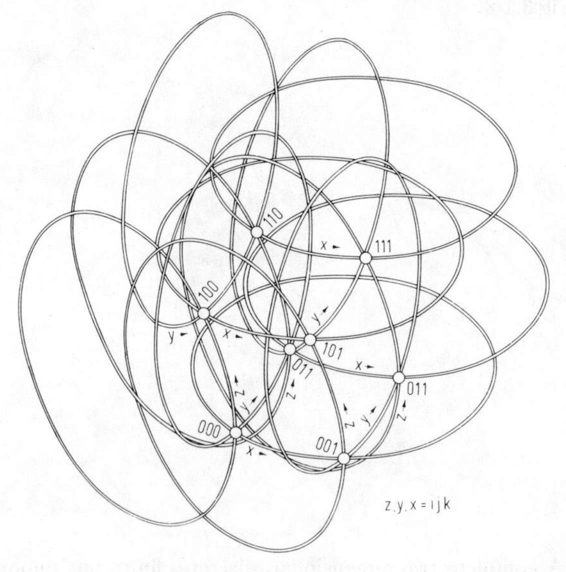

FIG.3.1-9. The analogue of Fig.3.1-8, with the straight axes and grid lines replaced by rings.

section, while Fig.3.1-3 needed only four. The generalization of Fig.3.1-4 from two to three dimensions calls for 12 rings. This generalization for $2^N = 8$ is shown in Fig.3.1-9. Since this illustration is hard to comprehend, the important part of it with the eight points shown in Fig.3.1-8 is shown again in Fig.3.1-10. It is now readily understandable that for large values of 2^N—and correspondingly large values of the radius of the rings—the grid of Fig.3.1-8 is approached. Hence, locally a coordinate system based on rings is like a coordinate system based on straight axes, provided the radius of the rings, or the number 2^N of their marks, is sufficiently large.

For two dimensions we were able to draw a complete grid of coordinates

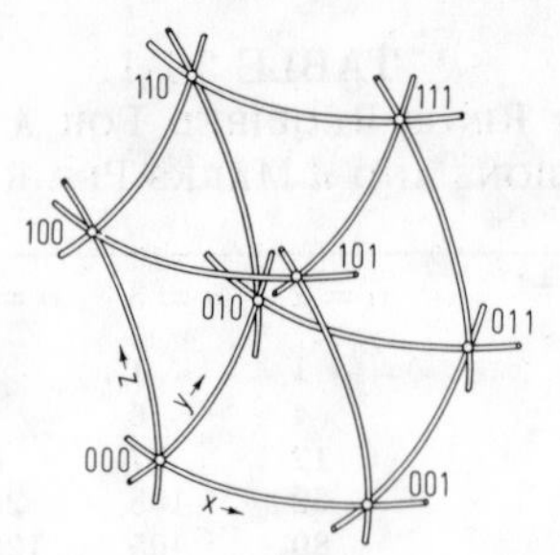

FIG.3.1-10. The section of Fig.3.1-9 containing the eight marks corresponding to the ones in Fig.3.1-8.

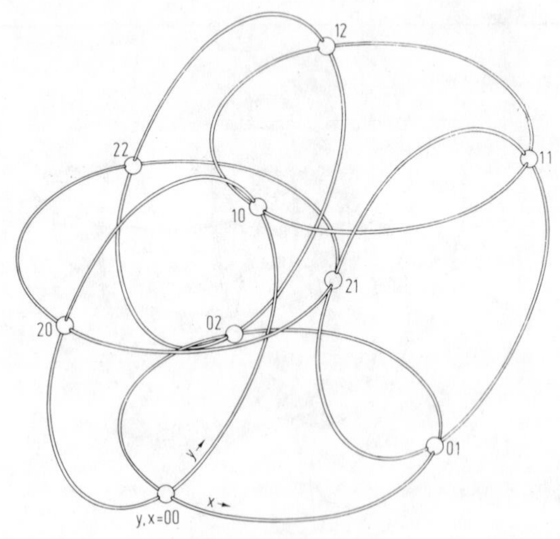

FIG.3.1-11. A complete two-dimensional, discrete, finite but unbounded coordinate system using the ring $n = 3$ of Fig.3.1-1.

for $2^N = n = 4$ marks on a ring in Fig.3.1-5. This drawing required 8 rings. According to Table 3.1-1 one would need 48 rings for the same grid in three dimensions. Such an illustration is too complex to draw or to comprehend. Hence, we choose instead the smaller number of marks $n = 3$. It requires 6 rings in two dimensions and 27 rings in three dimensions according to Table 3.1-1.

Figure 3.1-11 shows a two-dimensional coordinate system with three marks per ring ($r = 2$, $n = 3$). In order to simplify this illustration we substitute straight rods for the circular sections between the marks. This modified drawing is shown in Fig.3.1-12. Note that the use of circles in Fig.3.1-1 is only convenient but not necessary. Any closed curve could be

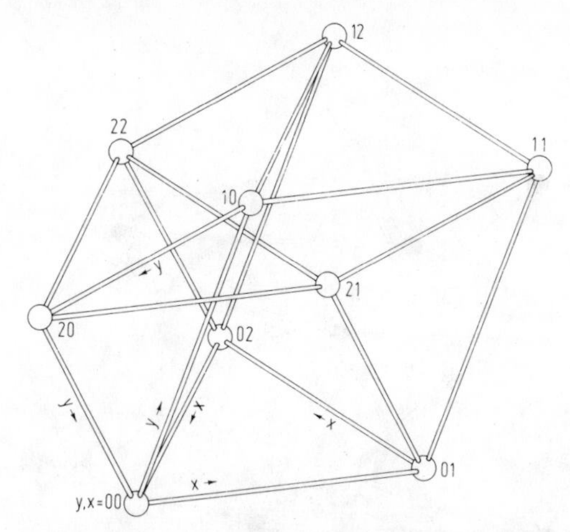

FIG.3.1-12. The coordinate system of Fig.3.1-11 with the circular rods replaced by straight rods.

used too, e.g., a triangle for $n = 3$, a quadrangle for $2^N = 4$, etc.

The extension of Fig.3.1-12 from $r = 2$ to $r = 3$ dimensions is shown in Fig.3.1-13. Consider first the part of the drawing with white spheres, all having the coordinate $z = 0$. This part of the drawing is equal to Fig.3.1-12. Added to it are two more parts distinguished by grey spheres ($z = 1$) and by black spheres ($z = 2$). They are again equal to Fig.3.1-12.

According to Table 3.1-1 there are 27 rings—or rather triangles—in Fig.3.1-13, with $3 \times 27 = 81$ rods and 27 spheres or marks.

A model actually constructed with 81 rods and 27 spheres is shown in Fig.3.1-14. This photograph is less lucid than the drawing of Fig.3.1-13, mainly because of the perspective distortions caused by the short focal length of the camera lens, but also because it is more difficult to see which rod is in front and which behind wherever the rods are crossing.

The structures shown in Figs.3.1-13 and 3.1-14 are three-dimensional, discrete, finite and unbounded coordinate systems. They can obviously be constructed in what we usually call the three-dimensional space. One may readily verify that six rods connect at each sphere, and that the rods between the spheres are all equally long. It is usual to claim that a two-dimensional unbounded surface—such as the surface of the Earth—must be imbedded in a three-dimensional space, and a three-dimensional unbounded "space" must be imbedded in a four-dimensional space. The two-dimensional unbounded coordinate systems of Figs.3.1-5, 3.1-6, 3.1-11, and 3.1-12 clearly require our usual three-dimensional space if we want to construct them with spheres, rings or rods. But the three-dimensional, unbounded structures of

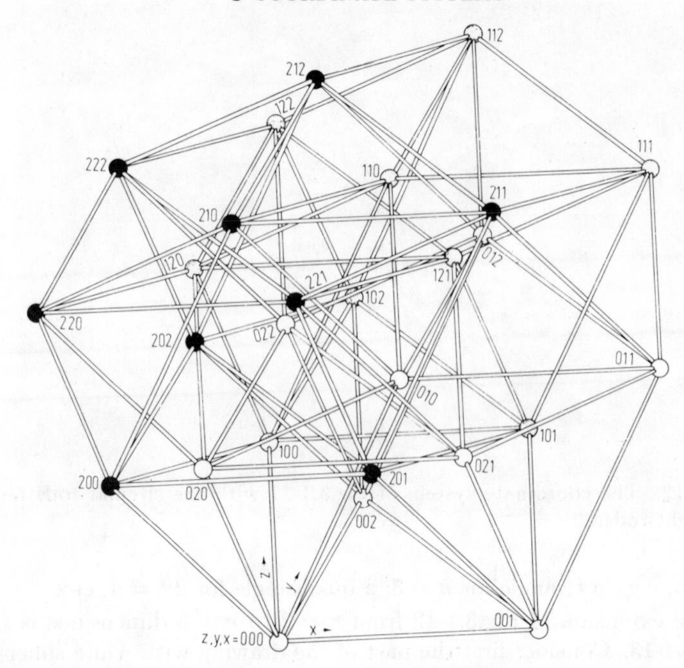

FIG.3.1-13. The generalization of the coordinate system of Fig.3.1-12 from $r = 2$ to $r = 3$ dimensions. The number of marks per ring equals $n = 3$. The marks for $z = 0$ are shown white, those for $z = 1$ grey, and those for $z = 2$ black. This is a complete, three-dimensional, discrete, finite but unbounded coordinate system.

Figs.3.1-13 and 3.1-14 fit into the very same "space". There are two reasons for this.

First, the numbers rings in Fig.3.1-1 show a finite number of points, marks, or numbers only. This finite number is important for the two-dimensional structure of Fig.3.1-5 as well as for the three-dimensional structure of Fig.3.1-13. There is no lucid way to generalize or—better—abstract these structures for denumerably or nondenumerably many points. The usual coordinate system used in connection with differential calculus assumes nondenumerably many points for each coordinate axis. Essentially all our concepts of space are based on the mathematical abstraction of nondenumerably many points and the continuum. The use of a finite number of discrete points leads to very different results.

A second reason is that space is not a physically observable thing like the rods and the spheres of Fig.3.1-14. Hence, its features are not derived from experience and observation but from definitions that came from abstract mathematics. For instance, it is often claimed that the space of our

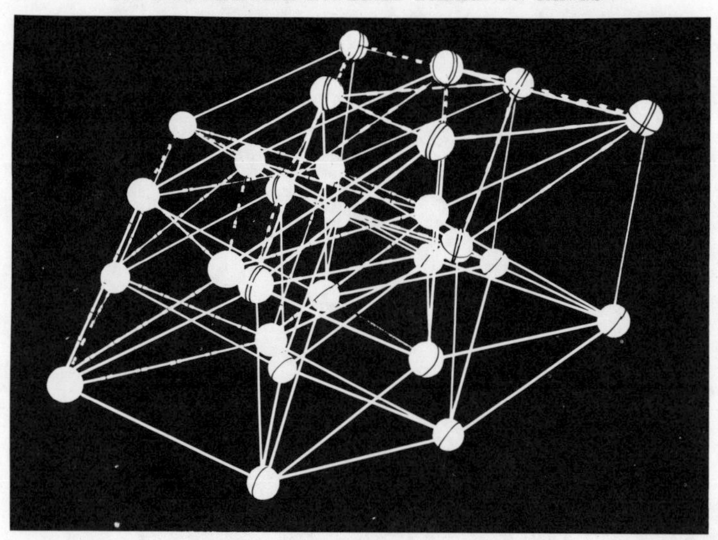

FIG.3.1-14. Photograph of a model of the coordinate system of Fig.3.1-13. The spheres for $z = 0$ have one black ribbon, those for $z = 2$ have two, while those for $z = 1$ have none.

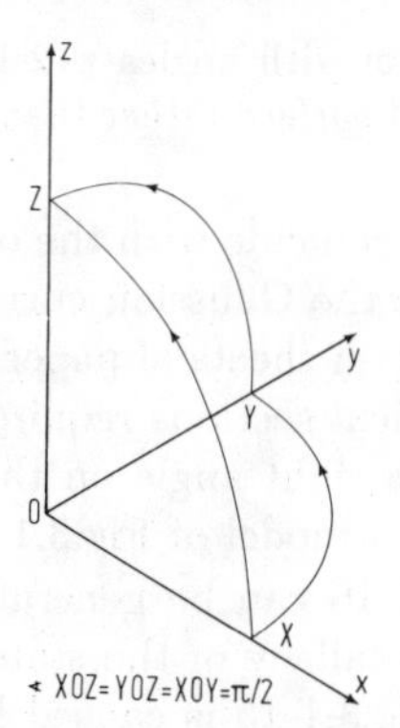

FIG.3.1-15. Three axes x, y, z with angles $\pi/2$ between them are often used to claim that "our space" is three-dimensional.

everyday experience has three dimensions since we can visualize only three vectors perpendicular to each other; a typical illustration for this claim is shown by Fig.3.1-15. Although this illustration looks very persuasive, it is readily disproved by Fig.3.1-16. This illustration shows four vectors or axes x, y, z, w. The right angle between x and z is measured in the plane XOZ, the right angle between y and w in the plane WOY. The right angles between x and y, y and z, z and w, as well as between w and x are measured on

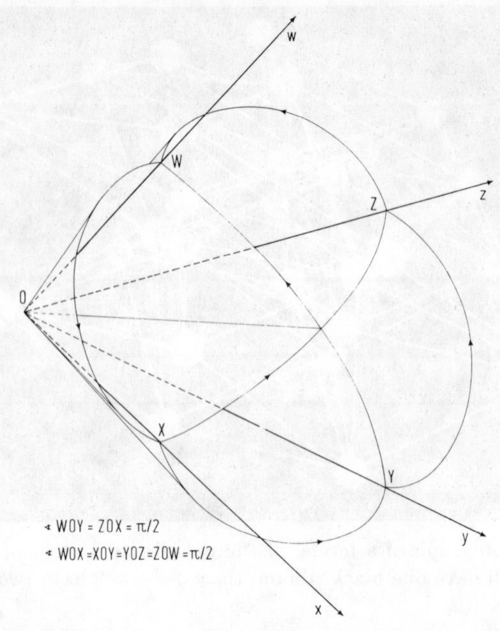

FIG.3.1-16. Four axes x, y, z, w with angles $\pi/2$ between them. Some of the angles are measured on a conical surface rather than on an Euclidean plane.

conical surfaces whose apexes coincide with the origin O of the axes x, y, z, w. Since conical surfaces have the Gaussian curvature zero like a plane, we can plot four right angles first on sheets of paper in a plane, and then bend the paper sheets into the conical sections required for Fig.3.1-16. There is thus no problem in defining a right angle on the conical surfaces. Figure 3.1-17 shows a photograph of a model of Fig.3.1-16 produced in that way.

It is evident that Fig.3.1-16 can be generalized to any finite number of axes or "dimensions". The fallacy of the statement "our space is three-dimensional" derived from Fig.3.1-15 is caused by *defining* first that "our space" must be Euclidean, which then leads to the conclusion that it must be three-dimensional.

We are accustomed from Riemann geometries to consider geometries that are Euclidean at short distances. How can Figs.3.1-16 and 3.1-17 be modified to be three-dimensional, Euclidean close to the origin O of the axes x, y, z, w but four-dimensional, non-Euclidean at some larger distance from O? Let the four axes in Fig.3.1-16 be implemented by rods. The distance between the origin O and the points X, Y, Z, W shall be R. The points X and Y, Y and Z, Z and W, as well as W and X are then connected by wires of length $\pi R/2$ each. These wires may be bent with constant curvature like

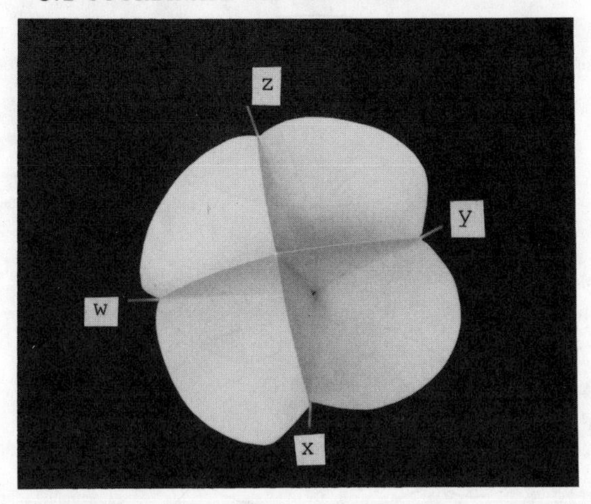

FIG.3.1-17. Photograph of a model with four axes with angles $\pi/2$ between them according to Fig.3.1-16.

a quarter section of the circumference of a circle with radius R, but this is not mandatory. Let now a soap bubble be placed over this structure of four rods and four circular sections of wires. The surface of this soap bubble close to the circular wires will consist essentially of four conical surfaces as in Figs.3.1-16 and 3.1-17. Close to the origin O, however, we will get planar surfaces between the rods x and y, y and z, z and w, as well as w and x. The angles between these rods close to the origin O will be $\pi/3$ or $60°$, while the angles between the rods x and z as well as between w and y will be $\pi/2$ or $90°$, which is what Euclidean geometry in three dimensions requires.

What are the angles measured along the quarter circles of wires between the points X and Y, Y and Z, Z and W, W and X? The distance of these points from the origin O is R. If R is sufficiently large, the surface of the soap bubble close to the quarter circles of wires will be curved like the conical surfaces in Figs.3.1-16 and 3.1-17. The length of the quarter circles is $\pi R/2$, and the angle inferred from this length as well as the distance R from the origin equals $(\pi R/2)/R = \pi/2$, since there is locally no difference between the models of Figs.3.1-16 and 3.1-17 and the soap bubble model close to the quarter circles between the points X and Y, Y and Z, Z and W, W and X. The angles between x and z as well as between y and w, measured along the quarter circles from X to Z and W to Y, yield also $(\pi R/2)/R = \pi/2$. Hence, we have a non-Euclidean geometry in four dimensions at a sufficiently large distance R from the origin O.

As another example for our ability to construct and visualize structures with more than three dimensions, let us construct a four-dimensional,

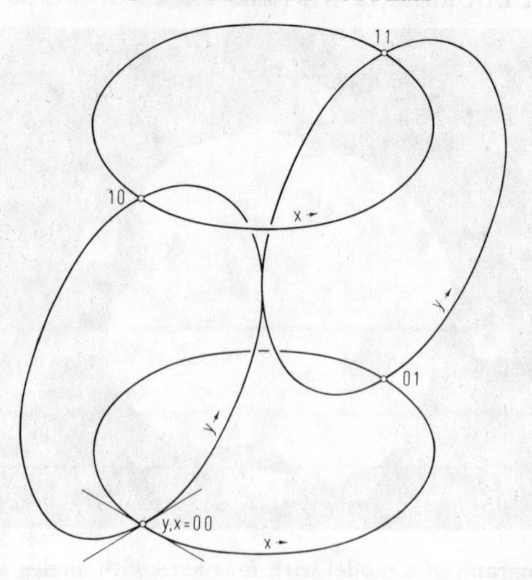

FIG.3.1-18. Two-dimensional dyadic coordinate system based on the ring $2^N = 2$ of Fig.3.1-1.

unbounded coordinate system. Since the complexity of multidimensional structures increases rapidly with the number of dimensions, we choose the simplest ring, which has only the points 0 and 1 according to Fig.3.1-1. Even then we need 32 rings for a four-dimensional coordinate system according to Table 3.1-1.

We proceed step by step. Figure 3.1-18 shows a two-dimensional coordinate system based on the ring with two points 0 and 1, or a *two-dimensional dyadic coordinate system.* We call it two-dimensional because two rings intersect at the points y, x = 00, 01, 10, 11. It is an unbounded coordinate system, because two rings intersect at *all* the points.

The extension to three dimensions is shown in Fig.3.1-19. This illustration is somewhat overwhelming. The reader may want to look first at the three rings of Fig.3.1-7, and associate them with the three rings intersecting at the point z, y, x = 000 in Fig.3.1-19. Then one may verify that there are three more rings parallel to the $x-$ring, three more parallel to the $y-$ring, and three more parallel to the $z-$ring. This adds up to the 12 rings shown for $r = 3$, $n = 2$ in Table 3.1-1. One may finally verify that the coordinate system is unbounded, since three rings intersect at each one of the 8 points 000 to 111. The term "parallel" may be dropped once the structure is understood, since bending and twisting would not change the fact that there are 8 coordinate points with 3 neighbors each.

To extend Fig.3.1-19 from 12 to 32 rings required for four dimensions appears impractical. A simpler representation is required. Figure 3.1-20a

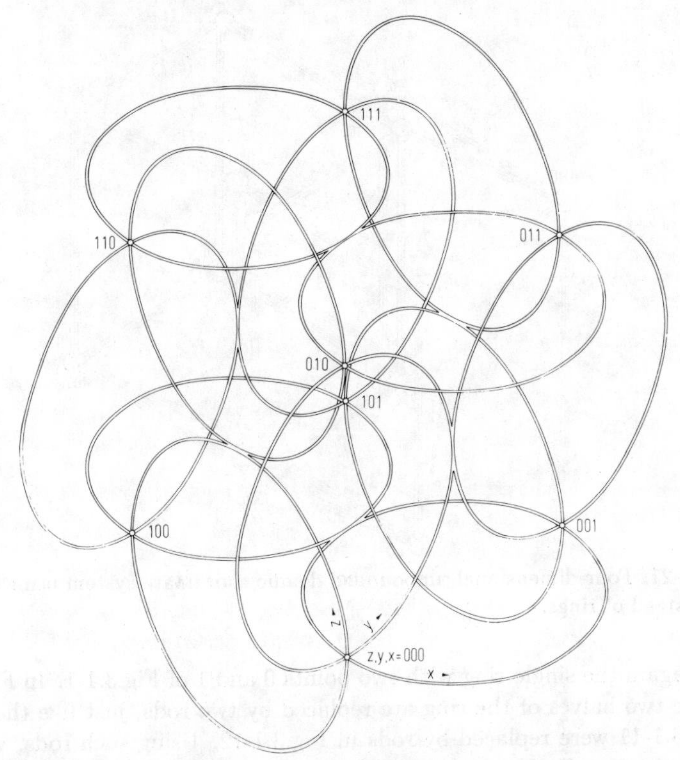

FIG.3.1-19. Three-dimensional dyadic coordinate system based on the ring $2^N = 2$ of Fig.3.1-1.

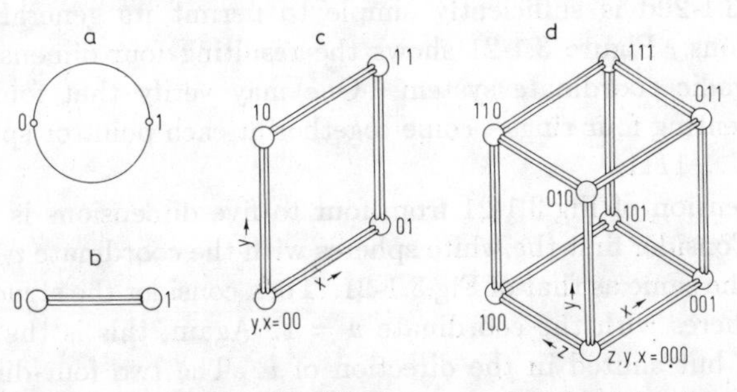

FIG.3.1-20. Development of a simplified representation of dyadic coordinate systems; (a) one dimension using a ring; (b) one dimension using double rods; (c) two dimensions as in Fig.3.1-18 but using double rods; (d) three dimensions as in Fig.3.1-19 but using double rods.

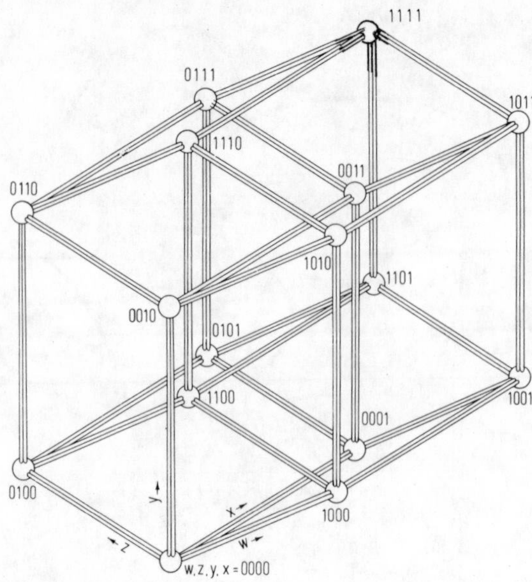

FIG.3.1-21. Four-dimensional, unbounded dyadic coordinate system using double rods instead of rings.

shows again the single ring with two points 0 and 1 of Fig.3.1-1. In Fig.3.1-20b the two halves of the ring are replaced by two rods, just like the rings of Fig.3.1-11 were replaced by rods in Fig.3.1-12. Using such rods, we can redraw the two-dimensional coordinate system of Fig.3.1-18 in the form of Fig.3.1-20c, while Fig.3.1-19 is redrawn into Fig.3.1-20d.

Figure 3.1-20d is sufficiently simple to permit its generalization to four dimensions. Figure 3.1-21 shows the resulting four-dimensional, unbounded, dyadic coordinate system. One may verify that four pairs of rods—representing four rings—come together at each point or sphere w, z, y, $x = 0000 \ldots 1111$.

The extension of Fig.3.1-21 from four to five dimensions is shown by Fig.3.1-22. Consider first the white spheres with the coordinate $v = 0$. This structure is the same as that of Fig.3.1-21. Then consider the structure with the black spheres with the coordinate $v = 1$. Again, this is the structure of Fig.3.1-21 but shifted in the direction of v. The two four-dimensional structures are connected by the double-rods v to produce a five-dimensional, unbounded dyadic coordinate system.

Note that 5 double rods come together at each sphere in Fig.3.1-22. In a more elaborate drawing we would have five rings intersect at each one of the 32 spheres or coordinate marks. At any coordinate mark one could proceed

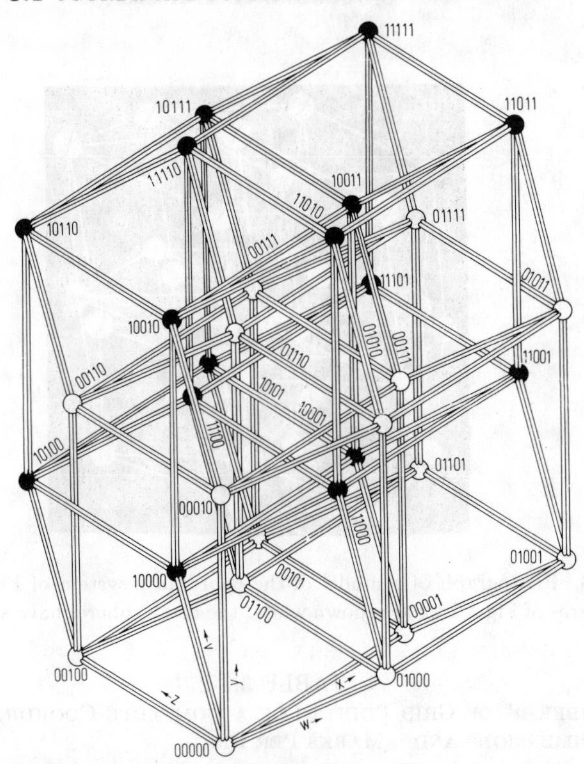

FIG.3.1-22. Five-dimensional, unbounded dyadic coordinate system. The spheres for $v = 0$ are shown white; this part of the coordinate system is equal to the one of Fig.3.1-21. The spheres for $v = 1$ are shown black.

in 10 directions to a neighbor, which is our definition $n/2 = 10/2 = 5$ for the dimension of a coordinate system. Note that there are always two ways to reach the same neighbor, since neighbors are double connected. This feature of dyadic coordinate systems is immediately evident from Fig.3.1-1, which shows that one can go clockwise or counter clockwise around the circle to reach 1 from 0 or 0 from 1.

A photograph of a model of Fig.3.1-22 is shown in Fig.3.1-23. The black spheres of Fig.3.1-22 are represented by the white spheres with a black ribbon. Only single rods are used instead of the double rods shown in Fig.3.1-22; according to Table 3.1-1, one saves 80 rods by using only one rather than two of them.

As a further example of a structure with more than three dimensions, let us design a four-dimensional Cartesian coordinate system. These coordinate systems hold for the limit $2^N \to \infty$ in Fig.3.1-1. Using rods and

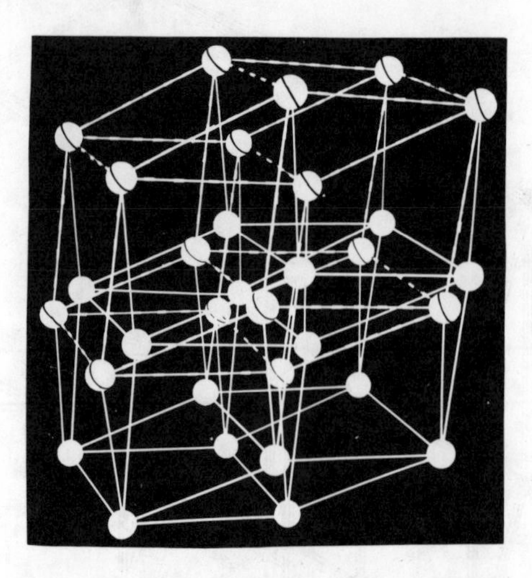

FIG.3.1-23. Photograph of a model of the coordinate system of Fig.3.1-22. The white spheres of Fig.3.1-22 are shown white, the black spheres have a black ribbon.

TABLE 3.1-2

THE NUMBER n^r OF GRID POINTS FOR A COMPLETE COORDINATE SYSTEM WITH r DIMENSIONS AND n MARKS PER RING.

r	n^r	$n=2$	$n=3$	$n=4$	$n=8$
1	n^1	2	3	4	8
2	n^2	4	9	16	64
3	n^3	8	27	64	512
4	n^4	16	81	256	4096
5	n^5	32	243	1024	32768

spheres we draw first the three-dimensional coordinate system with $3^3 = 27$ grid points in Fig.3.1-24. This is a readily understandable illustration. It shows three axes intersecting at the point $z, y, x = 1, 1, 1$, which is the justification for the term three-dimensional. Other points, like 1, 0, 1, show two intersecting axes and one ending axis; these are surface points. One intersecting axis and two ending axes distinguish edge points, e.g., 1, 0, 0. Finally, three ending axes distinguish corner points, e.g., 0, 0, 0 or 2, 0, 2. The number 3^3 of grid points used for this coordinate system is the smallest possible one to yield at least one point with three intersecting axes. The importance of this remark will become evident when we advance to four

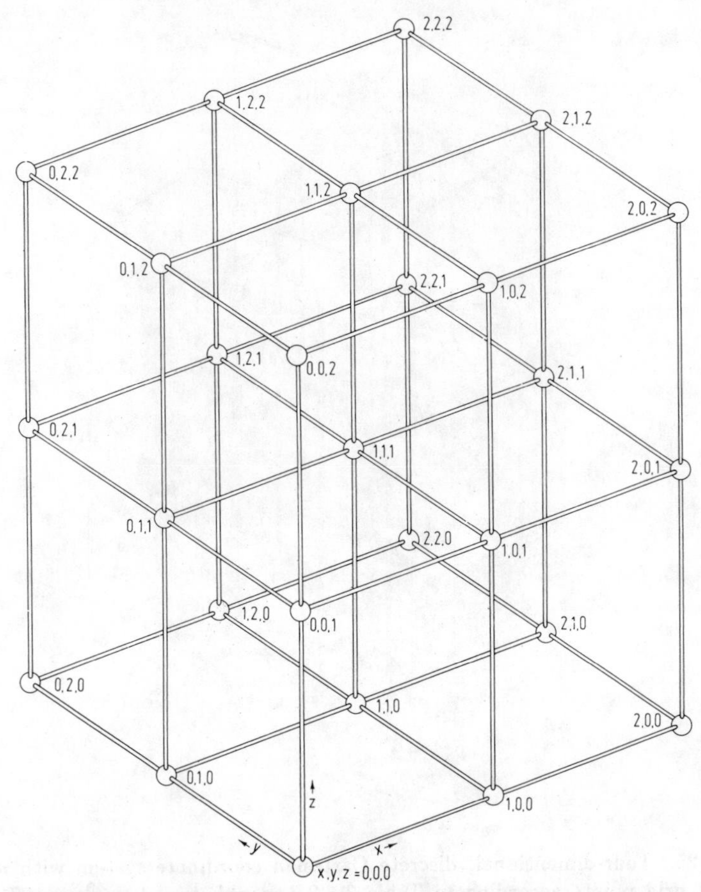

FIG.3.1-24. Three-dimensional, discrete Cartesian coordinate system with $n^r = 3^3 = 27$ grid points according to Table 3.1-2 and $rn^{r-1} = 3 \times 3^2 = 27$ rods according to Table 3.1-1.

dimensions[4].

The generalization of Fig.3.1-24 from three to four dimensions requires three structures according to Fig.3.1-24, each having 27 grid points and 27

[4]Figure 3.1-24 has caused misunderstandings. For instance, a reviewer wrote: " I am very puzzled by the idea of the 'number of rods coming together at a point' differing from point to point. Surely these spaces are topologically homogeneous? Should there not, for example, be a rod directly connecting points 1, 0, 2 and 1, 2, 2 in Fig.3.1-24?" The answer is that the structure of Fig.3.1-24 is a coordinate system constructed with rods. There is no space, homogeneous or otherwise. The structure is intended to show how we are forced to either extend it to infinity or to make it closed, to avoid the problem of a different number of rods coming together at various points.

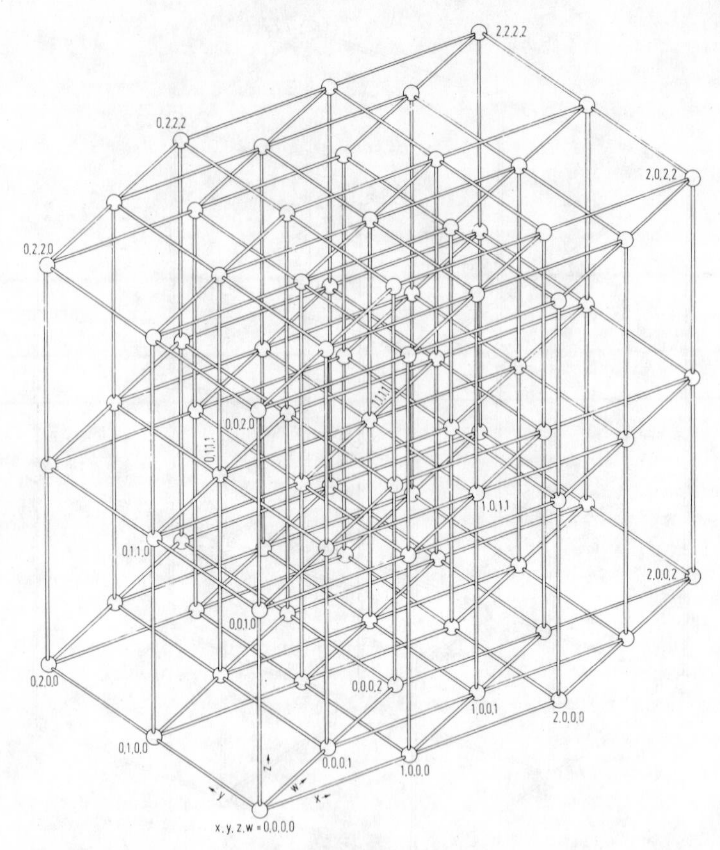

FIG.3.1-25. Four-dimensional, discrete Cartesian coordinate system with $n^r = 3^4 = 81$ grid points according to Table 3.1-2 and $rn^{r-1} = 4 \times 3^3 = 108$ rods according to Table 3.1-1.

rods. An additional 27 rods are needed in the direction of the new $w-$axis connecting the three structures. This adds up to a total of $3 \times 27 = 81$ grid points and $3 \times 27 + 27 = 108$ rods, in accordance with Tables 3.1-2 ($n^r = 3^4$) and 3.1-1 ($rn^{r-1} = 4 \times 3^3$). Such a coordinate system is shown in Fig.3.1-25. Since this is again a not-so-lucid structure, the following help is given:

a) One structure according to Fig.3.1-24 starts at the point $w, z, y, x = 0, 0, 0, 0$, and consists of the $x-$, $y-$, and $z-$axes.
b) A second such structure starts at the point 1, 0, 0, 0.
c) A third such structure starts at the point 2, 0, 0, 0.

In the center of Fig.3.1-25 one may see the point 1, 1, 1, 1 where four axes intersect, and which justifies the term four-dimensional coordinate system. Three intersecting axes and one ending axis distinguish body points (e.g., 1, 1, 1, 0); two intersecting axes and two ending axis distinguish surface points (e.g., 0, 1, 1, 0); one intersecting axis and three ending axis distinguish edge points (e.g., 0, 0, 0, 1); finally, four ending axes distinguish corner points (e.g., 0, 0, 0, 0; 2, 2, 2, 2).

If we had not used the required minimum of $n = 3$ points per axis but $n = 4$ points, the structure of Fig.3.1-25 would have increased to $4^4 = 256$ grid points and $4 \times 4^3 = 256$ rods, but we would have obtained 16 points with four intersecting axes rather than the one point 1, 1, 1, 1 in Fig.3.1-25.

Since we are so used to see a coordinate system "on a two-dimensional paper surface" or "in our three-dimensional space" it often helps one's thinking to look at the coordinate system alone without reference "on what" or "in what" it exists. Let the coordinate system of Fig.3.1-24 be welded together from wires that represent the rods, and let a little ant creep along the rods. The ant can observe local features of the coordinate system only. In particular, it can count the number of rods coming together at any point. The maximum number of rods is found to come together at the point 1, 1, 1. Since this number is 6, the coordinate system is classified as three-dimensional. In Fig.3.1-25, on the other hand, there are 8 rods intersecting at the point 1, 1, 1, 1, and the ant will conclude that this must be a four-dimensional coordinate system.

Let the ant be located at the point 1, 1, 1 in Fig.3.1-24. To keep it imprisoned there one may put traps or glue on the 6 rods leading away from this point. These 6 barriers would be insufficient for an ant located at the point 1, 1, 1, 1 in Fig.3.1-25. If only 6 rods are barred that correspond to the 6 rods leading away from the point 1, 1, 1 in Fig.3.1-24—which are the ones leading to the points with coordinate $w = 1$—there would still be two passable rods leading to the points $x, y, z, w = 1, 1, 1, 0$ and $x, y, z, w = 1, 1, 1, 2$. Hence, the ant can escape the three-dimensional prison by using the fourth dimension.

The example of a prisoner escaping a prison cell through the fourth dimension is often cited to show our inability to visualize four-dimensional phenomena. Obviously, the example achieves this goal only if one thinks in terms of a continuum. It is not really surprising that a deduction based on a nonobservable mathematical abstraction is not understandable—or visualizable—in physical terms. Many more things are thinkable than observable. The struggle to distinguish between thinkable and observable concerned once religious concepts, but at the current time it concerns mathematical concepts.

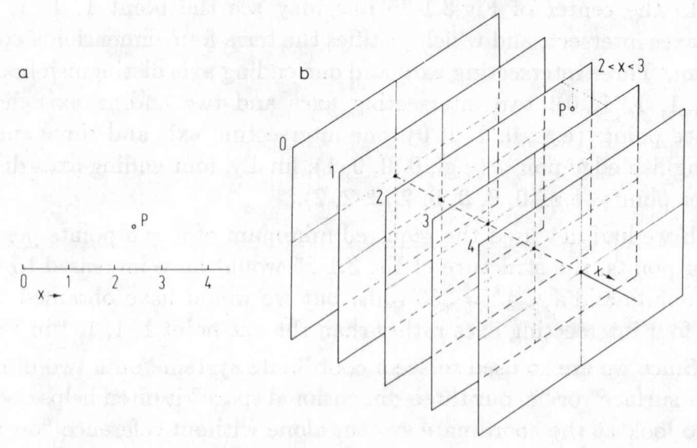

FIG.3.2-1. A ruler defines a one-dimensional interval $2 < x < 3$ in which the point P is located.

3.2 THE BELIEF IN THREE-DIMENSIONAL SPACE

Consider a ruler as shown in Fig.3.2-1 with marks $x = 0, 1, 2, \ldots$. A point P is said to be in the interval $2 < x < 3$ if it is at the location shown. Since P is not actually *on* the ruler, we visualize planes perpendicular to the axis x at the location of the marks $x = 0, 1, \ldots$ as shown in Fig.3.2-1b. A point between the planes at $x = 2$ and $x = 3$ is then in the interval $2 < x < 3$.

Instead of saying that P is in the interval $2 < x < 3$ we could just as well say that P is closer to the point $x = 2.5$ than to any other of the points $x = 0.5, 1.5, 2.5, 3.5, \ldots$. The marks of the ruler should in this case be at the locations $x = 0.5, 1.5, 2.5, \ldots$ rather than at $x = 0, 1, 2, \ldots$, but nothing else is changed. The statements "P is in the interval between two marks m and $m + 1$" and "P is closest to the mark $m + 1/2$" are so far equivalent.

Although the location of P in Fig.3.2-1 has been restricted to the interval $2 < x < 3$, the distance of P from the x-axis may be arbitrarily large, if we follow the usual mathematical abstraction and extend the planes in Fig.3.2-1b to infinity. In readily understandable terminology one may say that the point P in Fig.3.2-1b is located in a volume of the order ∞^2, if it is known to be in the interval $2 < x < 3$. In an experimental science one cannot extend these planes to infinity quite so readily, and one cannot simply assume that the volume between two planes is infinite. If the volume of the universe is finite, the volume between two planes in Fig.3.2-1b must be finite too. We have particularly excluded the infinitely short and

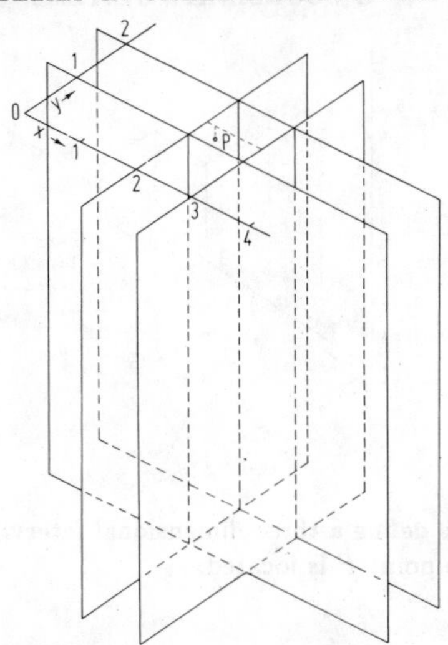

FIG.3.2-2. Two rulers define a two-dimensional interval $2 < x < 3$, $1 < y < 2$ in which the point P is located.

the infinitely long as unobservable, and that prohibits the assumption of the planes in Fig.3.2-1b extending to infinity. The problem is completely avoided by stating "P is closest to the mark $m + 1/2$" rather than "P is in the interval between two marks m and $m+1$", since the concept *interval* introduces the planes in Fig.3.2-1b and the volume between two planes. Only the concept of a finite distance is required by the statement "P is closest to the mark $m + 1/2$".

Let us advance from one ruler to two rulers or a two-dimensional coordinate system. Figure 3.2-2 shows on top the two coordinate axes x and y with the marks 0, 1, 2, ... The two planes at $x = 2$ and $x = 3$ are shown as in Fig.3.2-1b. In addition, two planes perpendicular to the y−axis at the points $y = 1$ and $y = 2$ are shown. The location of the point P may now be specified as the two-dimensional interval $2 < x < 3$, $1 < y < 2$ with a volume of the order ∞. Alternately, one may state that P is closest to the mark $x = 2.5$, $y = 1.5$ of a two-dimensional coordinate system, and avoid thus the nonobservable infinite volume.

The advance from a two-dimensional to a three-dimensional coordinate system is shown in Fig.3.2-3. The infinitely long cylinder with square cross section of Fig.3.2-2 is intersected by two planes perpendicular to the z−axis at the points $z = 3$ and $z = 4$. The resulting three-dimensional interval or cube $2 < x < 3$, $1 < y < 2$, $3 < z < 4$ is shown in Fig.3.2-3. The volume of this cube is finite. Hence, if we assume infinite planes through

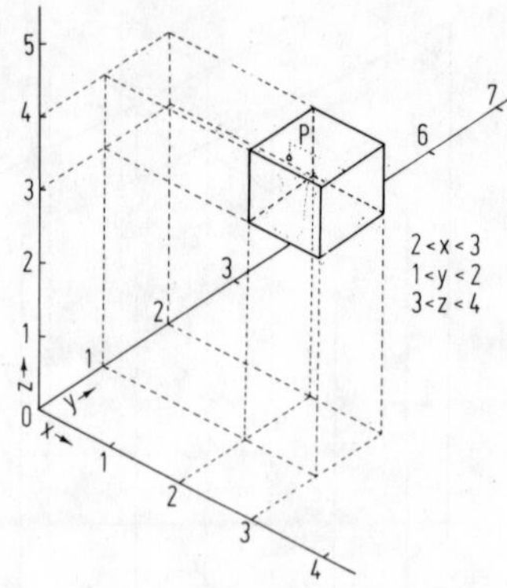

FIG.3.2-3. Three rulers define a three-dimensional interval $2 < x < 3$, $1 < y < 2$, $3 < z < 4$ in which the point P is located.

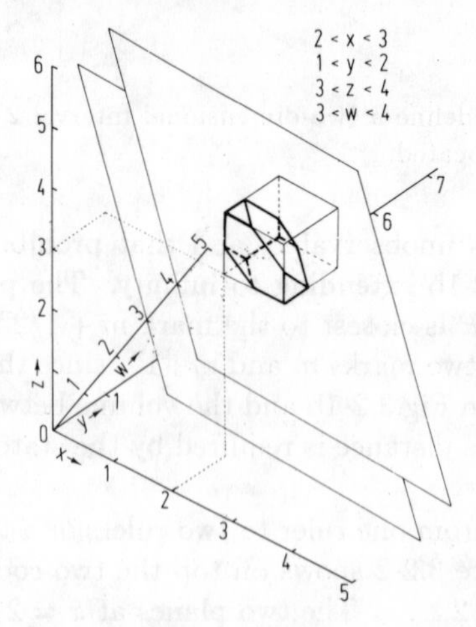

FIG.3.2-4. Four rulers define a four-dimensional interval $2 < x < 3$, $1 < y < 2$, $3 < z < 4$, $3 < w < 4$ in which the point P is located.

the marks of the rulers in Figs.3.2-1, 3.2-2, and 3.2-3, the three-dimensional coordinate system is distinguished, since it is the one of lowest order that permits to specify the location of the point P within a finite volume. On the other hand, we may state that "P is closest to the mark $x = 2.5$, $y = 1.5$, $z = 3.5$ of a three-dimensional coordinate system", and the distinction of the three-dimensional system disappear.

Let us go on to a coordinate system with four dimensions, as shown in Fig.3.2-4. The axis w is added. Two planes perpendicular to this axis and intersecting it at the points $w = 3$ or $w = 4$ are shown. These two planes

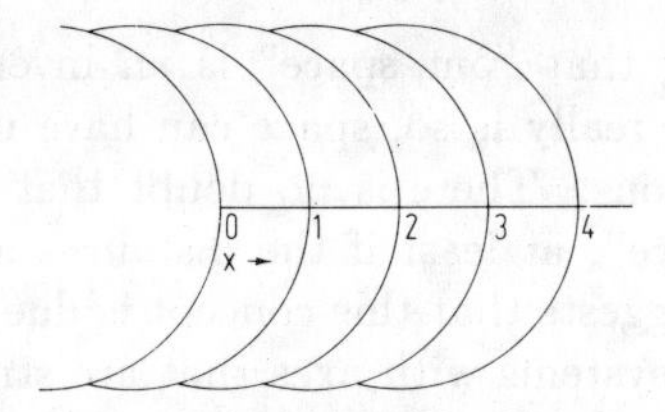

FIG.3.2-5. Replacement of the infinite planes in Fig.3.2-1b by the surfaces of spheres.

cut a section out of the cube of Fig.3.2-3. The point P is now specified in the four-dimensional interval $2 < x < 3$, $1 < y < 2$, $3 < z < 4$, $3 < w < 4$. The volume of this four-dimensional interval is smaller than that of the three-dimensional interval in Fig.3.2-3, but the reduction is no longer by a factor $1/\infty$ as it was in the case of three or fewer dimensions. Hence, three dimensions are highly distinguished if we base the discussion on intervals, planes, and volumes. On the other hand, we can say that P in Fig.3.2-4 is closest to the mark $x = 2.5$, $y = 1.5$, $z = 3.5$, $w = 3.5$, and there is no distinction of three dimensions. The distinction of three dimensions comes from our using Cartesian coordinate systems with straight axes and planes in Fig.3.2-1, 3.2-2, and 3.3-3. These are elements of Euclidean geometry. In connection with Figs.3.1-15 to 3.1-17 we had already pointed out that the assumption of Euclidean geometry leads to the conclusion that we live in a three-dimensional "space". We have now here a second example of Euclidean geometry distinguishing three dimensions, while the distinction disappears if we avoid the concepts of Euclidean geometry.

To show more clearly how the distinction of the three-dimensional "space" depends on Euclidean geometry, let us replace the planes in Fig.3.2-1b by surfaces of large spheres, as shown in Fig.3.2-5. The volume between two spherical surfaces is finite. Hence, if we substituted spherical surfaces for the planes in Figs.3.2-2, 3.2-3, and 3.2-4 the conspicuous reduction of the volume in which the point P was located from infinite in Fig.3.2-2 to finite in Fig.3.2-3 would not occur.

Euclidean geometry has exerted an incredibly strong influence on human thought. We have already discussed in Section 1.2 its influence before Bolyai and Lobachevskii. Not so well known is that Kepler devoted much time to linking the number of planets with the five regular solids of Euclidean geometry[1] (Kepler, 1596). In Section 1.4 we had quoted Einstein

[1] There are five regular solids in Euclidean geometry. Listed according to their number of faces they are the tetrahedron, the cube, the octahedron, the dodecahedron, and the icosahedron. There were also five intervals between the six planets Mercury, Venus, Earth, Mars, Jupiter, and Saturn known at the time of Kepler. Attempts to find

and Schlick as saying that "our space" is an invention without physical reality (p. 15). If this really is so, space can have no such observable feature as three dimensions. There is no doubt that we assign this feature generally to "our space", at least if the distances are not too large. Our discussion strongly suggests that this concept is due to the universal use of Cartesian coordinate systems with axes that are straight in the Euclidean sense, are infinitely long, and have the distance dx between adjacent marks[2]. Such a coordinate system with three axes permits us to describe many observed phenomena in a satisfactory way, and it assumed an identity well beyond that of a mathematical tool as "our space". It is a common experience that mathematical concepts become physical concepts, although we are aware that mathematics can only provide models for the description of physical observations, and only experience derived from observations can decide whether a mathematical model is useful for physics[3]. Nevertheless, Euclidean geometry defined the metric of "our space" for a long time, the mathematical concept of the continuum is believed almost generally to define the topology of "our space" at the present time, and the Cartesian coordinate system with Euclidean straight axes and distances dx between adjacent marks appears to define "our space". Reversing the sequence of this sentence, the definition of "our space" by such a coordinate system leads logically to the topology of the continuum and the Euclidean metric.

A group of phenomena, usually referred to as *mirror images*, is often believed to provide us with observations of the concept of dimension. For instance, the letter L and its mirror image, which is the same letter facing left, cannot be transformed into each other by shifts and rotations in the two-dimensional paper plane, but this transformation is possible if a movement through the third dimension is permitted. We shall discuss this group

---

a link between the number of regular solids and the number of intervals between the planets came to an end with the discovery of Uranus in 1781. See the chapters "The Cosmic Mystery" (Koestler, 1968) or "Weltgeheimnis" (Zinner, 1951).

[2]The expression "Distance dx between adjacent marks" implies that any finite distance is divided into nondenumerably many intervals. We have to add that in addition to these features of the coordinate system one also has to assume the measurement of distances according to the Pythagorean concept $s^2 = x^2 + y^2 + z^2$. This concept and its connection with the rotation of the coordinate system was discussed briefly in connection with Fig.1.3-1, and will be discussed more fully in Section 3.4. In connection with Fig.2.4-4 we have discussed how a coordinate system constructed of rods becomes a surface with coordinate lines painted on it, if one uses nondenumerably many rods with diameter dx. We have here the three-dimensional analogue.

[3]An excellent example is a paper by Penney (1965). It points out that Maxwell's equations, Dirac's equations, and Einstein's equations for empty space favor four dimensions for space-time. The question then arises, whether this fact proves that these mathematical models are good for an assumed four-dimensional space-time (for which the equations were written), or whether it proves that there is a space-time and that it is four-dimensional.

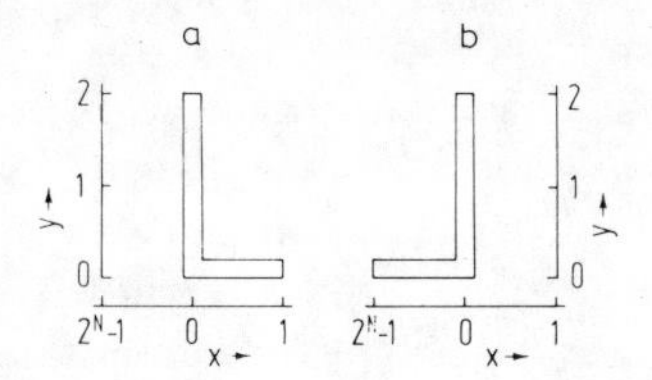

FIG.3.3-1. The letter L and its mirror image in a two-dimensional Cartesian coordinate system.

of phenomena in the following section.

3.3 Right and Left Handed Structures

One of the more conspicuous features of a *mathematical* space with dimension $n+1$ is that it permits to transform an object in a space with n dimensions into its mirror image. Consider the letter L in the two-dimensional coordinate system of Fig.3.3-1a. Its mirror image is shown in Fig.3.3-1b. We cannot transform L by any rotations or shifts in the $xy-$plane into its mirror image, However, we can perform this transformation by "flipping L through the third dimension".

Consider now the same process in a two-dimensional, finite but unbounded coordinate system according to Fig.3.1-2. We show this coordinate system again in Fig.3.3-2, but with the $y-$coordinate ring through the point $x = 2^N/2 = 4$ added. The letter L at the location $x = 0$, $y = 0$ extends along the $x-$coordinate ring from $x = 0$, $y = 0$ to $x = 1$, $y = 0$; along the $y-$ring it extends from $x = 0$, $y = 0$ to $x = 0$, $y = 2$.

One of the letters L in Fig.3.3-2 shall be moved along the $x-$ring to the location of the other letter L. These two letters will coincide. On the other hand, if we transport one of the letters "through the three-dimensional space" to the location of the other, they will be each other's mirror image. The situation is thus very much as in Fig.3.3-1.

We turn to a three-dimensional, finite but unbounded coordinate system according to Fig.3.1-7. Its essential parts are shown in Fig.3.3-3. The $x-$coordinate ring is the same as in Fig.3.1-7; only sections of the $y-$ and $z-$coordinate rings through the point $x = 0$, $y = 0$, $z = 0$ are shown. Also shown are sections of the $y-$ and $z-$rings through the point $x = 2^N/2 = 4$, $y = 0$, $z = 0$.

Instead of the letter L we use now a three-dimensional tripod. It's center is at $x = 0$, $y = 0$, $z = 0$, while the ends of its three legs are at the points (a) $x = 1$, $y = 0$, $z = 0$; (b) $x = 0$, $y = 1$, $z = 0$; (c) $x = 0$, $y = 0$, $z = 1$. The more familiar axes of a three-dimensional, right handed, Cartesian coordinate system are shown in Fig.3.3-3 close to the point $x = 0$, $y = 0$, $z = 0$ to help with the understanding of the curved tripod.

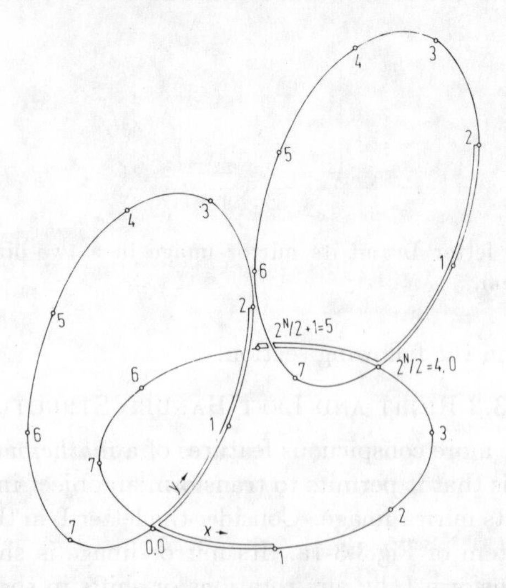

FIG.3.3-2. The letter L and its "mirror image" in a two-dimensional coordinate system based on rings.

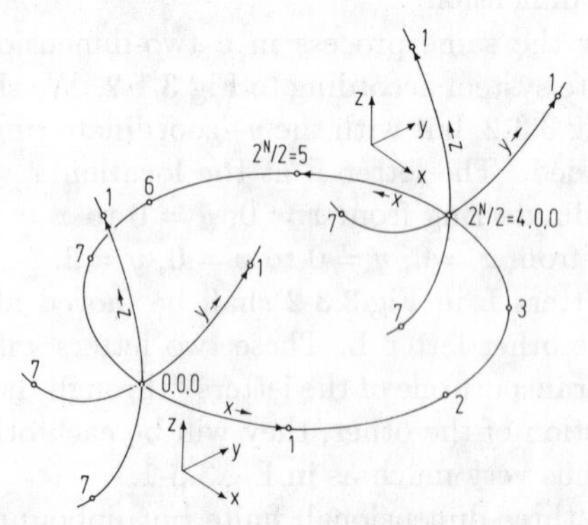

FIG.3.3-3. A right-handed tripod at the location $x = 0$, $y = 0$, $z = 0$ and $x = 2^{N/2} = 4$, $y = 0$, $z = 0$ of a three-dimensional coordinate system based on rings.

A second, equal tripod is constructed at the point $x = 2^N/2 = 4$, $y = 0$, $z = 0$. The ends of the three legs are now at the points (a) $x = 2^N/2+1 = 5$, $y = 0$, $z = 0$; (b) $x = 2^N/2 = 4$, $y = 1$, $z = 0$; (c) $x = 2^N/2 = 4$, $y = 0$, $z = 1$.

If one of the two tripods in Fig.3.3-3 is transported along the coordinate ring x to the location of the other, they will coincide. Both are right-handed tripods from this point of view. On the other hand, let the tripod with apex

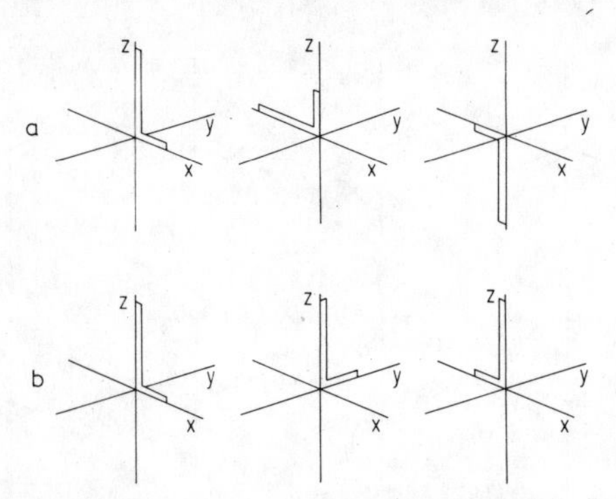

FIG.3.3-4. Rotation of a letter L in a three-dimensional Cartesian coordinate system.

at $x = 0$, $y = 0$, $z = 0$ be transported "through space" to the location of the other tripod. To make this more clear, we show the axes of the three-dimensional, right-handed, Cartesian coordinate system shifted "through space" from the vicinity of the point $x = 0$, $y = 0$, $z = 0$ to the vicinity of the point $x = 2^N/2 = 4$, $y = 0$, $z = 0$. It is evident that the tripod shifted "through space" is the mirror image of the tripod shifted along the $x-$coordinate ring. If the one is right-handed, the other is left-handed.

It is generally claimed that the transformation of a right-handed coordinate system—or a right-handed glove—into a left-handed one, requires transportation through a four-dimensional space. Hence, if one builds a model with rods and spheres of Fig.3.3-3 and performs the shifts just discussed, one must call four-dimensional space what is usually called "our three-dimensional space". We will interpret Fig.3.3-3 as a lucid example that the number of dimensions is defined by the coordinate system used, while the nonobservable concept "our space" or "physical space" does not have the feature *dimension*.

For a further elucidation of the generation of mirror images in multidimensional coordinate systems refer to Fig.3.3-4. Row a shows the letter L, using the $x-$ and $y-$coordinate of a three-dimensional coordinate system; this may either be a Cartesian coordinate system or one using rings with large diameter instead of straight axes. The letter L is rotated around the $y-$axis. No mirror image is produced.

Row b shows the letter L rotated around the $z-$axis. The mirror image of L is obtained by rotating the short bar of L first from the $x-$ to the $y-$axis, and then from the $y-$ to the negative $x-$axis.

Let us now turn to the tripod. It is shown in row a of Fig.3.3-5. Originally, the three legs 1, 2, 3 point in the direction of the positive $x-$,

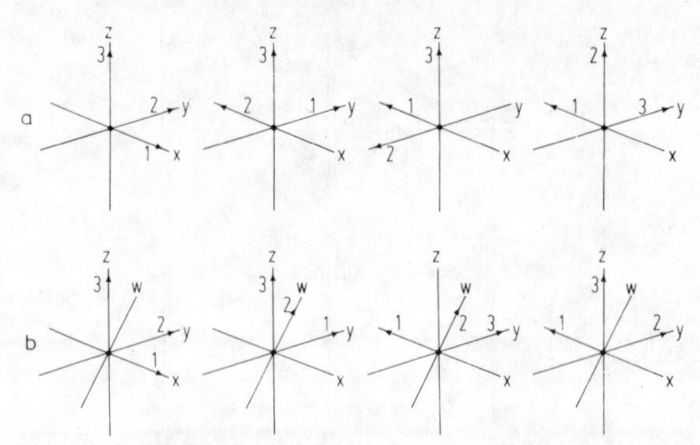

FIG.3.3-5. Rotation of a tripod in a three-dimensional, Cartesian coordinate system (a), and in a four-dimensional Cartesian coordinate system (b).

$y-$, and $z-$axes. A rotation around the $z-$axis makes leg 1 point in the direction $+y$, and leg 2 in the direction $-x$. A second rotation around the $z-$axis makes leg 1 point in the direction $-x$, and leg 2 in the direction $-y$. Leg 1 has now the desired direction opposite to its original direction $+x$. A rotation around the $x-$axis maintains the direction of leg 1, but legs 2 and 3 point now in the directions $+z$ and $+y$, which means we still have a right-handed tripod.

Row b shows how a left-handed tripod is obtained if there is a fourth coordinate axis or ring w. Originally, the three legs point in the directions $+x, +y, +z$. A rotation around the $z-$axis leaves leg 3 unchanged, while legs 1 and 2 are rotated in the directions $+y$ and $+w$. A second rotation around the $w-$axis leaves leg 2 unchanged, but brings legs 1 and 3 in the directions $-x$ and $+y$. Leg 1 has now the wanted direction. A further rotation is thus made around the $x-$axis. This leaves leg 1 unchanged, while legs 2 and 3 rotate into the directions $+y$ and $+z$. This is a left-handed tripod.

The problem of left- and right-handedness is frequently discussed in terms of left- and right-handed gloves. Let us try this approach. A manufacturer of gloves sets up two factories at opposite points of the equator, represented by the points 0, 0, 0 and 4, 0, 0 in Fig.3.3-3. Digitally controlled machines are used for all cutting and sewing operations. Only one program is used to control the machines; they only produce right-handed gloves. To do so, the machines and the program that controls them must use a three-dimensional coordinate system. Ordinarily, the coordinate system is built into the machines (sideways right, forward, upward), and the programs are written for this built-in coordinate system. However, the glove manufacturer uses the coordinate system based on the directions East ($x-$axis), North ($z-$axis), and direction to the Sun at 00:00 hours solar time on 21

March at the location 0, 0, 0—which is Noon at the location 4, 0, 0 (y–axis). Both factories turn out right-handed gloves in terms of their coordinate system. Shipping one half of the produced gloves to the other factory provides each with pairs of left- and right-handed gloves.

Consider the transportation of a rigid tripod along the coordinate ring x in Fig.3.3-3. Initially, the tripod is at the location 0, 0, 0. Its legs have the directions x, y, z as shown, which correspond to the directions of the coordinate rings in this point. Let now the tripod be transported along the x–ring to the point 4, 0, 0 in such a way that the directions of the three legs correspond to the directions of the coordinate rings in the points $x = 1, 2, 3, 4$ along the path of transport. It is evident that this causes no problem for the leg of the tripod denoted z, but either the leg denoted x or the one denoted y must be bent. Hence, forces are needed to make the right-handed tripod right-handed when transported around the coordinate system of Fig.3.3-3. Alternately, one may claim that mysterious forces bend a right-handed tripod into a left-handed one during its transportation.

Let us note that the two-dimensional letter L in Fig.3.3-1 can be moved around or flipped through the third dimension into its mirror image without need of forces for either process. However, when the letter is flipped into its mirror image, one actually sees its "underside". In three dimensions, we may transform a right-handed glove into a left-handed glove by turning it inside out. We see then the "inside" of the glove rather then the "underside" of the letter L, but if we do not distinguish between underside and upperside for the letter L we should not distinguish between inside and outside of a glove either. This transformation of a glove is quite impressive if done with a glove of thin material without any conspicuous difference between inside and outside, such as surgical gloves, electrician's rubber gloves, or a cheap type of kitchen gloves. Instead of turning a glove inside out, one may do the same with the arms of a coat or a shirt, or with the legs of trousers. Right- and left-handed are reversed in these operations, which shows that transformations that are supposed to require four dimensions are a matter of everyday experience.

3.4 Distance in Discrete Coordinate Systems

Consider a Cartesian coordinate system with two axes. A point P_0 has the coordinates x_0, y_0 and a second point P_1 has the coordinates x_1, y_1. We may define the coordinate difference or coordinate distance D_c:

$$D_c = |x_1 - x_0|,\ |y_1 - y_0| \tag{1}$$

The coordinate distance consists here of two numbers, or generally of one number per variable. One can derive from D_c distances that consist of one number only. The best known one is the Pythagorean distance D_{py}:

$$D_{py} = \left(|x_1 - x_0|^2 + |y_1 - y_0|^2\right)^{1/2} \quad (2)$$

The Pythagorean distance is well suited for Euclidean geometry. It can generally not be used for discrete coordinate systems, as will be discussed later on. Two other distances that can be derived from D_c are the absolute distance D_{ab},

$$D_{ab} = |x_1 - x_0| + |y_1 - y_0| \quad (3)$$

and the minimum distance D_{min}:

$$D_{min} = (|x_1 - x_0| + |y_1 + y_0|)_{min} \quad (4)$$

The meaning of D_{min} is that the coordinate system is rotated so that D_{ab} assumes a minimum value, as discussed in connection with Fig.1.3-1. One may readily see from this illustration that D_{min} yields the same value as the Pythagorean distance D_{py} in a plane with a Cartesian coordinate system, but it also yields the distance on the surface of the Earth along a great circle, if the location of the points P_1 and P_2 is characterized by the coordinate system of meridians and parallel circles.

Consider now the coordinate difference or coordinate distance D_c on a ring with 2^N points according to Fig.3.1-1 instead of the distance along an axis. One obtains:

$$\begin{aligned} D_c &= |x_1 - x_0| && \text{for } |x_1 - x_0| \leq 2^N/2 \\ &= 2^N - |x_1 - x_0| && \text{for } |x_1 - x_0| \geq 2^N/2 \end{aligned} \quad (5)$$

Alternately, we may write this definition in the following form:

$$D_c = |x_1 - x_0|_R \quad (6)$$

where the index R indicates that the smaller one of the two values $|x_1 - x_0|$ and $2^N - |x_1 - x_0|$ shall be taken. The generalization of the coordinate distance D_c on rings for two or more variables is straightforward:

$$D_c = |x_1 - x_0|_R, \quad |y_1 - y_0|_R, \quad \ldots \quad (7)$$

Just as before one can derive an absolute distance[1] D_{ab} and a minimum absolute distance D_{min} from D_c:

[1] For $2^N = 2$ the distance D_{ab} defined here becomes the Hamming distance that was introduced during the development of coding theory (Hamming, 1950).

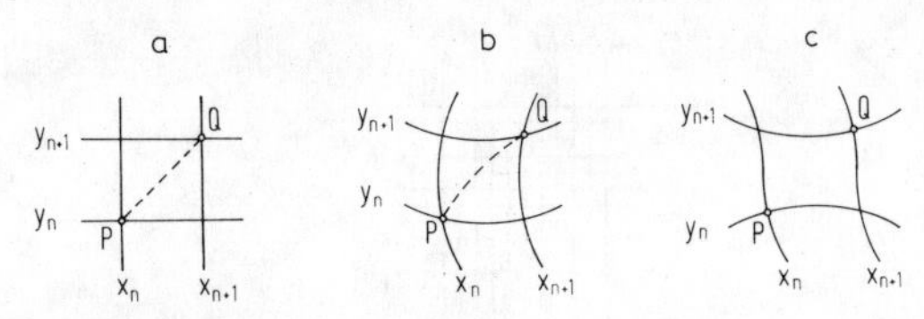

FIG.3.4-1. Distance between two points P and Q in coordinate systems that suggest an Euclidean plane (a), a sphere (b), and no particular surface (c).

$$D_{\mathrm{ab}} = |x_1 - x_0|_{\mathrm{R}} + |y_1 - y_0|_{\mathrm{R}} + \cdots \tag{8}$$

$$D_{\mathrm{min}} = (|x_1 - x_0|_{\mathrm{R}} + |y_1 - y_0|_{\mathrm{R}} + \cdots)_{\mathrm{min}} \tag{9}$$

Formally, one can also write the Pythagorean distance, but its physical meaning is not evident. For an explanation of this statement refer to Fig.3.4-1a. The distance between the points P and Q will almost automatically be stated to be the Pythagorean distance,

$$D_{\mathrm{py}} = \left[(x_{n-1} - x_n)^2 + (y_{n+1} - y_n)^2\right]^{1/2} \tag{10}$$

shown by the dashed line. This presumes, of course, that P and Q are located on a two-dimensional, continuous Euclidean plane. If we visualize the four coordinate lines to be plotted on a map, we immediately realize that the dashed line may cross a mountain or some other obstruction that makes it not the shortest distance. On a continuous, abstract mathematical surface, we may have one or more singularities on the dashed line.

The curved coordinate lines in Fig.3.4-1b prevent us from drawing an Euclidean straight line between P and Q, and give the Pythagorean distance of Eq.(10) as their distance. But the familiar curvature of the grid lines suggests to measure the distance along a great circle on a sphere. Only the unspecific curvature of the coordinate lines of Fig.3.4-1c does not suggest any specific continuous surface on which to measure the shortest distance between P and Q. If there is no surface but only the four coordinate lines or rods x_n, x_{n+1}, y_n, y_{n+1} one can obviously specify the coordinate distance between P and Q,

$$D_{\mathrm{c}} = |x_{n+1} - x_n|, \quad |y_{n+1} - y_n|,$$

or the absolute distance

$$D_{\mathrm{ab}} = |x_{n+1} - x_n| + |y_{n+1} - y_n|,$$

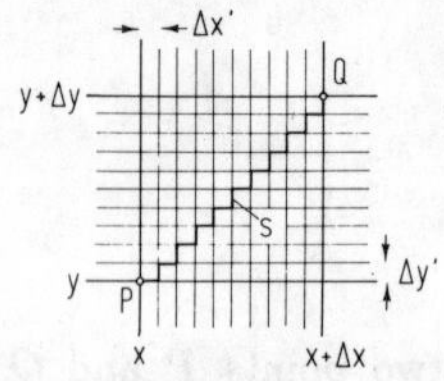

FIG.3.4-2. The reduction of the intervals Δx and Δy to $\Delta x' = \Delta x/10$ and $\Delta y' = \Delta y/10$ does not decrease the coordinate distance between P and Q; it always equals $s = 10\Delta x' + 10\Delta y' = \Delta x + \Delta y$.

but nothing more.

There is something very unsatisfactory about this way of defining distance. A physical object, be it the classical arrow or the more modern photon, will propagate from P to Q without any regard to the coordinate system. The usual way to overcome this problem is to assume, (a) that the coordinate lines can be drawn infinitesimally close together, and (b) that the coordinate lines are "on a surface" or "in a space" with certain properties. The source of the knowledge of these certain properties should be observations, but there are no observations that prove there is a continuous space, let alone a space with a more detailed list of features. That the Pythagorean distance requires more than an infinitesimal distance between coordinate lines is shown by Fig.3.4-2. Let there first be two pairs of coordinate lines at x_n, $x_n + \Delta x$, and y_n, $y_n + \Delta y$. The absolute distance between P and Q equals $\Delta x + \Delta y$. If the coordinate lines are now spaced with the smaller intervals $\Delta x'$, $\Delta y'$ one gets no reduction of the absolute distance. The length of the line s is always $\Delta x + \Delta y$, regardless of how small $\Delta x'$ and $\Delta y'$ are. This is, of course, the reason why Riemann geometries require the additional hypothesis for the length element $ds^2 = g_{ik} dx^i dx^k$.

There is a second way to overcome the problem of measurement of distance, that requires neither surfaces nor spaces. We can turn or bend the coordinate lines in Fig.3.4-1c so that either the line x_n or the line y_n passes through both points P and Q. Instead of having to define properties for a hypothetical surface or space, we have then to define only properties of a—mentally or actually—constructed coordinate system.

Consider the field equation of the general theory of relativity,

$$\mathbf{G} = 8\pi \mathbf{T} \tag{11}$$

where $\mathbf{G}$ is the Einstein tensor and $\mathbf{T}$ the stress-energy tensor of matter. The usual interpretation of this equation is that the stress-energy tensor $\mathbf{T}$ generates an average curvature $\mathbf{G}$ of space-time. The tensor exists "in complete absence of coordinates" (Misner *et al.*, 1973, pp. 32, 41, 42). The

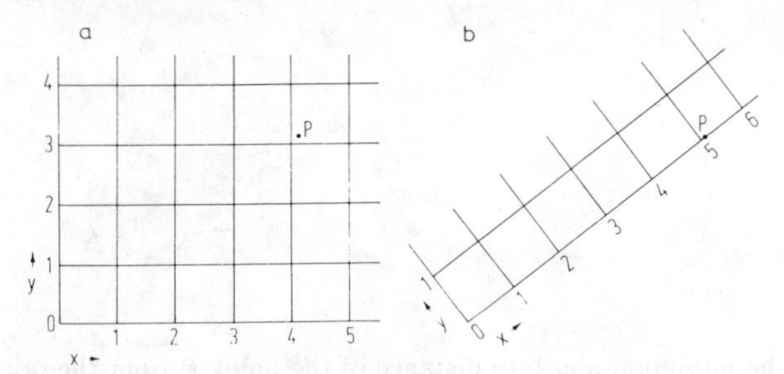

FIG.3.4-3. The Pythagorean distance of point P from the origin (a) equals the minimum absolute distance, which is found by rotating the coordinate system so that the origin and P lie on either the $x-$ or the $y-$axis.

point of view here is the exact opposite. The stress-energy tensor **T** generates a curvature tensor **G** of a coordinate system *in the complete absence of space*[2]. What is curved is not space but the coordinate lines, e.g., the lines or rods x_n and y_n in Fig.3.4-1c. If the coordinate lines are bent so that they are geodesics, the distance between any two points along the same coordinate line will be a stationary distance[3], and we have a satisfactory solution to the problem of distance in a discrete coordinate system. This is the minimum absolute distance measured with a coordinate system characterized by the stress-energy tensor **T**.

For a further study of distance refer to Fig.3.4-3. The point P is closest to the grid point $x = 4$, $y = 3$ of the Cartesian coordinate system. Its absolute distance from the origin $x = 0$, $y = 0$ equals $3+4 = 7$. Its minimum absolute distance equals the Pythagorean distance $(3^2 + 4^2)^{1/2} = 5$. The usual way to interpret the Pythagorean distance is to draw a straight line from $x = 0$, $y = 0$ to P in Fig.3.4-2a. However, we may just as well rotate the coordinate system so that $x = 0$, $y = 0$ and P are located either on the $x-$ or on the $y-$axis as shown in Fig.3.4-3b. This choice emphasizes the minimum absolute distance.

Consider now two points P_1 and P_2 on the surface of the Earth with the coordinates 0° longitude, 0° latitude and 10° E longitude, 10° N latitude. We can again rotate the coordinate system around the first point P_1 at 0°

[2]We use the word space rather than space-time since the concept of time has not yet been discussed. It will be taken up in Chapter 4.

[3]The term *stationary distance* instead of minimum distance is used because a geodesic does not only yield the minimum distance. For instance, a great circle between two points on the surface of the Earth yields two distances, one of which is the shortest distance. "A geodesic is a curve whose length has a stationary value with respect to arbitrary small variations of the curve, the end points being held fixed" (Synge and Schild, 1969, Sec. 2.4). The geodesic coordinate systems distinguished here by their physical properties are also distinguished in tensor analysis, since the absolute derivative of a tensor is reduced to the ordinary derivative in them (Synge and Schild, 1969, Sec. 8.2).

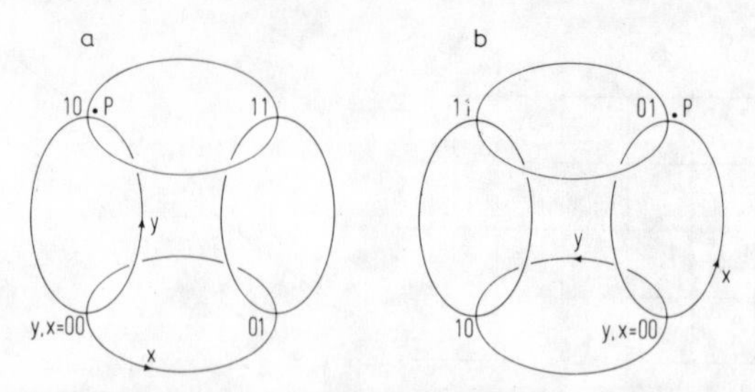

FIG.3.4-4. The minimum absolute distance of the point P from the origin y, $x =$ 00 (a) is found by rotating the coordinate system around 00 so that P lies either on the $x-$ or on the $y-$axis.

longitude, 0° latitude so that the second point P_2 lies either on the rotated meridian or the rotated equator. One of the distances between P_1 and P_2 along these great circles is the minimum absolute distance as well as the—generalized—Pythagorean distance on a spherical surface.

Let now P_1 be shifted to the coordinates 5° E longitude, 5° N latitude, while P_2 remains unchanged. Again one can rotate the coordinate system to make both P_1 and P_2 lie on the same rotated meridian, and thus obtain their minimum absolute or Pythagorean distance. However, one can no longer use rotation of the parallel circle 5° N, since this circle is not a great circle.

The question arises whether there are coordinate systems which do not generally permit to obtain a Pythagorean distance between two points by a rotation of the coordinate system. The answer is that coordinate systems based on the rings of Fig.3.1-1 are of this type. Refer to the simple dyadic coordinate system of Fig.3.4-4a. The point P located closest to the grid point $y = 1$, $x = 0$ has the coordinate distance $1, 0 = 10$ from the origin $y = 0$, $x = 0$ of the coordinate system. The absolute distance equals $1+0 = 1$. Let us now rotate the coordinate system around the origin $y = 0$, $x = 0$. The only other grid point that can come close to P is the one with the coordinates $y = 0$, $x = 1$ as shown in Fig.3.4-4b. The coordinate distance is now $0, 1 = 01$, and the absolute distance remains unchanged $0 + 1 = 1$. Hence, 1 is the minimum absolute distance. The concept of Pythagorean distance has no obvious meaning in this coordinate system; one would have to introduce the concepts of space and metric of space to give it a meaning.

Consider now a dyadic coordinate system with three variables x, y, z in the representation of Fig.3.1-20d. Such a coordinate system is shown in Fig.3.4-5a. A point P is located at the grid point $z = 0$, $y = 1$, $x = 1$. Its coordinate distance from the origin 000 is 011, its absolute distance is $0 + 1 + 1 = 2$. The coordinate system in Fig.3.4-5b is rotated around its

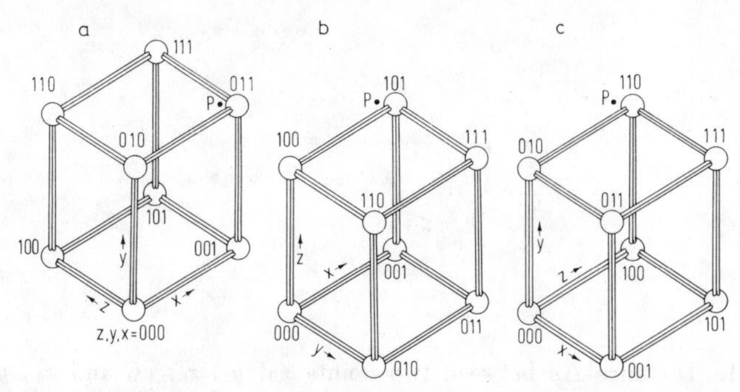

FIG.3.4-5. Generalization of Fig.3.4-3 to a three-dimensional coordinate system. Rotation of the coordinate system around the origin z, y, x = 000 brings the point P to the marks 011 (a), 101 (b), and 110 (c). The absolute distance equals $1 + 1 = 2$ in all three cases.

origin so the P is located at the grid point $z = 1$, $y = 0$, $x = 1$. The coordinate distance from the origin is now 101, but the absolute distance has remained unchanged $1+0+1 = 2$. In Fig.3.4-5c the coordinate system is rotated once more so that P is located at the grid point 110. The coordinate distance is changed to 110, but the absolute distance is again $1+1+0 = 2$. Hence, the absolute and the minimum absolute distance are equal for dyadic coordinate systems; a Pythagorean distance is not defined.

Let us generalize these results from dyadic coordinate systems to general coordinate systems based on rings. For large rings, $2^N \gg 1$, such coordinate systems are for short distances $k\Delta x$, $k \ll 2^N$, very similar to Cartesian coordinate systems, as shown for one dimension by Fig.3.1-1, and for three dimensions by Fig.3.1-10. Hence, for short distances—or locally—we can define a Pythagorean distance by means of a rotation of the coordinate system. However, we cannot do so if the absolute distance $|x_1 - x_0| + |y_1 - y_0| + |z_1 - z_0| + \ldots$ is not small compared with the circumference of the rings. For coordinate systems based on rings the Pythagorean concept of distance is a local feature that cannot be applied to large—or *global*—distances.

3.5 Coordinate Systems Defined by the Metric Tensor

We have claimed that the field equation $\mathbf{G} = 8\pi\mathbf{T}$ of the general theory of relativity means that the stress-energy tensor $\mathbf{T}$ generates a curvature tensor $\mathbf{G}$ of a coordinate system. For an elucidation of this statement we consider the following example. The two grid points x_0, y_0, z_0, w_0 and x_1, y_1, z_1, w_1 of a four-dimensional coordinate system have the coordinate distances $x_1 - x_0 = x^0$, $y_1 - y_0 = x^1$, $z_1 - z_0 = x^2$, $w_1 - w_0 = x^3$. The

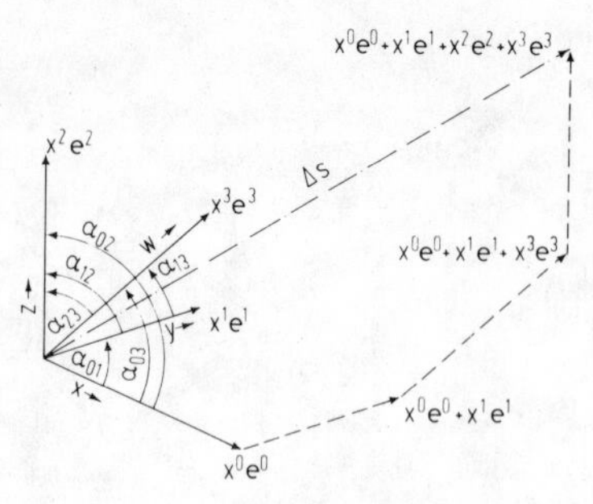

FIG.3.5-1. Distance Δs between two points x_0, y_0, z_0, w_0 and x_1, y_1, z_1, w_1 in a four-dimensional slanted coordinate system. $x^0 = x_1 - x_0$, $x^1 = y_1 - y_0$, $x^2 = z_1 - z_0$, $x^3 = w_1 - w_0$.

unit vectors $\mathbf{e}^0$, $\mathbf{e}^1$, $\mathbf{e}^2$, $\mathbf{e}^3$ are used to characterize the directions of the coordinate axes x, y, z, w in the points x_0, y_0, z_0, w_0. "Unit vector" means that $\mathbf{e}^0$ extends from x_0 along the x-axis to the adjacent mark $x_0 + \Delta x$, etc.

The Pythagorean distance between the points x_0, y_0, z_0, w_0 and x_1, y_1, z_1, w_1 may then be represented by the four vectors $x^0\mathbf{e}^0$, $x^1\mathbf{e}^1$, $x^2\mathbf{e}^2$, $x^3\mathbf{e}^3$ shown in Fig.3.5-1. The square $(\Delta s)^2$ of the length of Δs is defined by the following scalar product:

$$(\Delta s)^2 = (x^0\mathbf{e}^1 + x^1\mathbf{e}^1 + x^2\mathbf{e}^2 + x^3\mathbf{e}^3) \cdot (x^0\mathbf{e}^1 + x^1\mathbf{e}^1 + x^2\mathbf{e}^2 + x^3\mathbf{e}^3)$$

$$= \sum_{i=0}^{3} \sum_{k=0}^{3} g_{ik} x^i x^k$$

$$g_{ik} = \mathbf{e}^i \cdot \mathbf{e}^k = \cos\alpha_{ik} = \cos\alpha_{ki} = g_{ki} \tag{1}$$

The vector product assumes the validity of Euclidean geometry. In order to avoid the triviality of assuming Euclidean geometry for the whole coordinate system, we may restrict the validity of Eq.(1) in some way. We follow Riemann, who postulated Euclidean geometry for infinitesimal distances ds, and require that Eq.(1) shall only hold for sufficiently small but finite values of Δs. The components of the metric tensor for discrete coordinates are then nothing else but the cosines of the angles in Fig.3.5-1:

$$\begin{pmatrix} g_{00} & g_{01} & g_{02} & g_{03} \\ g_{10} & g_{11} & g_{12} & g_{13} \\ g_{20} & g_{21} & g_{22} & g_{23} \\ g_{30} & g_{31} & g_{32} & g_{33} \end{pmatrix} = \begin{pmatrix} 1 & \cos\alpha_{01} & \cos\alpha_{02} & \cos\alpha_{03} \\ \cos\alpha_{10} & 1 & \cos\alpha_{12} & \cos\alpha_{13} \\ \cos\alpha_{20} & \cos\alpha_{21} & 1 & \cos\alpha_{23} \\ \cos\alpha_{30} & \cos\alpha_{31} & \cos\alpha_{32} & 1 \end{pmatrix} \tag{2}$$

In the general theory of relativity one does not require Euclidean geometry for small distances but Lorentz geometry. This means the sign of $x^0\mathbf{e}^0$ in one of the expressions in Eq.(1) has to be reversed:

$$(\Delta s)^2 = (x^0\mathbf{e}^0 + x^1\mathbf{e}^1 + x^2\mathbf{e}^2 + x^3\mathbf{e}^3) \cdot (-x^0\mathbf{e}^0 + x^1\mathbf{e}^1 + x^2\mathbf{e}^2 + x^3\mathbf{e}^3)$$
$$= \sum_{i=0}^{3}\sum_{k=0}^{3} g'_{ik}x^i x^k \tag{3}$$

The coefficients g'_{ik} are again defined by Eq.(2), but the signs of the first column have to be changed to yield -1, $-\cos\alpha_{10}$, $-\cos\alpha_{20}$, $-\cos\alpha_{30}$.

Since we have not yet discussed the concept of time, we must tacitly assume what time means to give an interpretation of the components g'_{ik} of the metric tensor for space-time. We cannot simply say that $\mathbf{e}^0$ in Fig.3.5-1 is the direction of the time axis, since we have consistently interpreted "spatial" axes as axes of a coordinate system implementable by rods, and we could hardly claim that a time axis can be implemented by a rod. The vector $x^0\mathbf{e}^0$ in Fig.3.5-1 does not represent the time t but a shift $\mathbf{v}t$, where $\mathbf{v}$ is the velocity with which the local section of the coordinate system moves relative to another coordinate system. There is no difficulty assigning the angles α_{01}, α_{02}, and α_{03} between the unit vectors $\mathbf{e}^1$, $\mathbf{e}^2$, $\mathbf{e}^3$ of the axes of the coordinate system and the unit vector $\mathbf{e}^0$ of the direction of movement of this coordinate system. It becomes immediately evident why one generally cannot separate space-time into space and time, since this would be equivalent to separating a moving coordinate system into a coordinate system and a "move". Only for $\mathbf{v} = 0$ is the moving coordinate system reduced to a coordinate system, or space-time reduced to space.

The equality of the components of the metric tensor with the cosines of the angles between the axes of a *local* coordinate system and the direction of its local movement in the point x_0, y_0, z_0, w_0 shows that we do not need the concept of space. The only things that have to be bent or curved are the axes or "rods" of the coordinate system and the path along which it moves. A moving coordinate system is clearly as man-made a structure as a ruler. We do not need to construct such a coordinate system by means of rods and spheres, as in Fig.3.1-10 or Fig.3.1-25, and a similar second coordinate system relative to which the movement of the first one is defined. We can substitute electromagnetic waves concentrated in a small angle for the rods—either radiated by a radar or a laser, or received by a telescope—and the intersection of such *beams* for the spheres. What is observable are the bent or curved beams of electromagnetic waves but not something called curved space.

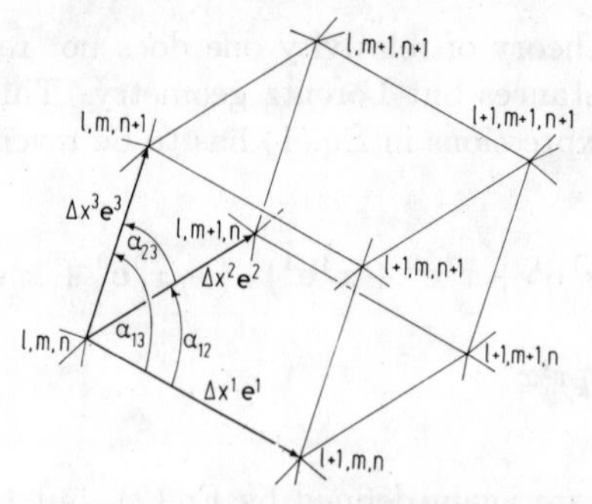

FIG.3.5-2. **The angles between the coordinate lines of a cell or a mesh of a coordinate system are defined by the components of the metric tensor $g_{ik} = g_{ik}(l, m, n) = \cos\alpha_{ik}$.**

A look at Fig.2.4-8 shows that in the absence of bending forces one can construct a bounded coordinate system with the same values of the angles between the axes at any grid point. The construction of an unbounded but finite coordinate system, on the other hand, requires bending, which means the angles cannot have the same value at every grid point characterized by the spheres in Fig.2.4-8. A look at Fig.3.1-11 confirms that the angles between the $x-$ and the $y-$ring at the points 00 and 01 are not equal.

Let us redraw Fig.3.5-1 to show more clearly that it represents a cell or a mesh of a coordinate system as shown in Fig.3.1-10 for a three-dimensional system. For simplicity, we use only three variables x^1, x^2, x^3. Figure 3.5-2 shows such a mesh with its origin at mark l for the variable x^1, at mark m for the variable x^2, and at mark n for x^3. The unit vectors $\mathbf{e}^1$, $\mathbf{e}^2$, $\mathbf{e}^3$ give the direction to the adjacent grid points $l + 1$, m, n or l, $m + 1$, n or l, m, $n + 1$, and Δx^1, Δx^2, Δx^3 give the distance to these points. The Pythagorean length element $(\Delta s)^2$ at the grid point l, m, n follows:

$$(\Delta s)^2 = (\Delta x^1\mathbf{e}^1 + \Delta x^2\mathbf{e}^2 + \Delta x^3\mathbf{e}^3) \cdot (\Delta x^1\mathbf{e}^1 + \Delta x^2\mathbf{e}^2 + \Delta x^3\mathbf{e}^3)$$

$$= \sum_{i=1}^{3}\sum_{k=1}^{3} g_{ik}\Delta x^i \Delta x^k$$

$$g_{ik} = g_{ki} = \mathbf{e}^i \cdot \mathbf{e}^k = \cos\alpha_{ik} \tag{4}$$

The metric tensor thus gives the angles between the coordinate lines at every grid point l, m, n of the coordinate system. Note that we do not have to specify the distance of the grid point l, m, n from the arbitrary origin of the coordinate system by any lengths along the x^i-axes, but only by the

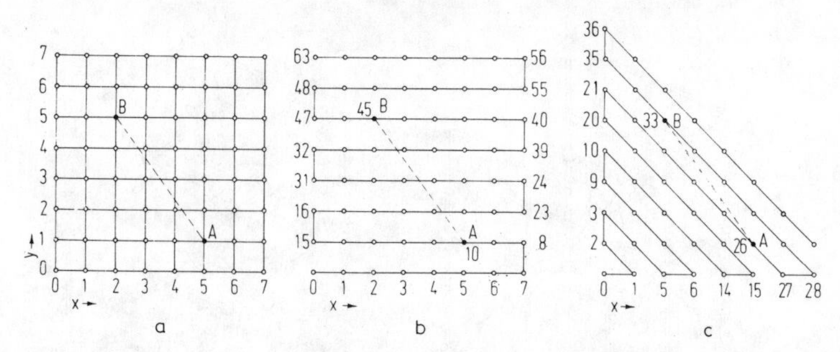

FIG.3.6-1. Two-dimensional, discrete Cartesian coordinate system (a), its replacement by an one-dimensional coordinate system (b), and the generalization of the principle to coordinate systems with denumerable many points.

number l, m, n of marks along the x^1-, x^2-, and x^3-axis. One could not do so if one used the continuum and differentials dx^i instead of a coordinate system with discrete marks.

3.6 COORDINATE SYSTEMS WITH ONE VARIABLE

We have seen that one may construct coordinate systems with more than three variables "in our three-dimensional space". For completeness sake we show now that one never needs more than one space variable if one uses discrete coordinate systems. Figure 3.6-1a shows a typical discrete Cartesian coordinate system with the two variables x and y; it is called a two-dimensional coordinate system because of the use of two variables.

If x and y assume only a finite number of values, say the eight values $0 \ldots 7$ shown in Fig.3.6-1a, one may connect the resulting 64 coordinate points as shown in Fig.3.6-1b. Only the one variable x is now needed to characterize any one of the 64 coordinate points. A short reflection shows that this result must hold for any finite number of coordinate points, and that there are many different ways in which these coordinate points can be characterized by one variable.

This result can be generalized to denumerable many coordinate points. Figure 3.6-1c shows how the points of a Cartesian coordinate system according to Fig.6.3-1a, but without any upper limit on the values of x and y, can be characterized by only one variable x.

The principle of Fig.3.6-1c cannot be generalized to nondenumerable many values of the variables x and y in Fig.3.6-1a. We see that the assumption of a continuum for the variables thus leads to the conclusion that one must use two variables, which in turn implies two dimensions.

The distance between the points A and B along an Euclidean straight line follows readily for Fig.3.6-1a from the Pythagorean relation

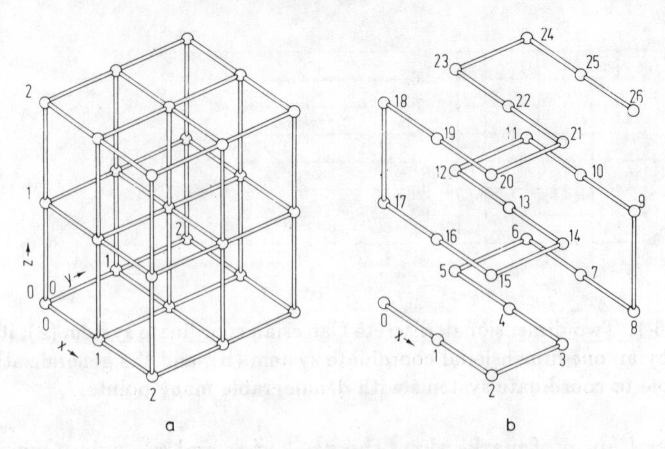

FIG.3.6-2. Three-dimensional, discrete Cartesian coordinate system (a), and its replacement by an one-dimensional coordinate system (b).

$$\left[(x_B - x_A)^2 + (y_B - y_A)^2\right]^{1/2} = (3^2 + 4^2)^{1/2} = 5$$

but this relation has no meaning for Figs.3.6-1b or c. The minimum absolute distance obtained by a rotation of the coordinate system around either point A or point B yields 5 for Figs.3.6-1a and b, but not for Fig.3.6-1c. The problem is caused by not having the same distance between any two adjacent points in Fig.3.6-1c—this distance being either 1 or $\sqrt{2}$—while this distance is always 1 in Figs.3.6-1a and b. We are here only interested in showing that one variable is sufficient to characterize the coordinate points of discrete coordinate systems, and do not advocate the use of only one variable; hence, there is no need to investigate distance measurements in a coordinate system according to Fig.3.6-1c.

Let us turn to the discrete Cartesian coordinate system with three variables x, y, z shown in Fig.3.6-2a. Figure 3.6-2b shows how the same coordinate points can be characterized by one variable x according to the principle of Fig.3.6-1b.

The replacement of three variables with denumerable many values by one variable according to Fig.3.6-1c is shown in Fig.3.6-3. To make this illustration easier to understand, we show with dashed lines three coordinate axes ξ, η, ζ with the same directions as the axes x, y, z in Fig.3.6-2a. Furthermore, the dotted lines between the points 1 and 3, 4 and 9, 10 and 19, 20 and 34, 13 and 15, as well as 24 and 28 , help to recognize planes with triangular boundaries that intersect a coordinate system according to Fig.3.6-2a to obtain the sequence in which the variable x runs through the coordinate points in Fig.3.6-3. As in the case of Fig.3.6-1c, one cannot

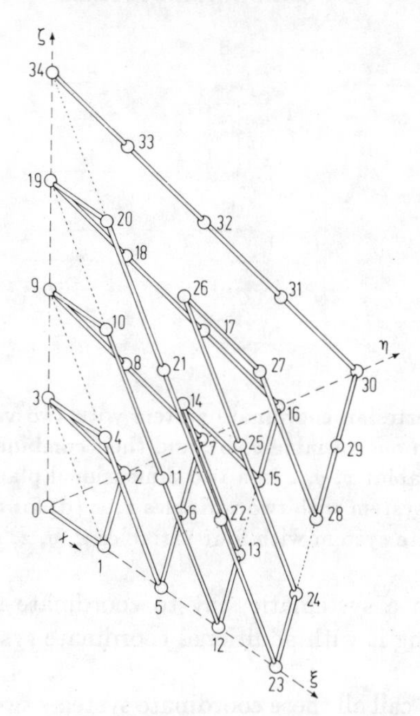

FIG.3.6-3. Replacement of a three-dimensional, discrete Cartesian coordinate system with denumerable many points according to Fig.3.6-2a by an one-dimensional coordinate system.

generalize Fig.3.6-3 to nondenumerable many coordinate points. This illustration is thus a very lucid proof that the need for three variables and the concept of a three-dimensional space is a result of the assumption of a continuum.

We have already shown that one can use more than three variables "in our three-dimensional space", but we give another example here that is easier for drafting. Figure 3.6-4a shows a discrete Cartesian coordinate system with the two variables x and y. A coordinate system with the one variable z is shown in Fig.3.6-4b. The combination of both coordinate systems yields the coordinate system of Fig.3.6-4c with three variables x, y, z. Note that z must have a finite range of values, while the range of x and y may be either finite or denumerable infinite.

If one uses the two-dimensional coordinate system with the variables z, w of Fig.3.6-4d rather than the one-dimensional one of Fig.3.6-4b in conjunction with the two-dimensional system of Fig.3.6-4a one obtains a four-dimensional coordinate system with the variables x, y, z, w. This coordinate system has the same structure as the one of Fig.3.6-4a, and one

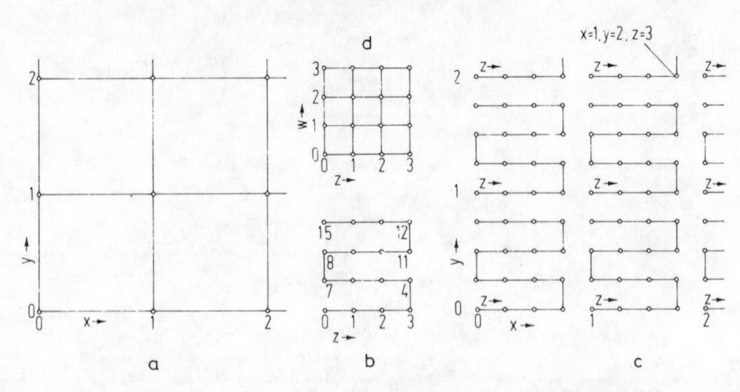

FIG.3.6-4. Discrete Cartesian coordinate system with two variables x and y (a), coordinate system with one variable z (b), and their combination to a coordinate system with three variables x, y, z in a two-dimensional plane (c). The substitution of the coordinate system with two variables z, w (d) for the one with variable z (b) yields a coordinate system with four variables x, y, z, w.

may thus advance in a systematic way to coordinate systems with more variables by combining it with additional coordinate systems of the type of Figs.3.6-4b or d.

It is tempting to call all these coordinate systems two-dimensional since they are plotted on the surface of a sheet of paper. However, if we eliminate the paper by constructing the coordinate systems of Fig.3.6-4 with wires, and bend the resulting grid structures at will, we would be hard put to decide whether the resulting structures are *on* a curved surface or *in* a space. The simplest solution is not to use the terms surface and space but only the term coordinate system.

4 Time and Motion

4.1 Time and Time Differences

It is usual to think of "space" as something provided by nature. A coordinate system constructed in this space permits to specify the location of points in the space. According to our discussion, this is not so. A coordinate system permits to specify the location of observable points relative to the marks of the coordinate axes or rings, or relative to the grid points of the coordinate system, but there is no need to say that either the observable points or the coordinate system are "in space". The term space is actually never needed, although it may be used as a convenient short word when not too much precision of expression is required. Without a coordinate system of observable points there is nothing observable that one could call space or—more precisely—physical space. Indeed, we may say that physical space is another term for coordinate system, or that a coordinate system defines or generates a physical space. We have previously pointed out that it is hard to see any difference between the usual concept of space and a three-dimensional coordinate system with infinitely long axes, infinitesimal distances between the marks on the axes, and the Pythagorean definition $s^2 = x^2 + y^2 + z^2$ for the concept of distance[1].

The relationship between time and clocks is essentially the same as between space and coordinate systems. We usually believe that clocks measure time, just as coordinate systems are believed to measure space (e.g., distances in space). In reality, clocks or—more generally—observable changes do not measure time but *generate* time. This statement sounds very provocative, but it becomes quite simple once we explain it in more detail. It is also in line with recent writings on time, from which we quote (Zwart, 1976)[2]:

---

[1] Whether this coordinate system was derived from concepts about the features of physical space, or whether the features of space were derived from this coordinate system, is a chicken-or-egg question, since a more specific description of the concept of space requires mathematical concepts, but these mathematical concepts were developed in order to define the concept of space more precisely.

[2] Other recent publications on time are due to Gibbs (1979) and Whitrow (1980).

> In physics time is not defined at all or at most only as that quantity which is measured by clocks. But as to what time really is, and what quantity is really measured by a clock, physics does not and need not offer an opinion (p. 10).
>
> ... where nothing happens and nothing changes it is if time were standing still. The concept of time could only have arisen where something happened, where one continually observed new and different phenomena (p. 27).
>
> We must realize clearly that the totality of events does not occur *in* time, but that it itself *constitutes* time (p. 43).
>
> However, it is inconceivable that out of points and moments without magnitude finite spatial and temporal intervals could be built up[3] (p. 191).

Our concept of time comes from our observation of changes. The most important change we observe is our own aging process[4]. This is a very crude clock, since it usually takes several years for the aging process to produce an observable physical effect. A much faster personal means to register changes is provided by our memory. This memory permits us to distinguish between yesterday and today, and to be aware of even faster changes. If there were no aging process and no memory, we would hardly have developed our concept of time. The rising of the Sun or the seasons of the year are periodically repeated processes that are quite different from our concept of time. We do not expect to reexperience memorized previous experiences periodically, and few of us expect to be reborn periodically. Our clocks and calendars are designed to replace our aging process and memories of change by something less personal.

The statement "clocks and observable changes generate time[5]" refers thus first of all to the observable change provided by our aging process—which is a sequence of observable facts—and our memory—which permits to

---

The concept of time generated by observable changes or clocks is either very close or equal to the concept of *relational time* used by philosophers.

[3]The content of this last quote differs from the three other quotes. It is included here, since we have emphasized that adjacent marks on a coordinate system must have a finite distance Δx, and will soon claim that time marks generated by a clock must have a finite distance Δt. Zwart's statement clearly rejects infinitesimal intervals dx and dt. Let us observe that in abstract mathematics one can build finite intervals from infinitesimal points by taking nondenumerably many of them, but one cannot carry over such a result from abstract mathematics to physics without showing how this can be done by measurement.

[4]Some philosophers, among them Hume and Locke, believe that our concept of time came from the succession of ideas in our mind. This is no real difference, since we would not be aware of the aging process without a memory. We will turn to the memory within a few sentences.

[5]The statement "clocks and observable changes *measure* time" is specifically avoided since it would suggest that there were something called time in the absence of clocks and observable changes.

register a sequence of observed facts. The statement refers to our wristwatch and other more conventional clocks only in the sense that these devices are designed to reproduce more objectively the time generated subjectively by our aging process and memory.

When we refer to time we mean a time t_1 relative to a reference time t_0, just as a grid point $x_1, y_1, z_1, w_1, \ldots$ of a coordinate system is understood relative to the reference point $x_0, y_0, z_0, w_0, \ldots$. The problem of distance in a coordinate system becomes the problem of difference between two times t_1 and t_0 generated or displayed by a clock. We take the time difference $t_1 - t_0$ for granted, but this fact is based on our concept of time derived from our aging process and our memory. For a periodic process with period T, such as the rising of the Sun, the proper concept of difference is "$t_1 - t_0$ modulo T", which means the smallest of the differences $t_1 - t_0$, $t_1 - t_0 - T$, $t_1 - t_0 - 2T, \ldots$, that is not negative, is the time difference. Hence, this type of time should be measured by something with the structure of the rings in Fig.3.1-1, and not by something with the structure of the numbers axis, which the difference $t_1 - t_0$ implies. This is usually done. The clock face is a ring with 12 or 24 marks for the hours, and 60 marks for the minutes and seconds. The days of the week are measured by rings with seven marks, the months by rings with 12 marks. Only the years are measured by us according to the marks or numbers of the numbers axis[6].

Let a clock face be divided into 2^N intervals. The distance T_c between two time points t_1 and t_0 is then given by Eq.(3.4-5), if D_c, x_1, and x_0 are replaced by T_c, t_1, and t_0. Instead of the normalized notation used for that equation one may use a non-normalized one, based on a period T and 2^N marks with distance $\Delta T = T/2^N$. The time difference between two time points $k\Delta T$ and $m\Delta T$ then becomes:

$$\begin{aligned} T_c &= |m-k|\Delta T \qquad \text{for } |m-k|\Delta T \leq T/2 \\ &= T - |m-k|\Delta T \quad \text{for } |m-k|\Delta T \geq T/2 \end{aligned} \tag{1}$$

We have a special name for such "time differences modulo T". We call them *phase differences.* The phase difference between two sinusoidal oscillations

[6] The Mayans used rings in a complicated fashion to keep track of the years. Their time keeping method led to a situation when time either had to start over again or had to come to an end. This unprecedented problem confronted and troubled the last adherents of Mayan culture on the island of Tayasal in Lake Petén-Itzá, Guatemala. Andrés de Avedaño, a Franciscan friar, delivered them from this evil in 1697 by bringing the Christian calendar and the Spanish sovereignty that came with it (Thompson, 1959, p. 156). The example shows how strongly one can believe in a concept of time different from ours. Tayasal is now the town of Flores, with a Christian church over the Mayan ruins, but the name Tayasal lives on for a place to the north of Flores across a bay of the lake.

with period T is the time difference between two distinguishable points, e.g., zero crossings with positive slope, measured modulo T. The periodicity of a sinusoidal oscillation makes it senseless to distinguish between the periods. We can only distinguish time points within a period; this is the phase difference or time difference modulo T. Instead of Eq.(1) we may introduce the sign $\underset{n}{-}$ for a modulo n subtraction. For instance, 2 o'clock minus 11 o'clock yields the time difference

$$T_c = 2 \underset{12}{-} 11 = 3\,\text{hours},$$

which is the smaller of the two values $|2 - 11| = 9$ and $12 - |2 - 11| = 3$.

Accepting a coordinate system based on rings is not too difficult for us. We are used to alternate between Cartesian, polar, spherical, cylindrical, etc. coordinate systems. There are also a variety of times, such as Mean Solar Time, Navigator's Time, Ephemeris Time, and Coordinated Universal Time, but we conceive of these times as "Real Time" measured by different methods with various accuracies. This Real Time starts at some arbitrarily defined moment, and increases from there on forever. Such a time cannot be measured, since any counter that registers the ticks and tacks of any kind of clock eventually reaches the highest number it can record, and then starts over again like the odometer of a car. Hence, the physically measurable—or generable—time behaves like the numbers on a very large ring according to Fig.3.1-1, and not like the numbers on the mathematical numbers axis. This is not easily recognized, because the limit of the largest measurable time is beyond one's comprehension. We write at this moment the year 1981, and the existing clocks can register the time generated by them since the beginning of 1981 with an ambiguity of about one microsecond. Ten billion years from now we will write 10 000 001 981 for the year, and the accumulated ambiguity of the time since 1981 will be about one hour, unless better clocks are invented. It is evident that the writing space required to keep track of the year is getting longer, but it is also evident that other processes will interfere with our generation and recording of time before the length of the number to be recorded becomes a problem. Time exists only as long as it is generated by a clock or observable changes, and as long as there is a memory or record of previous states. Time is as much a human invention or construction as a coordinate system. Clocks and coordinate systems enable us to order our observations and connect them by mathematical formulas. Time generated by a clock helps us describe planetary orbits just like a coordinate system centered at the Sun—or rather a point close to the Sun—helps, but this description requires no more a "Real Time" than it requires a space.

4.2 Shifts and Propagation

We have discussed coordinate systems that permit to specify a location for observable objects, and clocks that permit to record changes of these locations. Physical events are described in this coordinate-clock system; me may use the conventional term space-time instead of coordinate-clock system, as long as we keep in mind that space stands for coordinate system and time for clock. Having a coordinate system and one or more clocks one can introduce the concepts of propagation and velocity.

Consider the shifting of a function $f(x, y, z)$ in a three-dimensional coordinate system. Let the specific functional value $f(x_0, y_0, z_0)$ be shifted from the point x_0, y_0, z_0 to the point x_1, y_1, z_1. The following equation must be satisfied:

$$f(x_0, y_0, z_0) = f(x_1 - \Delta X, y_1 - \Delta Y, z_1 - \Delta Z)$$
$$\Delta X = x_1 - x_0,\ \Delta Y = y_1 - y_0,\ \Delta Z = z_1 - z_0 \qquad (1)$$

Let now the Cartesian coordinate system be replaced by one based on rings with n marks. One might be inclined to replace the ordinary subtraction in Eq.(1) by modulo n subtraction, but this would generally be wrong. The values ΔX, ΔY, ΔZ represent in essence coordinate distances according to Eq.(3.4-1). However, the coordinate distance D_c was defined to have always a positive sign, in order to comply with the requirement $s(A, B) = s(B, A)$ used in Eq.(1.3-1) for the definition of metric. The distances ΔX, ΔY, ΔZ in Eq.(1) are *signed* coordinate distances or *directional* coordinate distances D_{cd}:

$$D_{cd} = x_1 - x_0,\ y_1 - y_0,\ z_1 - z_0 \qquad (2)$$

On a ring, the (absolute) coordinate distance according to Eq.(3.4-7) equals

$$D_c = |x_1 - x_0|_R,\ |y_1 - y_0|_R,\ |z_1 - z_0|_R, \qquad (3)$$

where the subscript R indicates that the smaller of the two values $|x_1 - x_0|$ and $2^N - |x_1 - x_0|$ is to be used. In order to introduce a direction, we define as positive direction the advancement from 0 to 1, from 1 to 2, ..., from $2^N - 1$ to 0 in Fig.3.1-1. Negative direction is defined by advancement in the opposite direction. For large values of 2^N this definition assumes the same sense as the positive and negative direction along the numbers axis. The directional coordinate distance on a ring becomes thus:

$$\begin{aligned} D_{\text{cd}} = \Delta X &= x_1 - x_0 &&\text{for} \quad -2^N/2 \leq x_1 - x_0 \leq +2^N/2 \\ &= x_1 - x_0 + 2^N &&\text{for} \quad x_1 - x_0 \leq -2^N/2 \\ &= x_1 - x_0 - 2^N &&\text{for} \quad +2^N/2 \leq x_1 - x_0 \end{aligned} \tag{4}$$

As shorter notation we write

$$D_{\text{cd}} = \Delta X = (x_1 - x_0)_{\text{R}} \tag{5}$$

for the directional coordinate distance in analogy to Eq.(3.4-6).

Tables 4.2-1 to 4.2-4 show addition, subtraction, coordinate distance, and directional coordinate distance for a ring with $n = 8$ marks. These relations may readily be verified with the help of Fig.3.1-1. Note that the directional coordinate distance introduces negative numbers even though the subtraction modulo 8 uses positive numbers only.

Let us return to Eq.(1). The directional coordinate distances ΔX, ΔY, ΔZ must be computed according to Eq.(4). The differences $x_1 - \Delta X$, $y_1 - \Delta Y$, $z_1 - \Delta Z$, on the other hand, are obtained by modulo n subtraction, if the rings have $n = 2^N$ marks:

$$\begin{gathered} f(x_0, y_0, z_0, \dots) = f(x_1 \underset{\text{n}}{-} \Delta X, y_1 \underset{\text{n}}{-} \Delta Y, z_1 \underset{\text{n}}{-} \Delta Z, \dots) \\ \Delta X = (x_1 - x_0)_{\text{R}},\ \Delta Y = (y_1 - y_0)_{\text{R}},\ \Delta Z = (z_1 - z_0)_{\text{R}}, \dots \end{gathered} \tag{6}$$

A new problem arises. Since only non-negative numbers occur on rings according to Fig.3.1-1, the modulo n addition and subtraction is so far defined for positive numbers only. For instance, Tables 4.2-1 and 4.2-2 contain only the numbers 0, 1, ... , 7 for x and x_0. The directional coordinate distances $D_{\text{cd}} = \Delta X$ may be positive or negative. Hence, addition and subtraction modulo n must be extended to negative numbers.

Consider the rings of Fig.3.1-1. Addition of a positive number ΔX modulo n to a number x_1 means that one advances from x_1 in the mathematical positive sense ΔX marks, while subtraction of a positive number ΔX means that one advances from x_1 in the negative sense. For negative numbers we reverse the direction of advancement, which means modulo n addition becomes a modulo n subtraction and vice versa[1]:

---

[1] The implicit assumption is made that one has as many symbols as there are marks on the ring. For instance, the symbols 0, 1, ... , 9, A, B, C, D, E, F are used for the ring $2^N = 16$ in Fig.3.1-1, which plays a prominent role in the programming of microprocessors in hexadecimal code. There is, of course, a practical problem in finding enough different symbols for large rings, but we will be primarily interested in dyadic rings with only two symbols 0 and 1.

TABLE 4.2-1

ADDITION MODULO 8 OF x_0 AND x; $x_0 \underset{8}{+} x$.

x_0	$x=0$	$x=1$	$x=2$	$x=3$	$x=4$	$x=5$	$x=6$	$x=7$
0	0	1	2	3	4	5	6	7
1	1	2	3	4	5	6	7	0
2	2	3	4	5	6	7	0	1
3	3	4	5	6	7	0	1	2
4	4	5	6	7	0	1	2	3
5	5	6	7	0	1	2	3	4
6	6	7	0	1	2	3	4	5
7	7	0	1	2	3	4	5	6

TABLE 4.2-2

SUBTRACTION MODULO 8 OF x_0 FROM x; $x \underset{8}{-} x_0$.

x_0	$x=0$	$x=1$	$x=2$	$x=3$	$x=4$	$x=5$	$x=6$	$x=7$
0	0	1	2	3	4	5	6	7
1	7	0	1	2	3	4	5	6
2	6	7	0	1	2	3	4	5
3	5	6	7	0	1	2	3	4
4	4	5	6	7	0	1	2	3
5	3	4	5	6	7	0	1	2
6	2	3	4	5	6	7	0	1
7	1	2	3	4	5	6	7	0

$$x_1 \underset{n}{-} \Delta X = x_1 \underset{n}{+} |\Delta X| \quad \text{for} \quad \Delta X < 0$$

$$x_1 \underset{n}{+} \Delta X = x_1 \underset{n}{-} |\Delta X| \quad \text{for} \quad \Delta X < 0 \qquad (7)$$

Let us give an example. On a ring with $n = 2^N = 8$ marks we have $x_1 = 6$ and $x_0 = 1$. Table 4.2-4 or Fig.3.1-1 yield:

$$\Delta X = (x_1 - x_0)_{\mathrm{R}} = (6-1)_{\mathrm{R}} = -3$$

The inverse operation $x_1 \underset{8}{-} \Delta X$ yields with the help of Table 4.2-1 and Eq.(7):

$$x_1 \underset{8}{-} \Delta X = 6 \underset{8}{-} |-3| = 6 \underset{8}{+} 3 = 1 = x_0$$

TABLE 4.2-3

COORDINATE DISTANCE D_c BETWEEN x AND x_0 ON A RING WITH $n = 8$ MARKS ACCORDING TO EQ.(3.4-6); $|x - x_0|_R$.

x_0	$x = 0$	$x = 1$	$x = 2$	$x = 3$	$x = 4$	$x = 5$	$x = 6$	$x = 7$
0	0	1	2	3	4	3	2	1
1	1	0	1	2	3	4	3	2
2	2	1	0	1	2	3	4	3
3	3	2	1	0	1	2	3	4
4	4	3	2	1	0	1	2	3
5	3	4	3	2	1	0	1	2
6	2	3	4	3	2	1	0	1
7	1	2	3	4	3	2	1	0

TABLE 4.2-4

DIRECTIONAL COORDINATE DISTANCE D_c OF x_0 RELATIVE TO x ON A RING WITH $n = 8$ MARKS; $(x - x_0)_R$.

x_0	$x = 0$	$x = 1$	$x = 2$	$x = 3$	$x = 4$	$x = 5$	$x = 6$	$x = 7$
0	0	1	2	3	4	−3	−2	−1
1	−1	0	1	2	3	4	−3	−2
2	−2	−1	0	1	2	3	4	−3
3	−3	−2	−1	0	1	2	3	4
4	4	−3	−2	−1	0	1	2	3
5	3	4	−3	−2	−1	0	1	2
6	2	3	4	−3	−2	−1	0	1
7	1	2	3	4	−3	−2	−1	0

TABLE 4.2-5

ADDITION MODULO 2 OF x AND x_0, SUBTRACTION MODULO 2 OF x_0 FROM x, COORDINATE DISTANCE D_c BETWEEN x AND x_0 ON A DYADIC RING, AND DIRECTIONAL COORDINATE DISTANCE D_{cd} OF x_0 RELATIVE TO x ON A DYADIC RING.

$x \oplus x_0$			$x \ominus x_0$			$D_c = \|x - x_0\|_R$			$D_{cd} = (x - x_0)_R$		
x_0	$x = 0$	$x = 1$	x_0	$x = 0$	$x = 1$	x_0	$x = 0$	$x = 1$	x_0	$x = 0$	$x = 1$
0	0	1	0	0	1	0	0	1	0	0	1
1	1	0	1	1	0	1	1	0	1	1	0

Similarly, $x_1 = 1$ and $x_0 = 6$ yield:

$$\Delta X = (1-6)_{\mathrm{R}} = +3$$
$$x_1 \underset{8}{-} \Delta X = 1 \underset{8}{-} 3 = 6 = x_0$$

According to Fig.3.1-1 there are two distinguished rings. The one, $2^N \to \infty$, is the numbers axis of the integer numbers; it leads to Cartesian coordinate systems with finite distance Δx between adjacent marks. The other is the smallest ring, $2^N = 2$; it leads to the dyadic coordinate systems. The case $2^N \to \infty$ is of interest, since it eliminates only the infinitesimal distance dx between marks, but leaves otherwise the Cartesian and, by implication, the other conventional coordinate systems unchanged. The dyadic coordinate systems are of interest since they differ most radically from the conventional ones. Furthermore, dyadic coordinate systems simplify the operations just discussed.

Refer to Table 4.2-5. From left to right the table shows addition modulo 2, subtraction modulo 2, coordinate distance D_{c} according to Eqs.(3.4-5) and (3.4-6), as well as directional coordinate distance D_{cd} according to Eqs.(4) and (5). All four operations yield the same result. Hence, Eq.(6) can be rewritten for $n = 2^N$ in the following simpler form[2]:

$$f(x_0, y_0, z_0, \dots) = f(x_1 \oplus \Delta X, y_1 \oplus \Delta Y, z_1 \oplus \Delta Z, \dots)$$
$$\Delta X = x_1 \oplus x_0,\ \Delta Y = y_1 \oplus y_0,\ \Delta Z = z_1 \oplus z_0,\ \dots \qquad (8)$$

Shifting or propagating along the x–axis, $f(x - \Delta X)$, in a Cartesian coordinate system permits one to reach points arbitrarily far away from $x = 0$ for sufficiently large values of ΔX. This is not so for rings. In particular, on the dyadic ring $2^N = 2$ in Fig.3.1-1 one cannot get away any further from 0 then the mark 1. If one wants larger distances or more grid points, one must use more-dimensional coordinate systems according to Figs.3.1-18 to 3.1-22.

Consider the shifting of a point located nearest the grid point 0110 in Fig.3.1-21 to the neighborhood of the grid point 1010. This means a shift from the grid point with the coordinates $w_0 = 0$, $z_0 = 1$, $y_0 = 1$, $x_0 = 0$ to the grid point with coordinates $w_1 = 1$, $z_1 = 0$, $y_1 = 1$, $x_1 = 0$. To find the required changes ΔW, ΔZ, ΔY, ΔX of the coordinates we must subtract 0110 modulo 2 from 1010:

[2] We use the notation $\oplus$ for addition modulo 2. A dot under or over the addition sign + is also often used.

$$\begin{array}{r} 1010 \\ \oplus\, \underline{0110} \\ 1100 \end{array} \tag{9}$$

Hence, the $w-$ and $z-$coordinates have to be changed by 1, which transforms w from 1 to 0 and z from 0 to 1. The $y-$ and $x-$coordinates remain unchanged. The coordinate distance between 1010 and 0110 equals

$$D_c = 1,\ 1,\ 0,\ 0\ = 1100, \tag{10}$$

and the absolute distance equals

$$D_{ab} = 1 + 1 + 0 + 0 = 2 \tag{11}$$

In four-dimensional Cartesian coordinates a shift from the grid point $w = 1$, $z = 2$, $y = 3$, $x = 4$ to the grid point $w = 9$, $z = 8$, $y = 7$, $x = 6$ requires the change of coordinates

$$\begin{array}{r} 9\,8\,7\,6 \\ -\,\underline{1\,2\,3\,4} \\ 8\,6\,4\,2 \end{array} \tag{12}$$

which means an increase of the coordinates w, z, y, x by 8, 6, 4, 2. The only formal difference compared with the dyadic coordinate system is the use of the ordinary subtraction sign in Eq.(12) instead of the modulo 2 addition sign in Eq.(9). Of course, this is so only as long as there are no coordinates larger than 9, and each digit of the subtrahend is smaller than the respective digit of the minuend, in order to avoid the problem of a borrow for the subtraction. The coordinate distance between 9, 8, 7, 6 and 1, 2, 3, 4 equals

$$D_c = 8,\ 6,\ 4,\ 2 \tag{13}$$

and the absolute distance equals

$$D_{ab} = 8 + 6 + 4 + 2 = 20 \tag{14}$$

The minimum absolute distance cannot be determined without additional assumptions. If we assume that the coordinate axes w, z, y, x are perpendicular to each other in the Euclidean sense—which means an abstract, nonrealizable coordinate system—one obtains

$$D_{\text{min}} = D_{\text{py}} = (8^2 + 6^2 + 4^2 + 2^2)^{1/2} = 10.95 \tag{15}$$

or rather 11, since there are marks for 0, 1, 2, ... only, and not for fractional numbers. The coordinate distance and the absolute distance require an arbitrary coordinate system relative to which the two points are located whose distance is wanted. The minimum absolute distance, on the other hand, requires a particular coordinate system, that is typically distinguished by the measuring process or by the motion of something observable that propagates from one point to the other. In Fig.1.3-1b the coordinate system simulates the ruler of Fig.1.3-1c. If this ruler is straight in the Euclidean sense, one can calculate the distance s with the help of the coordinate system of Fig.1.3-1a and the theorem of Pythagoras. In Fig.3.4-4a the definition of the minimum absolute distance of the point P from the origin 000 makes physical sense for something that propagates from 000 to P via the grid points 001 or 010. No such thing as a "metric of space" is required in either case. The metric of space, from this point of view, is nothing more than a means to define particular coordinate systems with the help of an arbitrarily chosen one. For instance, the measurement of the distance between A and B in Fig.1.3-1c is simulated by the particular coordinate system of Fig.1.3-1b, whose grid points and coordinate line can be defined with the arbitrary coordinate system of Fig.1.3-1a as reference; the mathematical assumptions that permit the definition of one coordinate system by another define the metric. Note that no assumptions about a space are required, but only assumptions about how one coordinate system can be defined in terms of another.

As an example of propagation in a coordinate system intermediate to the dyadic system $2^N = 2$ and the Cartesian system $2^N \to \infty$ consider the hexadecimal ring $2^N = 16$ in Fig.3.1-1 with the 16 symbols 0, 1, ... , F for its marks. In a four-dimensional coordinate system one has the rings w, z, y, x. A shift from an initial grid point $w_0 = E$, $z_0 = 9$, $y_0 = 0$, $x_0 = D$ to a final grid point $w_1 = 1$, $z_1 = D$, $y_1 = 9$, $x_1 = A$ requires the change of coordinates according to Eq.(4) with $2^N/2 = 8$:

$$\Delta X = (x_1 - x_0)_{\text{R}} = (A - D)_{\text{R}} = -3, \quad \Delta Y = (y_1 - y_0)_{\text{R}} = (9 - 0)_{\text{R}} = -7$$
$$\Delta Z = (z_1 - z_0)_{\text{R}} = (D - 9)_{\text{R}} = +4, \Delta W = (w_1 - w_0)_{\text{R}} = (1 - E)_{\text{R}} = +3$$

The coordinates w and z must be increased by 3 and 4, but the coordinates y and x decreased by -7 and -3. Positive numbers (ΔZ , ΔW) occur if the shortest distance between the initial grid point and the final grid point is in the mathematical positive direction on the ring, while negative numbers (ΔX, ΔY) occur, if the shortest distance is in the mathematical negative direction. The coordinate distance D_c and the absolute distance D_{ab} equal $D_c = 3, 4, 7, 3 = 3473$ and $D_{ab} = 3 + 4 + 7 + 3 = 17$.

4.3 Velocity

We have so far discussed shifting or propagation without regard to the time required to propagate from one grid point of a coordinate system to another. By introducing time one introduces the concept of velocity. Our investigation will be carried out for dyadic coordinate systems only, but it will be quite evident how the generalization to other rings would have to be carried out.

First we rewrite Eq.(4.2-8) into the shorter form

$$f(\xi_0) = f(\xi_1 \oplus \Delta\xi), \quad \Delta\xi = \xi_1 \oplus \xi_0$$
$$\xi = \ldots, z/Z, y/Y, x/X; \quad \Delta\xi = \ldots, \Delta z/Z, \Delta y/Y, \Delta x/X \tag{1}$$

The position of the digits 0 and 1 in the numbers ξ_0, ξ_1, $\Delta\xi$ suffices to indicate for which ring x, y, z, ... they hold. The normalization ($\xi = x/X$, ...) is emphasized. The distinction between normalized and nonnormalized quantities is more important for dyadic coordinate systems than usual, since the distributive law of multiplication and division does not apply to addition modulo 2, as will be discussed soon.

The concepts of time and velocity are introduced into Eq.(4.2-1) by the substitutions

$$\Delta X = v_x \Delta T = v_x(t_1 - t_0) = v_x t_1 - v_x t_0 = x_1 - x_0$$
$$\Delta Y = v_y \Delta T, \quad \Delta Z = v_z \Delta T \tag{2}$$
$$f(x_0, y_0, z_0) = f(x_1 - v_x \Delta T, y_1 - v_y \Delta T, z_1 - v_z \Delta T) \tag{3}$$

The transformation of Eq.(2) for a dyadic coordinate system offers two choices. We may choose a dyadic coordinate system for the variable ξ, but measure time in the usual way, which means ordinary addition and subtraction are used for the time. The second choice is to use addition modulo 2 or—generally—addition and subtraction modulo n for the time as well as for the coordinates ξ.

We write $\theta = t/T$ for the time and $\beta_b = v_b/c$ for the velocity to emphasize that these are normalized quantities. The index b is used to

BOX 4.3-1
DYADIC MULTIPLICATION AND ORDINARY MULTTIPLICATION.

dyadic multiplication	ordinary multiplication
$1010 \otimes 11.01$	1010×11.01
1010	1010
1010	1010
0000	0000
$\oplus$ 1010	$+$ 1010
11100.10	100000.10

emphasize that β_b and v_b are velocities measured in a dyadic or binary coordinate system.

Our first choice, dyadic coordinate system but time added or subtracted as usual, yields the following replacements for Eqs.(2) and (3):

$$\xi_1 \oplus \xi_0 = \Delta\xi = \beta_b \Delta\theta = \beta_b(\theta_1 - \theta_0) = \beta_b\theta_1 - \beta_b\theta_0 \tag{4}$$

$$f(\xi_0) = f(\xi_1 \oplus \beta_b \Delta\theta) \tag{5}$$

The second choice, dyadic coordinate system and time added modulo 2, yields:

$$\xi_1 \oplus \xi_0 = \Delta\xi = \beta_b \otimes \Delta\theta = \beta_b \otimes (\theta_1 \oplus \theta_0) = \beta_b \otimes \theta_1 \oplus \beta_b \otimes \theta_0 \tag{6}$$

$$f(\xi_0) = f(\xi_1 \oplus \beta_b \otimes \Delta\theta) \tag{7}$$

The sign $\otimes$ is introduced here for *multiplication modulo 2.* It is needed since the distributive law of multiplication holds for ordinary addition and subtraction,

$$\beta_b(\theta_1 + \theta_0) = \beta_b\theta_1 + \beta_b\theta_0, \quad \beta_b(\theta_1 - \theta_0) = \beta_b\theta_1 - \beta_b\theta_0,$$

but not for addition modulo 2:

$$\beta_b(\theta_1 \oplus \theta_0) \neq \beta_b\theta_1 \oplus \beta_b\theta_0$$

The multiplication modulo 2 or *dyadic multiplication* is best explained by means of an example. The number 1010 is multiplied as usual by 11.01, but the resulting four summands are added modulo 2 rather than according to the usual multiplication. Box 4.3-1 shows this dyadic multiplication alongside the ordinary multiplication of the same numbers. A proof that dyadic multiplication satisfies the distributive law

$$\xi \otimes (\eta \oplus \zeta) = \xi \otimes \eta \oplus \xi \otimes \eta \tag{8}$$

required by Eq.(6) is given in Section 12.2.

The velocity β_b to be used in Eq.(7) follows from the inversion of the dyadic multiplication in Eq.(6), which yields the dyadic division:

$$\beta_b \otimes (\theta_1 \oplus \theta_0) = \xi_1 \oplus \xi_0$$
$$\beta_b = (\xi_1 \oplus \xi_0) \odot (\theta_1 \oplus \theta_0) \qquad (9)$$

A practical method for the performance of this dyadic division is discussed in Section 12.3.

Let us return to Eqs.(4) and (5) for an example of dyadic coordinates with usual time. Consider the shift from $\xi_0 = 0110$ to $\xi_1 = 1010$ in Fig.3.1-21. A propagating point shall be observed at the time $\theta_0 = 1100$ at ξ_0 and at $\theta_1 = 10100$ at ξ_1. The velocity β_b could be defined according to Eq.(4) as follows:

$$\beta_b = (\xi_1 \oplus \xi_0)/(\theta_1 - \theta_0) = (1010 \oplus 0110)/(10100 - 1100)$$
$$= 1100/1000 = 1.1 \qquad (10)$$

Such an equation is senseless. The position of the digits of $\xi_1 \oplus \xi_0$ characterize the shift $\Delta X/X = \Delta Y/Y = 1$, $\Delta Z/Z = \Delta W/W = 0$. Hence, any operation that changes the position of the digits is inadmissible. Instead of β_b, we must calculate its four components β_{bx}, β_{by}, β_{bz}, β_{bw}:

$$\beta_{bx} = \beta_{by} = 1/1000 = 0.001 = 1/8$$
$$\beta_{bz} = \beta_{bw} = 0/1000 = 0 \qquad (11)$$

In order to obtain "the velocity" from these four components, let us see how the same problem is solved in a Cartesian coordinate system. In this case one divides the Pythagorean distance D_{py},

$$D_{py} = \left[(x_1 - x_0)^2 + (y_1 - y_0)^2 + (z_1 - z_2)^2\right]^{1/2},$$

by Δt:

$$v = D_{py}/\Delta t \qquad (12)$$

Due to the relations of Eq.(2) one may instead derive v from v_x, v_y, v_z:

$$v = \left(v_x^2 + v_y^2 + v_z^2\right)^{1/2} \qquad (13)$$

We have seen that the best distance in coordinate systems, that do not assume that there is a space with certain features, is the minimum absolute distance D_{min}. For dyadic coordinate systems the absolute distance and the minimum absolute distance are equal, as has been explained with the help of Figs.3.4-4 and 3.4-5. The coordinate distance $\xi_1 \oplus \xi_0 = 1100$ of Eq.(10) yields the absolute distance 2:

$$D_{ab} = 2 = 10 \tag{14}$$

The absolute velocity is then defined as the quotient of D_{ab} and the time difference $\theta_1 - \theta_0 = 1000$:

$$\beta_b = v_b/c = D_{ab}/(\theta_1 - \theta_0) = 10/1000 = 0.01 = 1/4 \tag{15}$$

This is the equivalent of Eq.(12). The equivalent of Eq.(13) follows by summing the components of the velocity in Eq.(11) according to the rules of ordinary addition:

$$\beta_{bx} + \beta_{by} + \beta_{bz} + \beta_{bw} = 0.001 + 0.001 + 0 + 0 = 0.01 = 1/4 \tag{16}$$

For an example of a dyadic coordinate system and time added modulo 2, we return to Eqs.(6) and (7). Again we consider the shift from $\xi_0 = 0110$ to $\xi = 1010$ in Fig.3.1-21. A propagating point shall be observed at the time $\theta_0 = 1100$ at ξ_0 and at the time $\theta_1 = 10100$ at ξ_1. The coordinate distance $\xi_1 \oplus \xi_0 = 1100$ and the time difference $\theta_1 \oplus \theta_0 = 11000$ yield the components $\beta_{bx} \ldots \beta_{bz}$ of the velocity with the help of Eq.(9):

$$\begin{aligned} \beta_{bx} = \beta_{by} &= 1 \odot 11000 = 0.0001111\ldots = 1/8 \\ \beta_{bz} = \beta_{bw} &= 0 \odot 11000 = 0 \end{aligned} \tag{17}$$

In order to define an absolute velocity we start again with the absolute distance $D_{ab} = 2$ between $\xi_1 = 1010$ and $\xi_0 = 0110$. Since the time variable θ is treated like the coordinate variable ξ in Eqs.(6) and (7), it is plausible to define an absolute time difference $\Delta\theta_{ab}$ in analogy to D_{ab}. From $\theta_1 = 10100$ and $\theta_0 = 1100$ follows

$$\theta_1 \oplus \theta_0 = 11000$$

and

$$\Delta\theta_{ab} = 1 + 1 + 0 + 0 + 0 = 2 = 10 \tag{18}$$

The absolute velocity β_b becomes now:

$$\beta_b = D_{ab} \odot \Delta\theta_{ab} = 10 \odot 10 = 1 \tag{19}$$

We may draw the conclusion that Eqs.(4) and (5) as well as Eqs.(6) and (7) permit an unambiguous definition of a velocity. Whether such definitions of velocity are more than an exercise in abstract mathematics cannot be said. We will return to this topic in Chapter 7, and also study time measurement by a dyadic clock in Section 5.7, but first we will investigate why we universally use three spatial variables but only one time variable.

4.4 Three Time Dimensions and One Space Dimension

We have shown—or at least attempted to show—that we *need* a clock and a ruler, which means one time dimension and one space dimension, but that we *may* construct coordinate systems which introduce the concept of several space dimensions. There is no theoretical limit to the number of dimensions that a discrete coordinate system may have, as long as the number is finite. The concept of three space dimensions is due simply because a mathematical space with the topology of the continuum—or the usual topology of real numbers—must have at least three dimensions to accommodate discrete coordinate systems with any finite number of dimensions. Hence, this mathematical space needs at least three but not more than three dimensions.

The distinction of a certain number of dimensions of a mathematical space is nothing unusual. A well know example is the colored map problem. Consider a political map of the world with countries shown in different colors. How many colors are needed so that two countries with common border are never shown in the same color? The answer is that one needs at least four colors. Figure 4.4-1 shows a simple map with countries A, B, C, and D shown in four shades of grey. It is evident that one needs four colors, but the proof that one never needs more is famous for its lengthiness[1]. If this four-color-problem had been known to the Greeks, they would have founded—no doubt—a philosophical school on it. Having become known at a time when people were more careful in making conclusions, it never became more than an interesting fact in map making and topology.

[1] In one dimension one needs only two colors. The proof is trivial, since one has to draw only a line, the first section in red and the second section in blue, to recognize that one needs two but never more than two colors. In three dimensions one needs infinitely many colors. Again, the proof is trivial, since one only has to think of many ropes with different colors that are twisted so that every rope touches somewhere every other rope. The result for two dimensions is correct only for single connected surfaces, such as the surface of a sphere. On the surface of a toroid one can readily construct examples that require five colors.

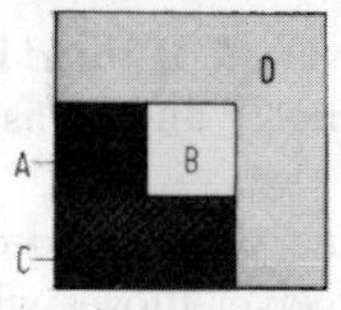

FIG.4.4-1. The four-color-map problem of topology on a single connected, two-dimensional surface.

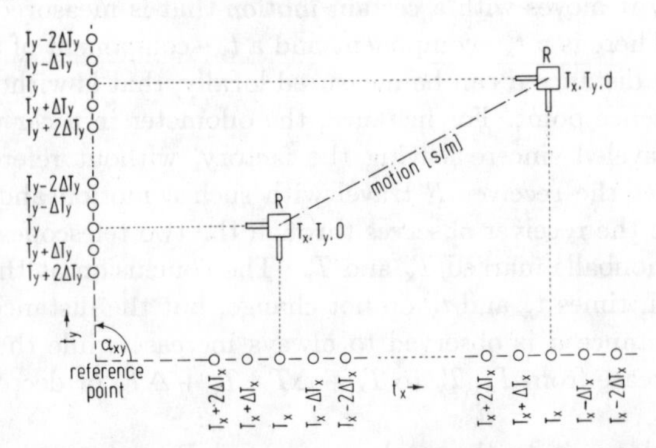

FIG.4.4-2. The use of two time variables t_x, t_y, and one space variable d, which never decreases, for the description of the movement of a vehicle R.

The assertion that our concept of three space dimensions is due only to the use of the overly abstract and nonobservable concept of the continuum runs so much counter to what we are used to, that it appears advisable to give some striking example that does not follow directly from what we have elaborated so far. Such an example would be, if we could show that the usual three space variables and one time variable could be replaced by three time variables and one space variable. We discuss such an arrangement with the help of Fig.4.4-2.

A coordinate system with the two axes t_x and t_y is constructed in the following way. Starting at a certain reference time $t_x = t_y = 0$, signals are sent out from the reference point in directions perpendicular to each other, $\alpha_{xy} = \pi/2$. Let us visualize these signals as streams of cannonballs having the marks $T_x - m\Delta T_x$, $T_x - (m-1)\Delta T_x, \ldots$ and $T_y - m\Delta T_y$, $T_y - (m-1)\Delta T_y$, $\ldots$ painted on them. A receiver R observes through telescopes, which are perpendicular to each other and to the propagating paths of the signals[2]. At a certain time the cannon balls marked T_x and T_y are observed. If this "certain time" was synchronized with the emission of the signals from the reference point, the observation of T_x and T_y specifies the two time

[2]The requirements stated here can be relaxed, but we do not want to go into too many details of something that is intended as an illustration of principles only.

coordinates t_x and t_y of the receiver R. If there is no such synchronization, the first observation will only define T_x and T_y as reference times for further observations.

Let the receiver R move along the dashed-dotted line the distance d. Usually we would say that the receiver moves with a certain velocity measured in meter per second, but in our example of space and time reversal we must say it moves with a certain *motion* that is measured in second(s) per meter. There is a t_x–component and a t_y–component of this total motion.

The distance d can be measured locally, that is without any help from the reference point. For instance, the odometer in a car measures the distance traveled sincere leaving the factory, without reference to anything else[3]. Let the receiver R travel with such a motion and in such a direction that the receiver observes through the two telescopes always the same two cannonballs marked T_x and T_y. The conclusion at the receiver will be that both times t_x and t_y do not change, but the distance d increases. Indeed, distance d is observed to always increase while the times t_x and t_y may increase from T_x, T_y to $T_x + \Delta T_x$, $T_y + \Delta T_y$ or decrease to $T_x - \Delta T_x$, $T_y - \Delta T_y$.

We leave it to the reader to extend Fig.4.4-2 from two to more dimensions and to investigate other effects, such as the replacement of the cannonballs by signals propagating with the speed of light. The whole scheme seems to be of academic interest only, provided we are on the surface of the Earth where rulers and rods are much easier to come by than streams of marked cannonballs. However, the situation is quite different for a space traveller, whose position and movement are more easily determined by streams of radar pulses than with the help of rulers, and who can measure locally primarily the travelled distance or the elapsed time since leaving the launch pad.

Despite this logical reasoning, many readers will object that we can return in space to our birthplace but not in time to our birthday. The reason why our personal time always advances is that we age and that we accumulate and store information. The storage of information is important since it makes us aware of the progress of time from one day to the next, which is too short a period to show the aging effect; also, we would not be aware of the aging process if we did not remember how beautiful we once looked. The reason why our clocks always advance is, of course, that we want them to represent more objectively our subjective experience.

---

[3]The distance is the integral of the velocity, and the velocity can be measured thanks to the rigid surface of the Earth on which the car travels. In a space ship one has no such rigid surface, but the acceleration can be measured and integrated twice to yield the distance d. Practical systems that are based on acceleration measurements have been developed under the names *inertial guidance* or *inertial navigation systems*.

Consider now an electron. It does not age, it has no memory to store information, and it does not carry a watch. When we talk about what an electron does as function of time, we mean what an observer with a clock is observing. Our usual concept of time is introduced as a means of description as arbitrarily as the coordinate systems in Chapter 3.

If we had a miracle drug that would change us physically to the state we were in at our birthday—and erase all stored information—we could indeed return in time to our birthday. This appears rather difficult, but it is no more difficult to envisage than a return to our birthplace. It seems rather easy to travel to our place of birth, but only if we consider this place with respect to the surface of the Earth. If we take the motion of the Earth around the Sun into account, we must return at an anniversary of the birthday to be at the previous location along the orbit. If we take the movement of the Sun relative to all other matter into account, we realize that a return to our place of birth would require that all matter has to be distributed as it was on our birthday. Indeed, if every atom would be where it was on our birthday, there would be no means to distinguish between our birthday and the day we returned to our birthplace; a return in space would imply a return in time. From this point of view, a return to the birthplace is much more of a feat than the development of a miracle drug that returns us—but not the rest of the world—to our birthday.

At this point, let us more modestly return to Eq.(4.3-1). We had there introduced the one space variable ξ for any number of space variables x, y, z, ... As a result, Eqs.(4.3-5) and (4.3-7) show formally one space and one time variable. This is not a mere matter of notation. A finite, bounded or unbounded, discrete coordinate system has a finite number of grid points or marks. Hence, we can number them 1, 2, ..., n. We need only one variable to characterize these marks. For an infinite, unbounded, discrete coordinate system, n becomes denumerably infinite, but we still need no more than one space variable. Only the transition to the continuum and coordinate axes with the usual topology of the real numbers introduce the need for more than one space variable.

5 Propagation in Unusual Coordinate Systems

5.1 Dyadic Metric

In Section 3.4 we had taken the view that a distinguished coordinate system is defined by the stress-energy tensor in the general theory of relativity. The coordinate rods or lines of such a system would be geodesics. The minimum absolute distance $D_{\min}$ between two coordinate points A and B is thus the distance actually travelled by a particle moving from A to B.

Since the general theory of relativity is based on Riemann geometries, all its results assume the topology of the continuum, either for "space-time" or for the coordinate system used; in addition, the generalized Pythagorean distance $ds^2 = g_{ik}dx^i dx^k$ is assumed. This means that only a certain class of coordinate systems is permitted, but the assumption of an infinitesimal resolution dx^i for these coordinate systems is not permissible according to information theory discussed in Section 2.2.

We will study the movement of test particles in discrete, unbounded coordinate systems that do not contradict the requirements of information theory. The axes, rings, or lines of these coordinate systems are assumed to be geodesics. Hence, the test particle always moves along the coordinate lines. This will sometimes appear very arbitrary, but we study purposely coordinate systems that differ strongly from the usual ones, since a radically different approach often makes one understand the conventional approach better. The development of non-Euclidean geometry by Bolyai and Lobachevskii showed that Euclidean geometry was not the only possible one, and thus made it easier to think about generalizations of Euclidean geometry. Since the concept of space is so deeply ingrained in our thinking, and since our concept of space is closely linked to Cartesian coordinate systems with coordinate axes having the usual topology of real numbers, we are interested in coordinate systems that are as different as possible from the Cartesian ones. A look at Fig.3.1-1 shows that the ring $2^N = 2$ differs the most from the conventional straight coordinate axis. Furthermore, we have seen in Figs.3.1-20 to 3.1-22 that the dyadic coordinate systems based on this ring can be plotted with manageable effort at least up to $n = 5$ dimensions. Finally, we have seen in Section 4.2, particularly from Table 4.2-5, that the coordinate distance and the directional coordinate distance on a dyadic ring are obtained by modulo 2 addition, while other rings require more complicated operations.

We had introduced the concept of metric in Section 1.3, Eq.(1.3-1), and that of coordinate distance in Section 3.4, particularly Eq.(3.4-5). Let us show first that the coordinate distance D_c, the absolute distance D_{ab}, and the minimum absolute distance D_{min} satisfy the four conditions of a metric as shown in Eq.(1.3-1).

For the coordinate distance, the first condition $s(A,B) = s(x_1,x_0) = 0$ for $A = B$ is obviously satisfied by the definition $D_c = |x_1 - x_0|_R$ of Eq.(3.4-6); the second condition, $s(x_1,x_0) > 0$ for $x_1 \neq x_0$, and the third condition $s(x_1,x_0) = s(x_0,x_1)$, are also evidently satisfied. The fourth condition, $s(x_2,x_1) + s(x_1,x_0) \geq s(x_2,x_0)$ is best recognized from Fig.3.1-1. As long as the sum $s(x_2,x_1) + s(x_1,x_0)$ is smaller or equal to $2^N/2$, it will equal $s(x_2,x_0)$; if the sum is larger than $2^N/2$, it will be larger than $s(x_2,x_0)$.

The generalization of the coordinate distance D_c from one to more dimensions by Eq.(3.4-7) shows that the conditions for a metric are satisfied for every variable. Hence, the coordinate distance D_c for any finite number of variables satisfies the condition of a metric.

The absolute distance D_{ab}, defined by Eq.(3.4-8) is zero if and only if $x_1 = x_0$, $y_1 = y_0$, and the condition $s(A,B) = 0$ for $A = B$ is thus satisfied. For $A \neq B$ at least one difference must be larger than zero, and the condition $s(A,B) > 0$ is satisfied. The condition $s(A,B) = s(B,A)$ is satisfied since only the absolute values $|x_1 - x_0|$, $|y_1 - y_0|$, ... enter Eq.(3.4-8). The condition $s(A,B) + s(B,C) \geq s(A,C)$ is satisfied since it is satisfied for every term $|x_1 - x_0|_R$, $|y_1 - y_0|_R$, ... as shown for the coordinate distance D_c, and the summation of several terms cannot change this fact.

The minimum absolute distance D_{min} satisfies the conditions for a metric, since it is a special case of the absolute distance D_{ab} obtained by rotating the coordinate system.

For dyadic coordinate systems the absolute and the minimum absolute distance are equal, as shown by Fig.3.4-4 for three dimensions and implied for more dimensions; they are called Hamming distance. The metric of dyadic coordinate systems is referred to as *dyadic metric*.

The feature metric is usually ascribed to mathematical spaces or surfaces, not to coordinate systems. Consider Fig.1.3-1a. The Pythagorean distance between the points A and B equals $s = (x^2 + y^2)^{1/2}$, which is shown to be also the minimum absolute distance in Fig.1.3-1b. Is the Pythagorean distance and the metric defined by it a feature of the coordinate system or of the paper plane on which it is drawn? We could construct the coordinate system of Fig.1.3-1a from steel wires welded together at the grid points. The Pythagorean distance and the minimum absolute distance would still hold, but there would be no observable plane. We could, of course, imagine some plane that is defined by the steel wire coordinate system, but if we deform the coordinate system according to Fig.1.3-2 we would obtain a different metric even though the imagined plane would not have to be changed. We are back to the view expounded before that the only things necessary and observable are the coordinate system and the objects observed with reference to it, and all features should be ascribed to the coordinate system and the objects. Our habit of seeing a coordinate system constructed on a surface or in a space with certain properties comes from abstract mathematics, and is reenforced by the fact that we implement coordinate systems by drawing them on a paper plane rather than welding steel wires together.

As long as we believed in absolute space or some observable thing called *physical space*, we were justified in ascribing features to this space. For example, the space was Euclidean or it was curved, it had the topology of the continuum or it had a discrete topology, and so on. Once we reconcile ourselves that "space" cannot be observed but only matter or energy, we are no longer justified to ascribe any features to this space. Space becomes a term like *ether* or *electron orbits of an atom* that is used for convenience and historical reasons. As an observable quantity, space is replaced by a coordinate system, which is typically implemented—in a very rudimentary form—by electromagnetic waves with a narrow beam; most of the time, the coordinate systems exists, of course, only in our mind. Finding and using coordinate systems that are particularly suited for a physical problem is a standard procedure. If a physical system has cylindrical symmetry, we will describe it automatically with cylindrical coordinates, since they lead to simpler mathematical manipulations and the results are more readily understood than with another coordinate system. Riemann geometries as well as the discrete coordinate system with Riemann-like[1] or dyadic metric discussed here may thus be viewed as a continuing effort to find coordinate systems that simplify mathematical manipulations and make results more lucid.

A coordinate system with m rings and n marks per ring defines n^m grid points. Hence, one can obtain many grid points either by using many marks and few rings (or dimensions), or by using many rings and few marks per ring. It is quite evident from Figs.3.1-18 to 3.1-22 that a dyadic coordinate system can extend over arbitrarily large distances even though the maximum distance along one ring is only 1 or Δx.

[1] Riemann-like means that the differentials in the equation $ds^2 = g_{ik}dx^i dx^k$ are replaced by finite differences.

TABLE 5.1-1
DYADIC GROUPS FOR $s = 1, 2, 3$.

	2^1	2^2	2^3
1	0	00	000
2	1	01	001
3		10	010
4		11	011
5			100
6			101
7			110
8			111

Dyadic coordinate systems are a representation of the dyadic group. This group consists of the 2^s numbers $0, 1, \ldots, 2^s - 1$, and has the modulo 2 addition as the group operation. The Hamming distance defines a distance in the dyadic group, which makes the dyadic group a topologic group. Since the Hamming distance also satisfies the conditions of a metric, we have a metric group; this explains the choice of the term dyadic metric for the heading of this section.

The dyadic group is best explained with the help of the three examples for $s = 1, 2, 3$ in Table 5.1-1. The modulo 2 addition of any two numbers yields another number of the group, for example, $010 \oplus 110 = 100$. The unit element is the zero, for example $010 \oplus 000 = 010$. The inverse element is the number itself, for example, $010 \oplus 010 = 000$. The groups with 2^1 and 2^2 elements in Table 5.1-1 are subgroups of the group with 2^3 elements.

In terms of the three-dimensional dyadic coordinate systems the addition $010 \oplus 110$ means a propagation or shift from the point $\xi_0 = 010$—or $z = 0$, $y = 1$, $x = 0$—to the point $\xi = 100$—or $z = 1$, $y = 0$, $x = 0$. The movement is in the mathematical positive direction for addition, but on a binary ring one cannot distinguish between mathematically positive or negative, which explains why subtraction and addition modulo 2 are identical.

Let us observe that there is nothing magic about the points 001, 010, and 100 having the same distance from the point 000 if one uses the three-dimensional coordinate systems of Figs.3.1-19 or 3.1-20d, whereas such a method of distance measurement would appear—at this point—surprising if the points 000, 001, 010, and 100 are plotted in the one-dimensional coordinate system of Fig.2.4-2a. A particle moving in Fig.2.4-2a from the point 000 to the point 001 would appear to move "continuously", whereas a move from 000 to 010 or 100 would appear as a jump. All three moves are equally continuous in Figs.3.1-19 or 3.1-20d.

It may appear somewhat abstract that the difference between a continuous and a jumpy movement should be due to the choice of a coordinate system, but it is actually experienced by anybody who travels around the Earth and crosses the international date line. The rotation of the Earth around its axis is described by the addition modulo 2 in Fig.3.1-20a. Let 0 denote a place on the Earth facing the Sun at noon. Let T be 12 hours, so that $t/T = 1$ is a place on Earth away from the Sun at midnight. The modulo 2 addition of $t/T = 1$ to either 0 or 1 will thus describe the transition from noon to midnight to noon, and so on. The jump at the date line occurs because our calendar is organized like the integer numbers on the numbers axis, that is, the days follow each other like the numbers 000, 001, ... in Fig.2.4-2a. The time shown by a clock with 24-hour division advances modulo 24—or modulo 2 if only noon and midnight are shown—and this time does not jump at the date line but advances like everywhere else.

For another example of time behaving like the points 0 and 1 in Fig.3.1-20a when 1 is added modulo 2, consider a primitive animal living in the tropics. It experiences day and night but knows nothing about absolute or relative time. All nights are equal and so are all days. Adding or subtracting a 12-hour period at any given moment will change day to night or night to day. The concept of time for this animal will be that of dyadic time. A distinction between past and future will not be made, except when it gets eaten by a bigger animal.

Let us study the concept of dyadic shifting with the help of the two-dimensional coordinate system of Fig.3.1-18, which is repeated in an easier-to-draw form in Fig.5.1-1a. In terms of computer programming we have four storage locations with addresses 00, 01, 10, and 11, which store the samples or numbers A, B, C, D. A dyadic shift can be performed by adding 01, 10, or 11 modulo 2 to the addresses and leaving the samples where they are. This is shown in Fig.5.1-1a. In terms of physics, the coordinate system is changed while the samples remain unchanged.

The drawing of circles according to Fig.5.1-1a is complicated. The same shifting is therefore repeated on a straight scale in row c. The samples A, B, C, and D remain in their locations, but the marking of the scale is changed by adding 01, 10, or 11 modulo 2 to the number of the scale at the extreme left. The unit of the scale is the normalized time t/T.

Figure 5.1-1b shows the shifting performed by leaving the address of the storage locations or the coordinate system unchanged and moving the samples A, B, C, D. One may readily verify that these four samples are at the same address in each column of rows a and b in Fig.5.1-1. Row d shows the shifting on a straight scale. The marking of the scale from 00 to 11 is the same in all four columns of row d, which corresponds to the usual increase of the time scale from left to right.

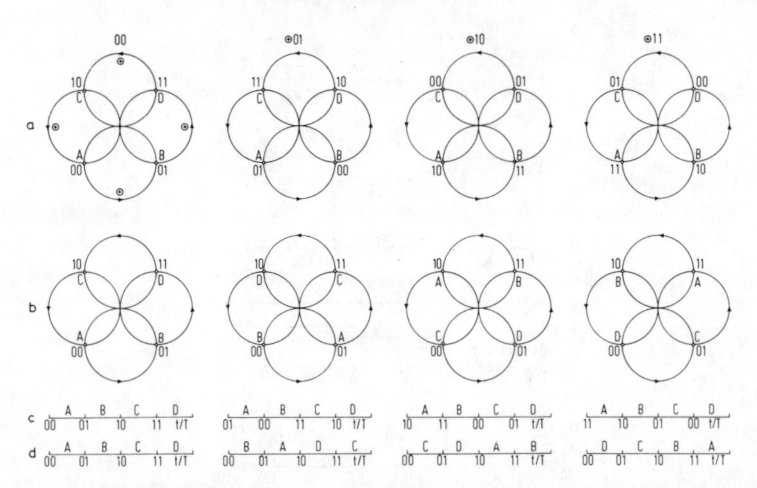

FIG.5.1-1. Dyadic shifting in two dimensions of the addresses of samples (a) and by moving the samples to various addresses (b). The first case represents a movement of the coordinate system relative to the samples A to D, the second case a movement of the samples A to D relative to the coordinate system. Rows (c) and (d) represent the same operations in one dimension.

In order to study shifting in three dimensions according to Fig.3.1-19 we use the simpler representation of Fig.3.1-20. It is shown again, in a slightly changed but readily understandable version, in Fig.5.1-2a. Let a sample A be at the location ξ_0. Equation (4.3-5) for the propagation of the function $f(\xi)$ in dyadic coordinates implies

$$\xi_1 \oplus \beta_b(\theta_1 - \theta_0) = \xi_0$$

$$\xi_1 = \xi_0 \oplus \beta_b(\theta_1 - \theta_0) \tag{1}$$

Let the velocity β_b equal 1 and let θ_0 equal 0. We then have the equation

$$\xi_1 = \xi_0 \oplus \theta_1 \tag{2}$$

For $\xi_0 = 0$—or $z/Z = y/Y = x/X = 0$—the sample A moves from $\xi_0 = 000$ to $\xi_1 = 001, 010, 011, \ldots$ as the time θ_1 increases from $\theta_0 = 0$ to $\theta_1 = 001, 010, 011, \ldots$. This is indicated in Fig.5.1-2a on the left by the lines with arrows and the numbers 1, 2, ... , 7. Sample A moves first in the direction x, then diagonally in the "xy–plane", again in direction x, diagonally through "xyz–space", and so on. Row b of Fig.5.1-2 shows this movement represented in one dimension. Furthermore, row c shows how

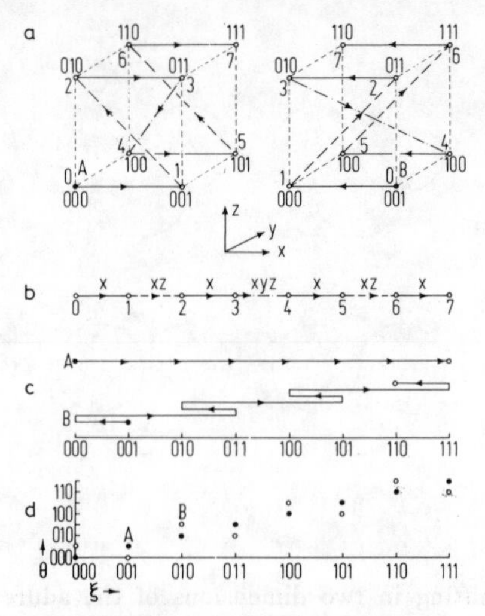

FIG.5.1-2. Various representations of dyadic shifting of samples A and B located at the points 000 and 001.

sample A moves if the points 000 to 111 are represented as points on an one-dimensional scale.

On the right in Fig.5.1-2a the movement of sample B originally located at the point $\xi_0 = 001$—or $z/Z = y/Y = 0$, $x/X = 1$—is shown. Equation (2) has now the form:

$$\xi_1 = 001 \oplus \theta_1 \tag{3}$$

The addition modulo 2 of $\theta_1 = 001, 010, 011, \ldots$ to 001 makes sample B move as indicated by the arrows. We observe that sample B moves first in the direction x, diagonally in the "xy–plane", again in direction x, diagonally through "xyz–space", and so on. This is exactly the same sequence as the one for sample A, and the one-dimensional representation of row b applies thus for sample B as well as for A.

The representation of Fig.5.1-2c, on the other hand, makes sample B move sometimes to the left and sometimes to the right, while sample A always moves to the right. The average movement of B is still from left to right. The jumpiness of the movement of B is strictly a result of this representation, which is borrowed from the real-number-line where a larger number is plotted to the right of a smaller number. There is no such distinction between a jumpy and a nonjumpy movement in the representations of Figs.5.1-2a and b.

Figure 5.1-2d shows movement of points A (black dot) and B (circle)

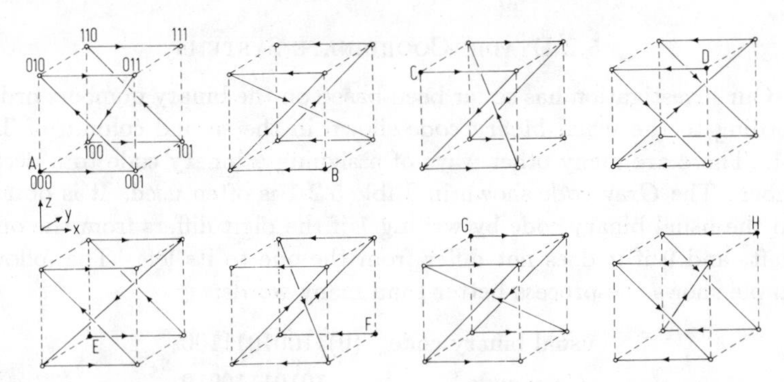

FIG.5.1-3. Dyadic shifting of each one of the eight samples A, B, ... , H located at the points 000, 001, ... , 111.

as function of time, according to Eqs.(2) and (3). This representation is the equivalent of the world line used to represent movement in a space-time with the usual real number topology—or in a Cartesian coordinate system with the variables x^0, x^1 having the usual topology of the real numbers.

The movement of sample B in Fig.5.1-2d is again jumpy, but we can interpret the jumps as movements from right to left or left to right, rather than forward and backward in time, which makes the representation acceptable. Figure 5.1-3 shows the dyadic shifting of all eight possible samples A, $B, \ldots, H$ with the initial location $\xi_0 = 000, 001, \ldots, 111$. One may verify that all samples are shifted in the same sequence shown by Fig.5.1-2b, first in direction x, then diagonally in the "$xy-$plane", and so on. The rules for movement in the world line representation of Fig.5.1-2d are difficult to recognize. We will have to return to this problem later.

The main objection against the coordinate systems of Figs.5.1-2 and 5.1-3 is that the samples do not always move along the coordinate lines but through the "$xy-$plane", the "$xyz-$space", etc. We had specifically said that we would use coordinate systems whose axes or coordinate lines would be geodesics, which implies that any samples would move along these geodesics. The result of not adhering to this principle is evident from Fig.5.1-2a. The time required by sample A to move from point 000 to 001, or from 001 to 010, or from 011 to 100, is always the same according to Eq.(2). However, the distances travelled with constant velocity in equal time intervals is not the same. The absolute distance D_{ab} between 000 and 001 equals 1, between 001 and 010 it equals 2, and between 011 and 100 it equals 3. We must modify our dyadic coordinate system so that constant velocity and equal time intervals imply equal absolute distance travelled.

5.2 Dyadic Coordinate Systems

Our investigation has so far been based on the binary numbers ordered according to the *usual binary code* shown in the second column of Table 5.2-1. There are many other ways of assigning a binary code to a decimal number. The *Gray code* shown in Table 5.2-1 is often used. It is obtained from the usual binary code by writing 1 if the digit differs from the one to its left, and 0 if it does not differ from the one to its left. The following example shows the process better than many words:

usual binary code	(0)11001011100
Gray code	10101110010

If we substitute the Gray code for the usual binary code in the dyadic group, we will not expect much of a change at first glance, since we are merely relabeling the elements. However, the minimum absolute distance—which is the Hamming distance in the case of the dyadic group—is changed, and this makes an investigation worthwhile. For instance, the Hamming distance between 8 and 12 in Table 5.2-1 equals 1 for the usual binary code, but 2 for the Gray code.

We discuss dyadic shifting using the Gray code with the help of Fig.5.2-1. This illustration is analogous to Fig.5.1-1 except for the change of the code. There are again the four storage or coordinate locations with the addresses 00, 01, 10, and 11 as in Fig.5.1-1, but the samples are stored according to the Gray code; this means C goes into location 11, which is a smaller number than 10 according to Table 5.2-1, and D goes into 10. A dyadic shift can be performed by adding 01, 11, or 10 modulo 2 to the address and leaving the samples where they are, which means the coordinate system is shifted. This is shown in row a of Fig.5.2-1.

The shifting of row a is shown without the help of circles in row c. The samples A, B, C, and D remain in their location, but the marking of the scale is changed by adding 01, 11, or 10 modulo 2 to the numbers on the scale at the extreme left. The unit of the scale is the normalized time t/T.

Row *b* shows the shifting performed by leaving the addresses of the coordinate system unchanged and moving the samples A, B, C, and D. One may readily verify that these four samples are at the same address in each column of rows a and b of Fig.5.2-1. Row d shows the shifting on a straight scale. The marking of the scale from 00 to 10 is the same for all columns of row d, which corresponds to the usual increase of the time scale from left to right.

In addition to the Gray code, there are many other possible codes that can replace the usual binary code. Table 5.2-1 shows in the fourth column the *minimized code*. The reason for this name will become clear later.

TABLE 5.2-1
THREE BINARY CODES FOR THE FIRST 32 INTEGERS $i = 0 \ldots 31$.

decimal	usual binary code	Gray code	minimized code
0	0	0	0
1	1	1	1
2	10	11	11
3	11	10	111
4	100	110	1111
5	101	111	11111
6	110	101	11110
7	111	100	11100
8	1000	1100	11000
9	1001	1101	1000
10	1010	1111	1001
11	1011	1110	1011
12	1100	1010	1010
13	1101	1011	1110
14	1110	1001	1100
15	1111	1000	1101
16	10000	11000	11101
17	10001	11001	11001
18	10010	11011	11011
19	10011	11010	11010
20	10100	11110	10010
21	10101	11111	10000
22	10110	11101	10001
23	10111	11100	10011
24	11000	10100	10111
25	11001	10101	10110
26	11010	10111	10100
27	11011	10110	10101
28	11100	10010	101
29	11101	10011	100
30	11110	10001	110
31	11111	10000	10

For the moment, let us note that any character of the minimized code or the Gray code has the Hamming distance 1 from the preceding as well as from the following character, a feature not shared by the usual binary code. For instance, character 16 of the usual binary code in Table 5.2-1 has the Hamming distance 5 from character 15, and the Hamming distance 1 from character 17, whereas character 16 of the Gray or the minimized code has the Hamming distance 1 from either character 15 or 17.

A listing of the binary code, the Gray code and the minimized code for 2, 4, 8, and 16 elements is shown in Table 5.2-2. One may see that the usual binary and the Gray code add new characters, but leave the old ones

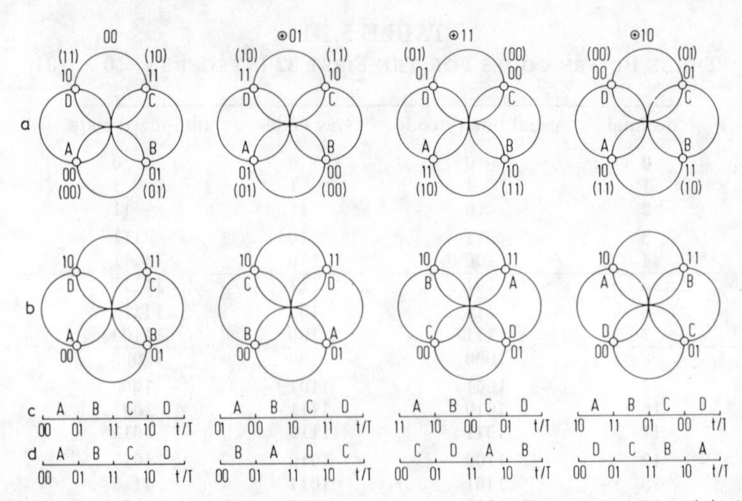

FIG.5.2-1. Dyadic shifting in two dimensions using the Gray code. Row (a) shows the shifting of the addresses (=coordinates) of the samples; row (b) shows the moving of the samples to various addresses. Rows (c) and (d) represent the same operations in one dimension. The numbers in parenthesis in row (a) show the usual binary code.

unchanged as the number of elements increases. The minimized code does not share this feature; for example, the first 8 characters of the code for 16 elements are not equal to the characters of the code for 8 elements. This code is characterized by a transition from the "all zeroes" character to the "all ones" character in a minimum of steps that permit a Hamming distance 1 between adjacent characters (0, 1, 11, 111, ...).

The dyadic shifting of two samples using the Gray code with eight elements is shown in Fig.5.2-2. This illustration is the equivalent of Fig.5.1-2, which holds for the usual binary code. Sample A is at the location $\xi_0 = 000$ in Fig.5.2-2a, left. Adding $\beta_b\theta_1$ modulo 2 to ξ_0 with $\beta_b = 1$ according to Eq.(5.1-2) yields the shifted positions ξ_1. For $\theta_1 = 001, 011, \ldots, 100$, sample A moves from 000 to 001, to 011, and so on. This is indicated by the lines with arrows and the numbers 0, 1, ... , 7. Sample A moves first in the direction x, then in the direction z, again in the direction x, then in the direction y, and so on. Row b of Fig.5.2-2 shows this movement represented in one dimension. Furthermore, row c shows how sample A moves if the corner points $\xi = 000$ to 100 of the cube are represented as points on the one-dimensional scale.

On the right of Fig.5.2-2a, the movement of sample B originally located at point 001 is shown. Adding 001, 011, 010, ... modulo 2 to 001 shows that

TABLE 5.2-2

BINARY CODE, GRAY CODE, AND MINIMIZED CODE FOR GROUPS WITH 2, 4, 8, AND 16 ELEMENTS.

decimal	usual binary code	Gray code	minimized code
0	0	0	0
1	1	1	1
0	0	0	0
1	1	1	1
2	10	11	11
3	11	10	10
0	0	0	0
1	1	1	1
2	10	11	11
3	11	10	111
4	100	110	101
5	101	111	100
6	110	101	110
7	111	100	10
0	0	0	0
1	1	1	1
2	10	11	11
3	11	10	111
4	100	110	1111
5	101	111	1110
6	110	101	1100
7	111	100	1000
8	1000	1100	1010
9	1001	1101	1011
10	1010	1111	1001
11	1011	1110	1101
12	1100	1010	101
13	1101	1011	100
14	1110	1001	110
15	1111	1000	10

sample B moves first in direction x, then in direction z, again in direction x, then in direction y, and so on. This is exactly the same sequence as the one for sample A, and the one-dimensional representation of row b applies thus for sample B as well as for A.

The representation of Fig.5.2-2c, on the other hand, makes sample B move sometimes to the left and sometimes to the right, whereas sample A always moves to the right. This representation, as well as the world line representation of Fig.5.2-2d, give the same result as the corresponding illustration for the usual binary code in Fig.5.1-2. Note that the samples

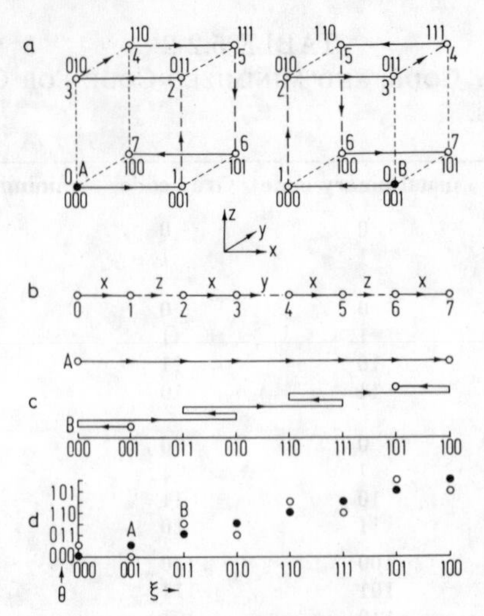

FIG.5.2-2. Various representations of dyadic shifting of samples A and B at points 000 and 001, using the Gray code.

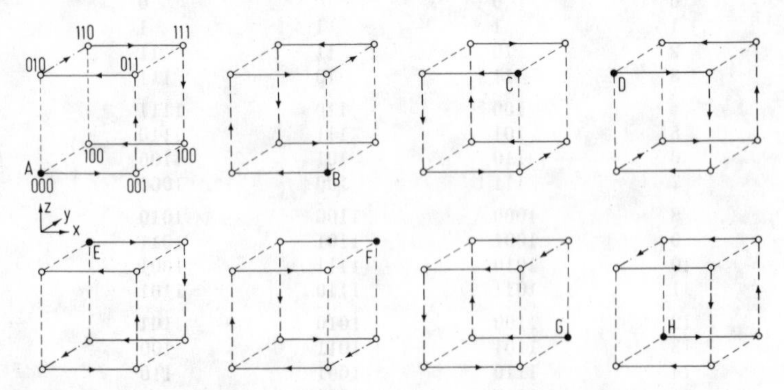

FIG.5.2-3. Dyadic shifting based on the Gray code of the samples A, B, ... , H located at the points 000, 001, ... , 100 according to the Gray code with eight elements in Table 5.2-2.

A and B always move the same distance X between successive points in Fig.5.2-2a, whereas this was not so in Fig.5.1-2a due to the movement along diagonals.

Figure 5.2-3 shows the movement of all possible eight samples A to H according to the Gray code. This illustration is the equivalent of Fig.5.1-3 for the usual binary code. In contrast to Fig.5.1-3, there is no movement on diagonals.

Let us turn to shifting according to the minimized code. We use eight

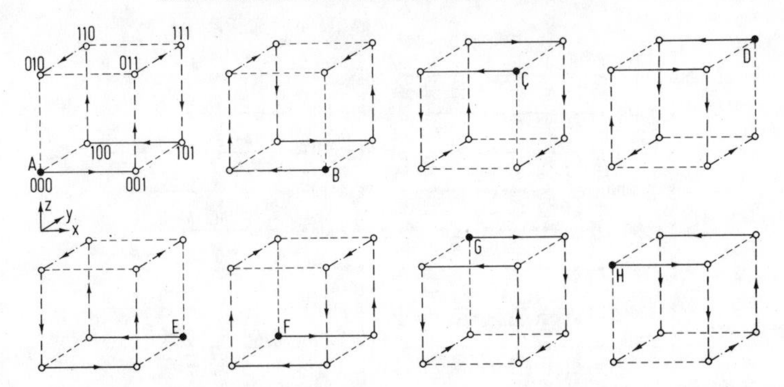

FIG.5.2-4. Dyadic shifting based on the minimized code of the samples A, B, ... , H located at the points 000, 001, ... , 010 according to the minimized code with eight elements in Table 5.2-2.

samples A to H and eight coordinate locations 000 to 111. The minimized code with eight elements in Table 5.2-2 has to be used. The eight corner points of a cube are plotted in the upper left corner of Fig.5.2-4 and labeled as in Figs.5.1-3 and 5.2-3. The sample A is at the location 000. According to Table 5.2-2 we must add the numbers 1, 11, 111, 101, 100, and 10 modulo 2 to 000, which leaves those numbers unchanged. Sample A thus moves in Fig.5.2-4 from 000 to 001, 011, 111, 101, 100, 110, and 010.

The samples B, C, ... , H in Fig.5.2-4 are located at 001, 011, ... , 010, which corresponds to the sequence of the eight elements of the minimized code in Table 5.2-2. The shifting of sample B requires that the numbers 1, 11, 111, 101, 100, 110, and 10 are added modulo 2 to the address 001 of B; and so on.

If we discuss shifting along a straight line in a three-dimensional Cartesian coordinate system, we rotate the coordinates so that the shifted sample moves along one of the axes. The shifting in three dimensions is thus reduced to a one-dimensional shift $f(x - vt)$ along the x–axis. If the shifting is not along a straight line but, e.g. along a circle, we use polar or spherical coordinates to obtain again a shift in one dimension, where the one dimension is a straight line that represents the actual circle. If we can find a one-dimensional representation for the shifting represented by Figs.5.2-3 and 5.2-4, we achieve a similar simplification as with the three-dimensional Cartesian or a spherical coordinate system reduced to one dimension.

Refer to Fig.5.2-5. The points 000, 001, ... , 100 that are at the corners of a cube in Fig.5.2-2a, are arranged along a straight line in Fig.5.2-5a, row 1. The samples A, B, ... , H are at the locations 000, 001, ... , 100 which correspond to their locations in Figs.5.2-2a and 5.2-3.

In order to obtain the locations of the samples A, B, ... , H when

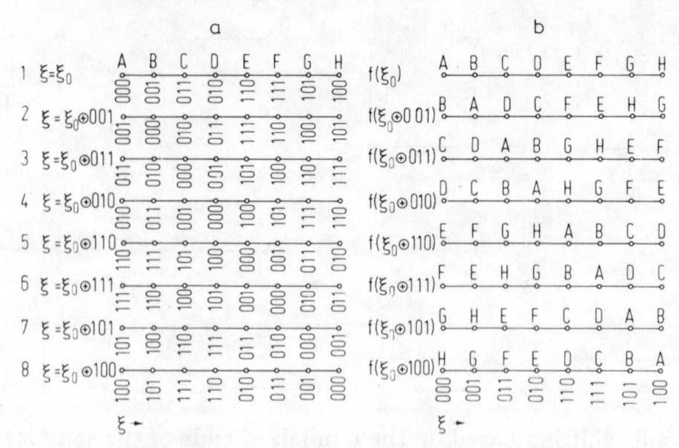

FIG.5.2-5. Dyadic shifting of the coordinate system with the samples A, B, ... , H remaining fixed (a), and dyadic shifting of the samples with the coordinate system remaining fixed (b).

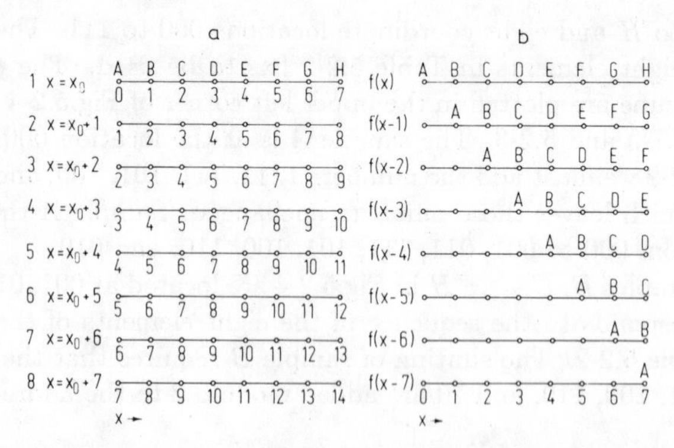

FIG.5.2-6. Usual shifting of the coordinate system in the direction of the x-axis with the samples A, B, ... , H remaining fixed (a), and usual shifting of the samples along the x-axis with the coordinate system remaining fixed (b).

shifted by 001, 011, ... , 100, we add 001, 011, ... , 100 modulo 2 to the numbers ξ_0 of Fig.5.2-5a, row 1. The resulting numbers are shown in rows 2 to 8. One may readily verify that the successive locations of samples A and B in Fig.5.2-5a are the same as in Fig.5.2-2a, while the same verification for samples C ... H in Fig.5.2-3 takes somewhat more time. Hence, we have found a one-dimensional representation for a dyadic shift in three dimensions; it is fairly evident, that this one-dimensional representation can be extended to more than three dimensions.

There is still a flaw in Fig.5.2-5a. It represents a shift of the coordinate points 000, 001, ... , 100 relative to the samples A, B, ... , H which are held fixed. We usually want the coordinate system fixed and the samples A

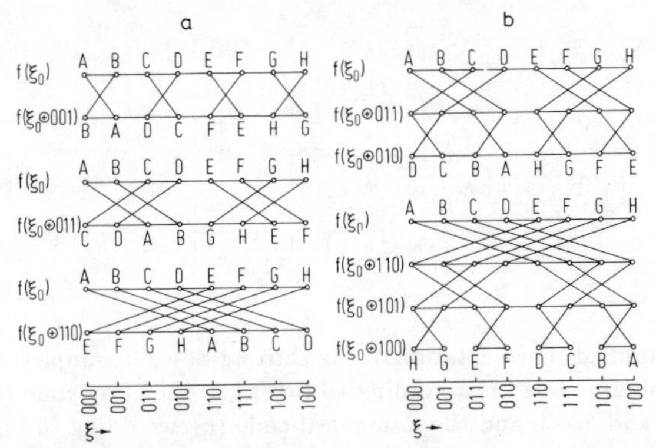

FIG.5.2-7. Exchange of neighbors, pairs of neighbors, and quadruples of neighbors by the dyadic shifts 001, 011, and 110 (a). The shifts $010 = 011 \oplus 001$ and $100 = 110 \oplus 011 \oplus 001 = 101 \oplus 001$ are composed from these basic shifts (b).

... H shifted relative to it. Such a representation is shown in Fig.5.2-5b. The coordinate marks 000, 001, ... , 100 are now fixed and the samples A, B, ... , H are moved so that for any shift 001, 011, ... , 100 they are located at the same coordinate points as in Fig.5.2-5a. For instance, row 2 shows in Fig.5.2-5a for $\xi = \xi_0 \oplus 001$ the coordinates 001, 000, 010, ... for the samples A, B, C, ... , while in Fig.5.2-5b for $f(\xi_0 \oplus 001)$ the samples B, A, D, ... are shown for the coordinates 000, 001, 011, ...

For a better understanding of the dyadic shift of the coordinate system relative to the samples A, B, ... in Fig.5.2-5a and the shift of the samples A, B, ... relative to the coordinate system in Fig.5.2-5b let us consider the equivalent operations in a Cartesian coordinate system. Figure 5.2-6a, row 1, shows the samples A, B, ... , H at the coordinate points 0, 1, ... , 7 of the $x-$axis. The shifts $x = x_0 + 1$, $x_0 + 2$, ... , $x_0 + 7$ add 1, 2, ... , 7 to the $x-$coordinates of the samples in rows 2 to 8. In Fig.5.2-6b the samples A, B, ... , H are shifted according to $f(x-1)$, $f(x-2)$, ... to increasing coordinate values.

The shifting according to Fig.5.2-6b is so familiar that we can write a particular shift, e.g. $f(x-5)$, without adding 5 to the coordinates x of the samples of the function $f(x)$, and writing A, B, ... at these new coordinate points. At first glance, no such simple way seems to exist for the dyadic shift according to Fig.5.2-5b. However, let us look at the shifts $f(\xi_0 \oplus 001)$, $f(\xi_0 \oplus 011)$, and $f(\xi_0 \oplus 110)$. Compared with $f(\xi_0)$ the first shift interchanges neighbors, the second pairs of neighbors, and the third quadruples of neighbors. This is shown more clearly in Fig.5.2-7a.

The Gray code numbers 001, 011, 110 are powers of 2. The general form of the number 2^n in Gray code is 110...0, with a total of $n+1$ digits.

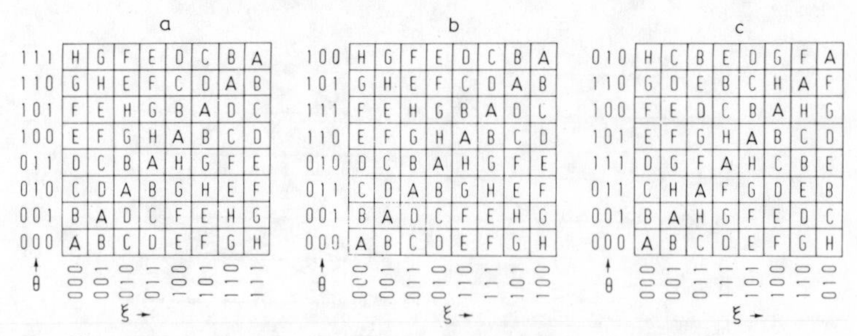

FIG.5.2-8. Simplified representation of the shifting of eight samples A through H for the usual binary code (a) according to Fig.5.1-3, the Gray code (b) according to Figs.5.2-5b and 5.2-3, and the minimized code (c) according to Fig.5.2-4.

One obtains, e.g. $2^0 = 1$, $2^1 = 11$, $2^2 = 110$, $2^3 = 1100$, $2^4 = 11000$. The dyadic shift by a power of two is clear from Fig.5.2-7a. Shifts that are not equal to a power of 2 can be decomposed into a sum of such powers. For instance, the number 010 is the modulo 2 sum of $2^1 = 11$ and $2^0 = 1$, $11 \oplus 01 = 10 = 010$. This shift is thus composed of a shift by 011 (exchange of pairs of neighbors) and a shift by 001 (exchange of neighbors) as shown on top of Fig.5.2-7b. The sequence in which these two shifts are performed is unimportant.

The number 100 can be decomposed into $2^2 \oplus 2^1 \oplus 2^0 = 110 \oplus 011 \oplus 001 = 101 \oplus 001$. This shift is shown as a further example in Fig.5.2-7b. The shift 110 means an exchange of quadruples of neighbors, the shift 011 one of pairs of neighbors, and the shift 001 one of neighbors. The sequence in which these shifts are performed is again unimportant.

Figure 5.2-8b shows once more the dyadic shifting of the samples $A \dots H$ of Fig.5.2-5b or 5.2-3 for the normalized velocity $\beta_b = v_b/c = 1$ according to the Gray code. The dyadic shifting according to the usual binary code of Fig.5.1-3 is represented by Fig.5.2-8a, while the shifting according to the minimized code of Fig.5.2-4 is represented by Fig.5.2-8c.

The samples in Fig.5.2-8a and b are both shifted according to the rules of dyadic shifting as shown by Fig.5.2-7. The reason is an automorphism of the dyadic group; the modulo 2 addition $a \oplus b = c$ remains unchanged if we use the Gray code instead of the usual binary code for a, b, and c, as may readily be verified with the help of Table 5.2-1. On the other hand, the samples A to H in Fig.5.2-8c do not follow the rules of dyadic shifting. However, sample A is still shifted proportionate to the normalized time θ as in Fig.5.2-8a and b. We say that A is shifted in its *eigencoordinates*. this means simply that a coordinate system is chosen so that A is at its origin $\xi = 0$ at the time $\theta = 0$. This causes no problem as long as there is only one sample, but we will have to investigate in more detail the shifting of more than one sample or particle.

Let us observe that the antipode is reached in Fig.5.2-4 in three steps: Sample A moves in three steps from 000 to 111, sample B from 002 to 110, sample C from 011 to 100, and so on. In Fig.5.2-3 each sample A to H is moved in five steps to its antipode, whereas seven steps are needed in Fig.5.1-3. This difference explains the choice of the term minimized code or minimized, Gray-like code.

5.3 Standing Waves and Topology

In theoretical work it is often quite evident which coordinate system one should use. Circular symmetry calls for a polar or cylindrical coordinate system, spherical symmetry for a spherical coordinate system, etc. In a practical situation it is often not obvious which coordinate system one should use, and one has to resort to observation. Should we use a Cartesian or some other coordinate system suitable for planar geometry on the surface of the Earth, or a coordinate system suitable for the surface of a sphere, such as circles of latitude and longitude? One way to answer this question is to observe the angles of large triangles, all having equally long sides, staked out on the surface of the Earth. If the sum equals π everywhere, one should use a coordinate system of planar geometry, while a sum larger than π and equal for all triangles calls for a coordinate system suitable for the surface of a sphere.

Can we come up with an observation that would tell us whether to use a (Cartesian) coordinate system with the usual topology of the real numbers for the marks along the axes, or some other coordinate system, for instance a dyadic coordinate system? For distances that are neither too large nor too small we know that a three-dimensional Cartesian coordinate system with the usual topology of the real numbers for the marks along the axes will do, since we regard this coordinate system as a precise mathematical representation of what we call "the three-dimensional continuous space". For very large distances we assume, under the influence of the general theory of relativity, that a closed coordinate system according to Fig.3.1-7 with a very large value of 2^N would be better. For very short distances we usually assume that the Cartesian coordinate system with real-number-topology will hold even for infinitesimal distances dx, but we have no physical theory that tells us this. It is strictly the authority of Aristotle and the success of differential mathematics that makes us use such coordinate systems for short distances, but it is worth remembering that the authority of Euclid and the success of Euclidean geometry were also once considered compelling.

It appears that standing waves permit us to decide by observation which coordinate system—not having the usual topological features—should be

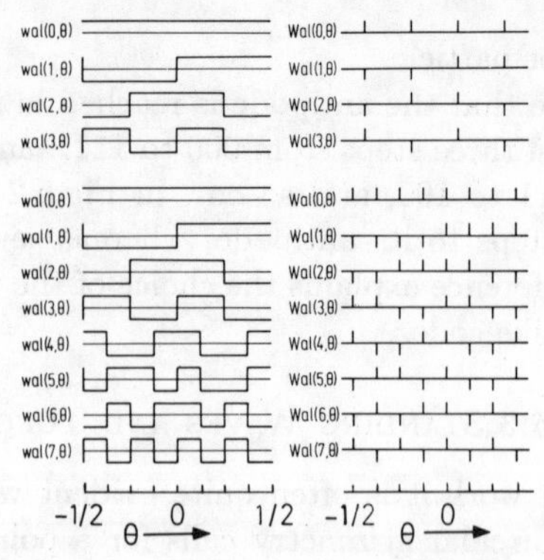

FIG.5.3-1. Walsh functions represented by almost everywhere continuous and differentiable functions $\mathrm{wal}(k,\theta)$, and by sampled functions $\mathrm{Wal}(k,\theta)$ in the interval $-1/2 \leq \theta \leq 1/2$. Note the difference in the notation $\mathrm{wal}(k,\theta)$ and $\mathrm{Wal}(k,\theta)$. The parameter k equals the number of jumps or zero crossings of the function $\mathrm{wal}(k,\theta)$ in the *open* interval $-1/2 < \theta < 1/2$.

used, and that they might do for (coordinate systems with) non-continuum topologies what the sum of the angles in a triangle did for non-Euclidean geometries.

Consider the following standing wave:

$$\cos(2\pi kx/X)\cos(2\pi kct/X) = \frac{1}{2}\{\cos[2\pi k(x+ct)/X] + \cos[2\pi k(x-ct)/X]\} \quad (1)$$

This formula shows that a standing sinusoidal wave can be produced by the superposition of a sinusoidal wave moving from right to left and another moving from left to right. The sum $x+ct$ and $x-ct$ show that addition and subtraction as defined for real numbers are required for this formula; in other words, the usual real-number-topology is implied for the variables x and t.

We have seen that the dyadic group has the addition modulo 2 as group operation; its inverse was the modulo 2 subtraction, which happens to yield the same result as the modulo 2 addition. In order to find the form of a standing wave for modulo 2 addition and subtraction, we replace $x/X+ct/X$ and $x/X-ct/X$ by $x/X \oplus ct/X$ and $x/X \ominus ct/X$. The functions distinguished by this dyadic shift will not be sinusoidal functions, but some yet to be determined functions $f(k,x/X)$ and $f(k,ct/X)$:

$$f(k, x/X)f(k, ct/X) = \frac{1}{2}[f(k, x/X \oplus ct/X) + f(k, x/X \ominus ct/X)]$$
$$= f(k, x/X \oplus ct/X) \qquad (2)$$

A known system of functions that satisfies this equation is the complete, orthonormal system of Walsh functions[1] $\mathrm{wal}(k,\theta)$ or $\mathrm{Wal}(k,\theta)$ shown in Fig.5.3-1. We prefer here the sampled representation of these functions, and replace Eq.(2) by the following equation:

$$\mathrm{Wal}(k, x/X)\,\mathrm{Wal}(k, ct/X) = \mathrm{Wal}(k, x/X \oplus ct/X) \qquad (3)$$

This equation shows that a standing Walsh wave will be obtained if we use the dyadic shift $x/X \oplus ct/X$ for a Walsh function.

For a sinusoidal function $\sin 2\pi k\theta$ it is obvious that k and θ play equal roles. Hence, a rule like

$$\sin[2\pi k(\theta + \theta')] = \sin(2\pi k\theta)\cos(2\pi k\theta') + \cos(2\pi k\theta)\sin(2\pi k\theta')$$

implies the rule:

$$\sin[2\pi(k + k')\theta] = \sin(2\pi k\theta)\cos(2\pi k'\theta) + \cos(2\pi k\theta)\sin(2\pi k'\theta)$$

The same equivalence of the variables k and θ applies to Walsh functions $\mathrm{Wal}(k,\theta)$. Hence, Eq.(3) written in the form

$$\mathrm{Wal}(k,\theta)\mathrm{Wal}(k,\theta') = \mathrm{Wal}(k, \theta \oplus \theta') \qquad (4)$$

implies the equation

$$\mathrm{Wal}(k,\theta)\,\mathrm{Wal}(k',\theta) = \mathrm{Wal}(k \oplus k', \theta) \qquad (5)$$

The Walsh functions are isomorphic to the dyadic group. The modulo 2 addition as group operation is replaced by the multiplication of Walsh functions. For instance $101 \oplus 011 = 110$ becomes:

$$\mathrm{Wal}(101,\theta)\,\mathrm{Wal}(011,\theta) = \mathrm{Wal}(110,\theta)$$
$$101 = 5,\ 011 = 3,\ 110 = 6,\ 101 \oplus 011 = 110 \qquad (6)$$

[1] Fine (1949, 1950); Harmuth (1977, p. 29); Harris and Wade (1971); Pichler (1967); Selfridge (1955); Vilenkin (1947); Wade (1969, 1971, 1975); Wallis *et al.* (1972, p. 437); Walsh (1923); Watari (1965).

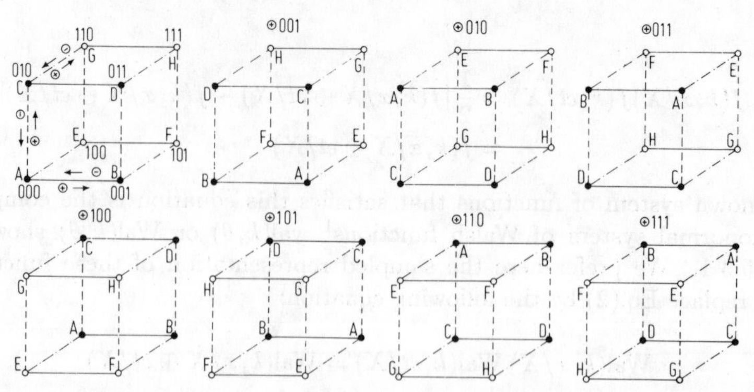

FIG.5.3-2. The standing wave $\mathrm{Wal}(1,\xi \oplus \theta)$ at the times $\theta = 000, 001, \ldots, 111$ at the coordinate points $\xi = 000, 001, \ldots, 111$.

We have just replaced the numbers 5, 6, and 3 by binary numbers using the usual binary code. One may just as well use the Gray code, due to the previously mentioned automorphism of the dyadic group for Gray code and usual binary code. Equation (6) reads in this case as follows:

$$\mathrm{Wal}(111,\theta)\,\mathrm{Wal}(010,\theta) = \mathrm{Wal}(101,\theta)$$
$$111 = 5,\ 010 = 3,\ 101 = 6,\ 111 \oplus 010 = 101 \tag{7}$$

The relation $\mathrm{Wal}(5,\theta)\,\mathrm{Wal}(3,\theta) = \mathrm{Wal}(6,\theta)$ may readily be verified from Fig.5.3-1 either for the functions $\mathrm{Wal}(k,\theta)$ or $\mathrm{wal}(k,\theta)$. The unit element is $\mathrm{Wal}(0,\theta)$; the inverse element of $\mathrm{Wal}(k,\theta)$ is the function $\mathrm{Wal}(k,\theta)$ itself. The shape of the functions $\mathrm{Wal}(2^n - 1,\theta)$ and $\mathrm{wal}(2^n - 1,\theta)$ is readily recognizable for $2^n - 1 = 1,\ 3,\ 7$, and holds generally. With these functions, the generalized relations of Eqs.(6) or (7),

$$\mathrm{Wal}(i,\theta)\,\mathrm{Wal}(k,\theta) = \mathrm{Wal}(i \oplus k,\theta) \tag{8}$$

and the definition of $\mathrm{Wal}(0,\theta)$ by Fig.5.3-1, one may produce functions $\mathrm{Wal}(k,\theta)$ or $\mathrm{wal}(k,\theta)$ with any value of k.

Let us see how the standing Walsh wave comes about, using the representation of Figs.5.1-3 or 5.2-3. Figure 5.3-2 shows the eight samples A, $B, \ldots, H$ located at the points $000, 001, \ldots, 111$; the usual binary code is used since it is more widely known than the Gray code. The samples represent the Walsh function $\mathrm{Wal}(k, x/X \oplus ct/X) = \mathrm{Wal}(k,\xi \oplus \theta)$ with $k = 1$. At the time $\theta = 0$ the samples A, B, C, and D have thus the value $+1$, while the samples E, F, G, and H have the value -1. The samples $+1$ and

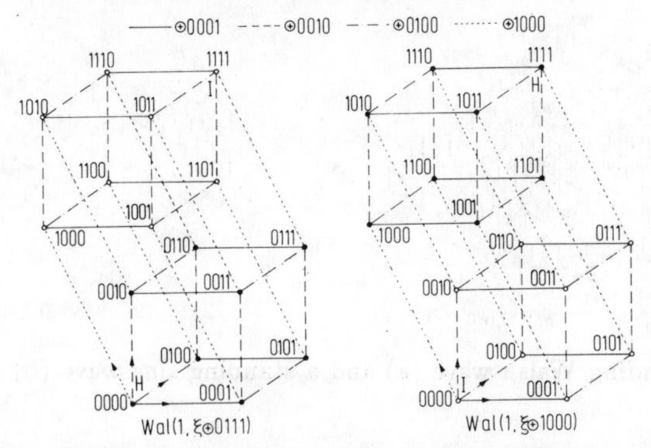

FIG.5.3-3. The standing Walsh wave Wal$(1,\xi\oplus\theta)$ with 16 samples in four dimensions; $\xi = 0000, 0001, \dots, 1111$; $\theta = 0111$ and 1000.

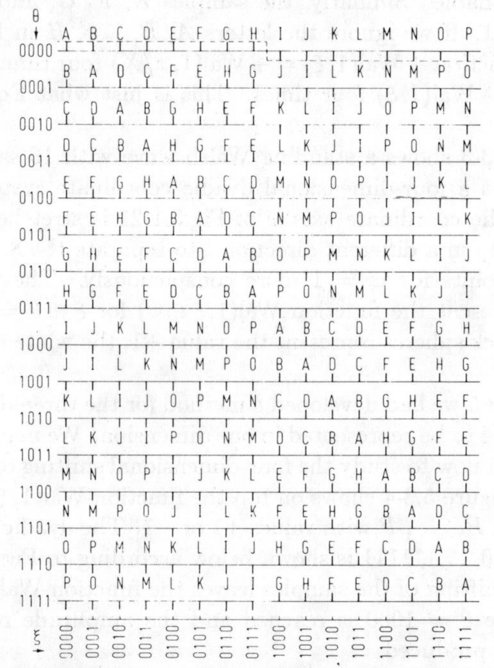

FIG.5.3-4. One-dimensional representation of the standing Walsh wave Wal$(1,\xi\oplus\theta)$ of Fig.5.3-3 for $\xi = 0000, 0001, \dots, 1111$ and $\theta = 0000, 0001, \dots, 1111$.

-1 are indicated by black and white spheres at the corners of the cube in the upper left of Fig.5.3-2.

Let us now perform the dyadic shifts $\oplus 001, \oplus 010, \dots, \oplus 111$ as shown in Fig.5.3-2. The eight samples $A, B, \dots, H$ are shifted as in Fig.5.1-3.

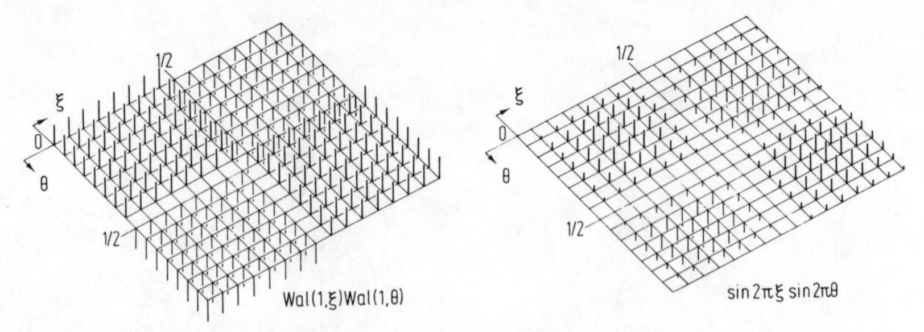

FIG.5.3-5. A standing **Walsh wave** (a) and a standing sine wave (b) represented by samples.

However, the samples A, B, C, and D all have the same value $+1$ and are not distinguishable. Similarly, the samples E, F, G, and H all have the same value -1. If we ignore the letters A, B, ..., H in Fig.5.3-2, we see on top the function $+\operatorname{Wal}(1,\xi) = +\operatorname{Wal}(1,x/X)$ four times and on bottom the function $-\operatorname{Wal}(1,\xi)$ four times. This is just what Eq.(3) requires for $k = 1$.

Figure 5.3-3 shows a standing Walsh wave with 16 samples. This example requires a four-dimensional dyadic coordinate system according to Fig.3.1-21. The coordinate axis w of Fig.3.1-21 is stretched in Fig.5.3-3—and also points in a different direction—to separate the 8 points for $w = 0$ from the 8 points for $w = 1$ more conspicuously. The two drawings of Fig.5.3-3 represent the function $\operatorname{Wal}(1, \xi \oplus \theta)$ for $\xi \oplus \theta = 0111$ and 1000; again, the black spheres represent the value $+1$, the white spheres the value -1.

In Fig.5.2-5 we had developed a method for the three-dimensional shifting of Fig.5.2-4 to be represented in one dimension. We may use this simpler representation now to study the four-dimensional shifting of Fig.5.3-3 in one dimension. Figure 5.3-4 shows on top the function $\operatorname{Wal}(1,\xi)$ represented by 16 samples A, B, ..., P with values $+1$ or -1. The dyadic shifting $\xi \oplus \theta$ for $\theta = 0001$, 0010, ..., 1111 is shown below according to Fig.5.2-5b. One can see how the shifting of the samples leaves the function $\operatorname{Wal}(1,\xi)$ unchanged until the time $\theta = 1000$ is reached and the amplitude reversed function $-\operatorname{Wal}(1,\xi)$ is produced.

Figure 5.3-5 shows a standing Walsh wave and a standing sine wave represented by samples on a discrete, two-dimensional grid produced by the space-variable ξ and the time variable θ. The scales for space- and time-axes are the same for both standing waves, since we have arbitrarily chosen this particular representation. What is not arbitrary is the fact that a dyadic shift $\xi \oplus \theta$ will produce a standing Walsh wave, while the usual shifts $\xi - \theta$ and $\xi + \theta$ will produce a standing sine wave. Hence, at least in these two cases

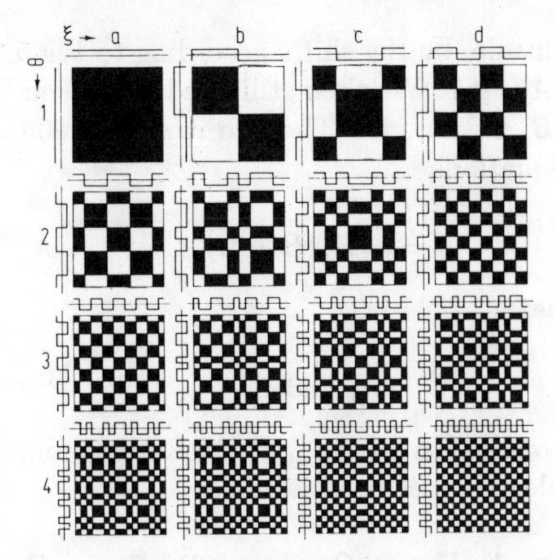

FIG.5.3-6. The patterns of standing waves produced by the continuous functions wal$(0,\xi)$ to wal$(15,\xi)$ in the interval $0 \le \xi < 1$, $0 \le \theta < 1$. Black represents $+1$, white represents -1.

one can infer the appropriate shift and from it the appropriate topology of the coordinate system from an observation of the standing waves produced.

Figure 5.3-6 shows patterns of standing waves produced by the Walsh functions shown above each pattern. Black stands for $+1$ and white for -1. These patterns are plotted for the functions wal(k,θ) according to Fig.5.3-1, since they can be represented by black and white, while the functions Wal(k,θ) would require an additional gray level; but there is no difficulty visualizing the patterns produced by the sampled functions Wal(k,θ).

Standing waves are usually associated with acoustic and electromagnetic oscillations, but the term should be interpreted here in a more general sense. For instance, the Walsh function wal$(1,\xi\oplus\theta)$ in column b, row 1 of Fig.5.3-6 might represent a proton (black) and a neutron (white). Sometime later, the proton and the neutron interchange positions, and this interchange is repeated periodically for larger time θ. Such an example is within the meaning of "standing wave" as used here.

5.4 Observed Shifts and Their Eigenfunctions

We call the sine and cosine functions the eigenfunctions of the shifts $\xi+\theta$, $\xi-\theta$, and the Walsh functions the eigenfunctions of the shifts $\xi\oplus\theta = \xi\ominus\theta$. The same Walsh function Wal$(k,\xi\oplus\theta)$ is obtained as standing wave whether we use the usual binary code or the Gray code for k; this is quite evident from Figs.5.2-8a and b. It is also evident from Fig.5.2-8c that the minimized code will *not* yield the Walsh functions as standing waves. Let us derive the eight functions that will remain unchanged, except perhaps for a reversal of amplitude, by the shifts according to Fig.5.2-8c.

We choose first $A = +1$. We then still need seven conditions to determine seven samples $B, C, \ldots, H$. The two rows $\theta = 000$ and $\theta = 001$ of Fig.5.2-8c yield the equations

$$A = \alpha_1 B, \; B = \alpha_1 A \tag{1}$$

with the two solutions:

$$\alpha_1 = \pm 1 \tag{2}$$

We can choose one value of α_1. The following four conditions are imposed on the samples B to H by this choice:

$$B = \alpha_1 A, \; H = \alpha_1 C, \; G = \alpha_1 D, \; F = \alpha_1 E \tag{3}$$

The conditions $E = \alpha_1 F$, $D = \alpha_1 G$, and $C = \alpha_1 H$, which also follow from the two rows $\theta = 000$ and $\theta = 001$, are automatically satisfied due to Eq.(2).

The rows $\theta = 000$ and $\theta = 011$ in Fig.5.2-8c yield a second set of equations,

$$A = \alpha_2 C, \; C = \alpha_2 A \tag{4}$$

with the solutions:

$$\alpha_2 = \pm 1 \tag{5}$$

Again, we can choose one value of α_2. Two additional conditions are imposed on the samples B to H by this choice:

$$C = \alpha_2 A, \; F = \alpha_2 D \tag{6}$$

The conditions $H = \alpha_2 B$ and $G = \alpha_2 E$ follow already from Eqs.(3) and (6),

$$H = \alpha_1 C = \alpha_1 \alpha_2 A, \; G = \alpha_1 D = (\alpha_1/\alpha_2) F \tag{7}$$

while the conditions $D = \alpha_2 F$, $E = \alpha_2 G$, and $B = \alpha_2 H$ are automatically satisfied due to Eq.(5).

A third set of equations is obtained from the rows $\theta = 000$ and $\theta = 111$ of Fig.5.2-8c:

$$D = \alpha_3 A, \; A = \alpha_3 D \tag{8}$$

Again, there are two solutions:

TABLE 5.4-1
THE 2^3 SETS OF SOLUTIONS FOR $\alpha_1, \alpha_2, \alpha_3$, AND THE VALUES OF THE SAMPLES $B, C, \ldots, H$ DEFINED BY THEM.

k	α_1	α_2	α_3	
0	+1	+1	+1	$A = 1$
1	+1	+1	−1	$B = \alpha_1 A = \alpha_1$
2	+1	−1	+1	$C = \alpha_2 A = \alpha_2$
3	+1	−1	−1	$D = \alpha_3 A = \alpha_3$
4	−1	+1	+1	$E = \alpha_1 F = \alpha_1\alpha_2\alpha_3$
5	−1	+1	−1	$F = \alpha_2 D = \alpha_2\alpha_3$
6	−1	−1	+1	$G = \alpha_1 D = \alpha_1\alpha_2$
7	−1	−1	−1	$H = \alpha_1 C = \alpha_1\alpha_2$

TABLE 5.4-2
THE EIGHT TWO-VALUED FUNCTIONS mgl(k, h) YIELDING STANDING WAVES FOR THE DYADIC SHIFT USING THE MINIMIZED CODE ACCORDING TO FIG.5.2-8c. THE SIGNS + AND − ARE WRITTEN FOR +1 AND −1.

	$h=0$	$h=1$	$h=2$	$h=3$	$h=4$	$h=5$	$h=6$	$h=7$
k	A	B	C	D	E	F	G	H
0	+	+	+	+	+	+	+	+
1	+	+	+	−	−	−	−	+
2	+	+	−	+	−	−	+	−
3	+	+	−	−	+	+	−	−
4	+	−	+	+	−	+	−	−
5	+	−	+	−	+	−	+	−
6	+	−	−	+	+	−	−	+
7	+	−	−	−	−	+	+	+

$$\alpha_3 = \pm 1 \tag{9}$$

Only one new condition is obtained,

$$D = \alpha_3 A \tag{10}$$

all the others, $G = \alpha_3 B$, to $E = \alpha_3 H$, are automatically satisfied due to Eqs.(2), (3), (5), (6), and (9).

The two choices permitted by each one of the three Eqs.(2), (5), and (9) yield 2^3 different sets of solutions for α_1, α_2, and α_3, which are shown in Table 5.4-1.

For the construction of the eight functions we choose first $A = +1$ in Table 5.4-2. The values for B, C, and D follow from Eqs.(3), (6), and (10) according to Table 5.4-1. The values of E, F, G, and H follow from Eqs.(3) and (6). The notation mgl(k, h) is used for this system of orthogonal functions (minimized Gray-code-like functions).

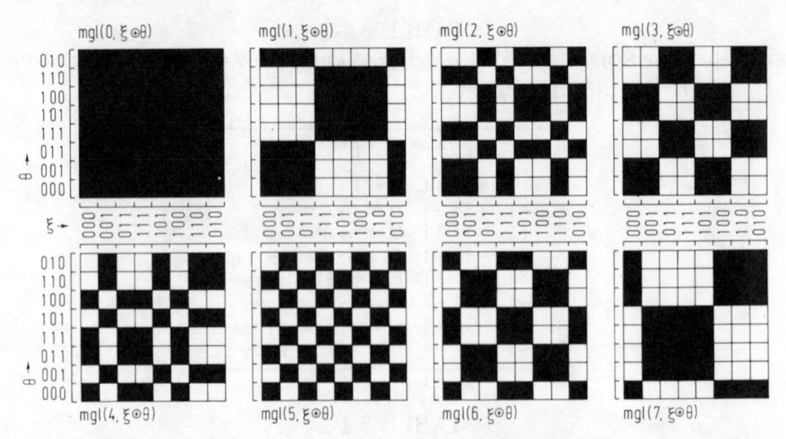

FIG.5.4-1. The patterns of standing waves produced by the functions $\mathrm{mgl}(0,\xi)$ to $\mathrm{mgl}(7,\xi)$. Black represents +1, white represents −1.

Figure 5.4-1 shows the patterns produced by the dyadic shifting of the functions $\mathrm{mgl}(k,h)$. It is evident that each function equals either $\mathrm{mgl}(k,\xi)$ or $-\mathrm{mgl}(k,\xi)$ for any value of $h = \xi \oplus \theta$, and that the relation

$$\mathrm{mgl}(k,\xi)\,\mathrm{mgl}(k,\theta) = \mathrm{mgl}(k,\xi \oplus \theta)$$

is satisfied in analogy to Eq.(5.3-3).

Figures 5.2-8a and b show that the automorphism between the usual binary code and the Gray code results in the same shifting pattern. An equivalent automorphism exists for the minimized code. The same process that transforms the usual binary code into the Gray code in Table 5.2-2 also transforms the minimized code into the *Gray-coded minimized code* as shown in Table 5.4-3. One may readily verify that the modulo 2 addition of two numbers of the minimized code, e.g. $1100 \oplus 1010 = 0110$, yields the same number as the modulo 2 addition of the respective numbers of the Gray-coded minimized numbers, $1010 \oplus 111 = 0101$.

Let us turn to the minimized code with 16 elements in Table 5.2-2. The dyadic shift of 16 samples $A, B, \ldots, P$ using this code is shown in Fig.5.4-2. One may see that the sample A is shifted along the main diagonal; a short reflection shows that this must be so for dyadic shifting using any code. The pattern of permutations of the other samples is not readily recognizable. A geometric representation of the shift of sample A is shown in Fig.5.4-3. The points $\xi = 0000, 0001, \ldots, 0010$ are located at the corners of a four-dimensional cube. Starting at 0000 sample A reaches the antipode 1111 in four steps.

The five-dimensional cube of Fig.5.4-4 represents the shifting of a sample A according to the minimized code with 32 elements according to Table

TABLE 5.4-3

MINIMIZED CODE AND GRAY-CODED MINIMIZED CODE FOR GROUPS WITH 2, 4, 8, AND 16 ELEMENTS.

decimal	minimized code	Gray-coded minimized code	decimal	minimized code	Gray-coded minimized code
0	0	0	0	0	0
1	1	1	1	1	1
			2	11	10
0	0	0	3	111	100
1	1	1	4	1111	1000
2	11	10	5	1110	1001
3	10	11	6	1100	1010
			7	1000	1100
0	0	0	8	1010	1111
1	1	1	9	1011	1110
2	11	10	10	1001	1101
3	111	100	11	1101	1011
4	101	111	12	101	111
5	100	110	13	100	110
6	110	101	14	110	101
7	10	11	15	10	11

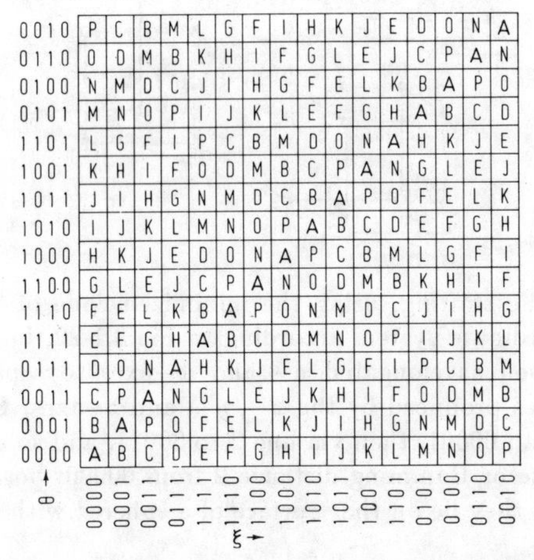

FIG.5.4-2. Dyadic shift of 16 samples $A \ldots P$ using the minimized code with 16 elements of Table 5.2-2. A certain sample, say M, is located at $\xi_0 = 0101$ at the time $\theta_0 = 0$. At a later time, say $\theta = 1100$, it must be located at $\xi \oplus \theta = \xi \oplus 1100 = \xi_0 + \theta_0 = 0101$, which yields $\xi = 0101 \oplus \theta = 0101 \oplus 1100 = 1001$.

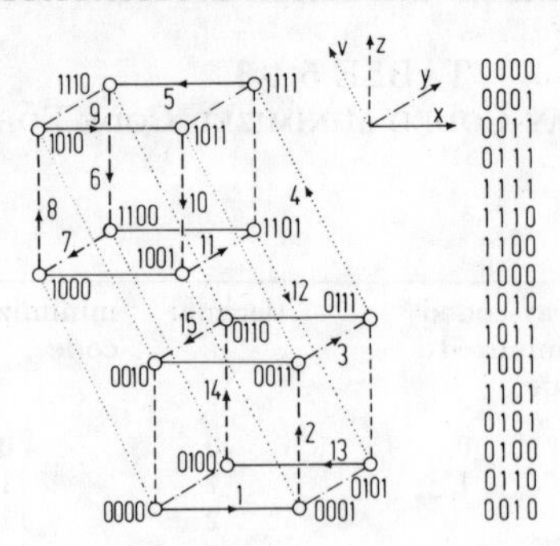

FIG.5.4-3. Shifting of the sample A of Fig.5.4-2 represented in a four-dimensional coordinate system according to Fig.3.1-21, but with differently oriented axes x, y, z, v, and an elongated $v-$axis to separate the two three-dimensional cubes produced by the $x-$, $y-$, and $z-$axes. Sample A moves from 0000 via line 1 to 0001, then via line 2 to 0011, via line 3 to 0111, reaches the antipode 1111 via line 4, continues via line 5 to 1110, and so on.

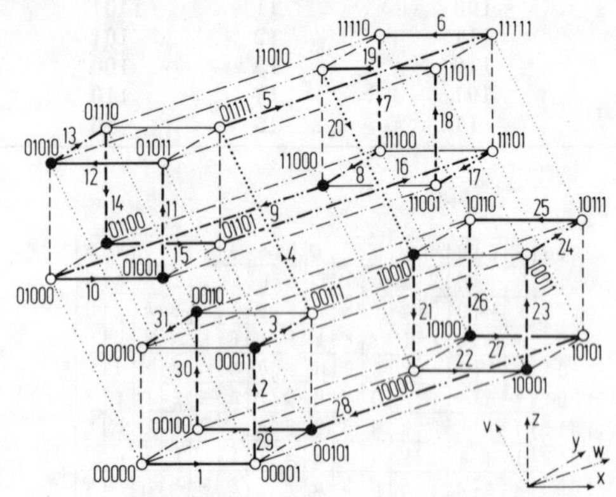

FIG.5.4-4. Dyadic shifting, using the minimized code of Table 5.2-1 in a five-dimensional coordinate system according to Fig.3.1-22, but with differently oriented axes as well as elongated $v-$ and $w-$axes to separate the four three-dimensional cubes produced by the $x-$, $y-$, and $z-$axes. Sample A moves from 00000 via line 1 to 00001, then via line 2 to 00011, and so on. The black spheres have the absolute or Hamming distance 2 from 00000; borrowing an expression, we may say that they lie on the "surface of a sphere" with Hamming radius 2.

5.2-1. The sample is shifted from 00000 via line 1 to 00001, via line 2 to 00011, and so on. The antipode 11111 is reached in five steps.

Consider the concept of the geodesic in a dyadic coordinate system. In Figure 5.4-4 there is one geodesic from the point $\xi = 00000$ to the point 00001; it is the solid line denoted 1. The length of the geodesic is the absolute or Hamming distance between its two ends. It is measured by counting the number of points or marks in Fig.5.4-4 along the geodesic, counting the point at the end but not the one at the beginning. Hence, the length of the geodesic between $\xi = 00000$ and $\xi = 00001$ equals 1. We can travel from 00000 to 00001 via 00010 and 00011; the length of this route equals 3, and it is thus no geodesic.

There are 2! geodesics between the points 00000 and 00011, one via 00010 and the other via 00001; the length of each equals 2. We find 3! geodesics between 00000 and 00111, each having the length 3; furthermore, 4! geodesics exist between 00000 and 01111 with length 4, and 5! geodesics exist between 00000 and 11111 with length 5. Hence, the Hamming distance n between two points equals the length of the geodesics—a property one would demand from an acceptable geodesic—while $n!$ is the number of geodesics between the two points.

The minimized code of Table 5.2-1 makes a sample shift from 00000 to the antipode 11111 along a particular geodesic. Other codes will shift the sample along another geodesic. We have just seen that there are 5! geodesics between 00000 and 11111, and there should thus be 5! codes defining these shifts. One obtains them from the minimized code of Table 5.2-1 by the 5! permutations of the 5 digits of the code[1]. The investigation of the codes is here equivalent to the determination of the functions of time $x = x(t)$, $y = y(t)$, $z = z(t)$ or $x^k = x^k(x^0)$ that define a geodesic in differential geometry.

Let us go on to six dimensions. In the dyadic group, this means that we have $2^6 = 64$ elements and the respective coordinate system has 64 points or marks. A three-dimensional, discrete Cartesian coordinate system defines $4^3 = 64$ coordinate marks. Figure 5.4-5 shows such a Cartesian coordinate system, with the 64 coordinate marks or grid points represented by small spheres. This connection of the 64 spheres corresponds to the usual topology of integer numbers.

Instead of connecting the 64 spheres in Fig.5.4-5 according to the usual topology of the integer numbers, we may connect them according to the topology of the dyadic group with Hamming distance, as shown in Fig.5.4-6. At first glance, this coordinate system does not look like the extension to six dimensions of the coordinate systems in Figs.3.1-20, 3.1-21, and 3.1-22, but it is indeed such a generalization. The rods between the spheres in Fig.5.4-6 are not assumed to be equally long, as was done for the coordinate systems with fewer dimensions, in order to permit the spheres

[1] The zeroes in front of the most significant digit 1 have been left out in Table 5.2-1. One must fill them in if one wants to make the permutations. For instance, 0 has to be replaced by 00000, 1 by 00001, etc.

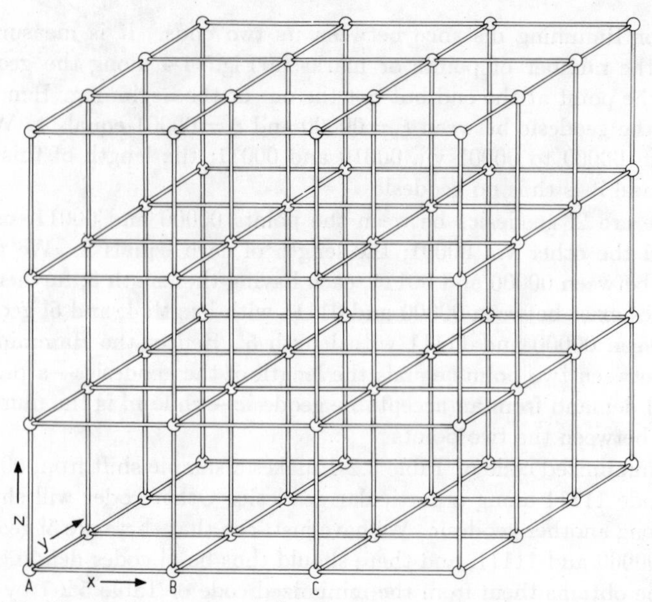

FIG.5.4-5. Connections between $4^3 = 64$ spheres or points according to the usual topology of a three-dimensional Cartesian coordinate system.

to be at the same locations as in Fig.5.4-5. Furthermore, the rods are shown as single rods rather than as double rods for simpler drafting; each rod in Fig.5.4-6 represents a double rod as in Fig.3.1-20b or a ring[2] as in Fig.3.1-20a.

The coordinate system of Fig.5.4-6 seems to be much more complicated than the Cartesian coordinate system of Fig.5.4-5. However, let us look at these two structures from the standpoint of local physics, or from the standpoint of a little ant that can crawl around the spheres and rods but cannot see the structure globally as we do. In Fig.5.4-6 there are always six rods or directions—or rather twelve since each rod shown represents two rods—leading away from each sphere or grid point; hence, all grid points are equal. On the other hand, the coordinate system of Fig.5.4-5 has grid points from which three rods or directions lead away (corner points), others with four rods (edge points), five rods (surface points), and finally there are points from which six rods lead away (interior points). From these local observations at all grid points, the structure of Fig.5.4-6 is classified as a six-dimensional, finite but unbounded coordinate system, while the structure of

[2]The two lines of which each rod in Fig.5.4-6 consists may be taken to be the circle in Fig.3.1-20a compressed into a narrow and very long loop. Of course, this visualization does not interpret correctly the crossings of bars in Fig.5.4-6.

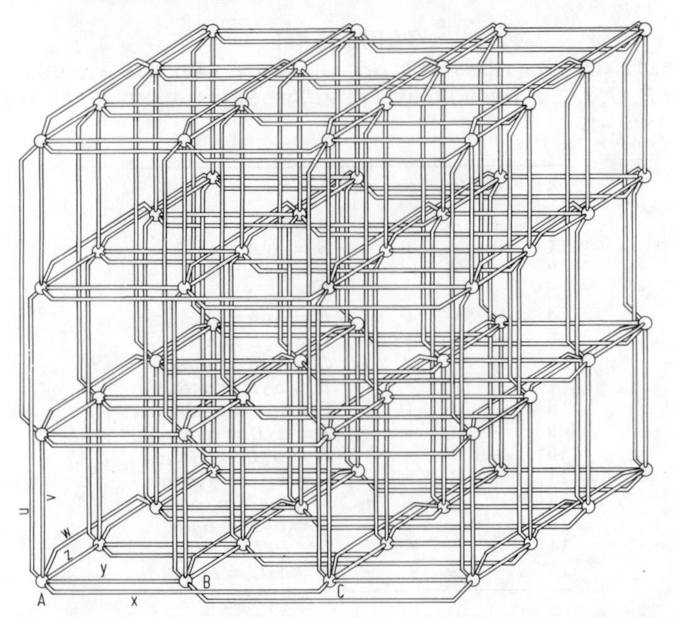

FIG.5.4-6. Connections between $2^6 = 64$ spheres or points according to the topology of a six-dimensional coordinate system representing the dyadic group with Hamming distance.

Fig.5.4-5 is classified as a three-dimensional, finite and bounded coordinate system.

The distance between two adjacent grid points in Fig.5.4-6, measured along the connecting rod, is not always the same. However, one can make these distances equal by using rods of fixed length and changing the position of the spheres accordingly, as done in Figs.3.1-21 and 3.1-22. Using bent rather than straight rods one can make the rods equally long and keep the pattern of Fig.5.4-6 for the grid points, but there are also ways to achieve equal distance within the limits of observability. Assume that the spheres or grid points in Fig.5.4-6 are as close together as we can resolve them with our measuring equipment. An (elementary) particle can be observed at either one point or another, but we cannot observe it travelling between adjacent points since this would require a resolution of more coordinate points. If we cannot observe "intermediate points" between two adjacent points, we cannot assign various values to the distance between two adjacent points, but we have to conclude that the distance has always the same value.

One may object that we can observe the time it takes an elementary particle to travel from one grid point to an adjacent one. This is not so. We can observe the particle at a certain time at one point and at a certain

TABLE 5.4-4

THE 2^4 SETS OF SOLUTIONS FOR α_1, α_2, α_3, α_4, AND THE VALUES OF THE SAMPLES $B, C, \ldots, P$ DEFINED BY THEM. THE SIGNS + AND − ARE WRITTEN FOR +1 AND −1.

k	α_1	α_2	α_3	α_4	
0	+	+	+	+	$A = 1$
1	+	+	+	−	$B = \alpha_1 A = \alpha_1$
2	+	+	−	+	$C = \alpha_2 A = \alpha_2$
3	+	+	−	−	$D = \alpha_3 A = \alpha_3$
4	+	−	+	+	$E = \alpha_4 A = \alpha_4$
5	+	−	+	−	$F = \alpha_1 E = \alpha_1 \alpha_4$
6	+	−	−	+	$G = \alpha_2 E = \alpha_2 \alpha_4$
7	+	−	−	−	$H = \alpha_3 E = \alpha_3 \alpha_4$
8	−	+	+	+	$I = \alpha_1 J = \alpha_1 \alpha_2 \alpha_3 \alpha_4$
9	−	+	+	−	$J = \alpha_2 H = \alpha_2 \alpha_3 \alpha_4$
10	−	+	−	+	$K = \alpha_1 H = \alpha_1 \alpha_3 \alpha_4$
11	−	+	−	−	$L = \alpha_1 G = \alpha_1 \alpha_2 \alpha_4$
12	−	−	+	+	$M = \alpha_2 N = \alpha_1 \alpha_2 \alpha_3$
13	−	−	+	−	$N = \alpha_2 D = \alpha_2 \alpha_3$
14	−	−	−	+	$O = \alpha_1 D = \alpha_1 \alpha_3$
15	−	−	−	−	$P = \alpha_1 C = \alpha_1 \alpha_2$

other time at an adjacent point, but whenever we do not observe it at the one point we must observe it at another. We cannot observe any travel time τ and derive from it a distance $c\tau$.

How can one decide whether the coordinate system of Fig.5.4-5 or 5.4-6 should be used[3]? Let us assume that we have measuring equipment set up that can resolve 64 points. These may either be the points in Fig.5.4-5 or in Fig.5.4-6, but there are of course no rods connecting them. Let an elementary particle be first at point A and then at point C. This is perfectly permissible in the six-dimensional structure of Fig.5.4-6, since point C is as adjacent to point A as point B; the Hamming distance is 1 between A and C as well as between A and B. In Fig.5.4-5, the elementary particle would have to pass through point B on its way from A to C. If it does not, we would say that the particle did not move continuously but jumped. Hence, the observation of jumps is an indication that a coordinate system like the one of Fig.5.4-6 may make motion look simpler than the coordinate system of Fig.5.4-5 with the usual topology of integer numbers in three dimensions.

We determine the eigenfunctions for the shifting according to Fig.5.4-2. The comparison of the rows $\theta = 0000$ and $\theta = 0001$ yields

[3]Putting it less precisely, one could say one wants to decide whether space has the topology of Fig.5.4-5 or 5.4-6.

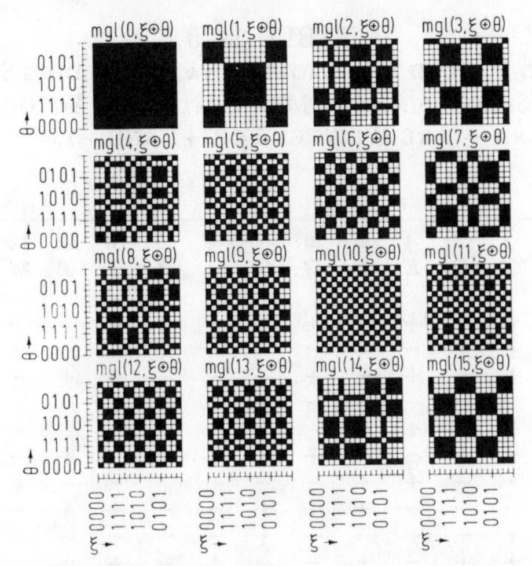

FIG.5.4-7. Patterns of the 16 eigenfunctions for the dyadic shifting according to Fig.5.4-2. For explanation consider the pattern mgl$(1, \xi \oplus \theta)$. For $\theta = 0$ one has the samples $A = B = C = D = M = N = O = P = +1$ (black) and $E = F = G = H = I = K = J = L = -1$ (white). For the times $\theta = 0001$, 0011, 0111 the pattern remains unchanged since samples with value $+1$ are interchanged only with each other. At the time $\theta = 1111$ the samples -1 are interchanged with the samples $+1$ (white squares and black squares interchanged); etc. The patterns mgl$(k, \xi \oplus \theta)$ produced in this way by dyadic shifting can clearly be represented by the product of Eq.(19).

$$A = \alpha_1 B,\ B = \alpha_1 A,\ \alpha_1 = \pm 1 \tag{11}$$

and the following relations:

$$\begin{aligned} &B = \alpha_1 A,\ P = \alpha_1 C,\ O = \alpha_1 D,\ F = \alpha_1 E, \\ &L = \alpha_1 G,\ K = \alpha_1 H,\ J = \alpha_1 I,\ N = \alpha_1 M \end{aligned} \tag{12}$$

The comparison of lines $\theta = 0000$ and $\theta = 0011$ yields:

$$A = \alpha_2 C,\ C = \alpha_2 A,\ \alpha_2 = \pm 1 \tag{13}$$

$$C = \alpha_2 A,\ N = \alpha_2 D,\ G = \alpha_2 E,\ J = \alpha_2 H \tag{14}$$

Lines $\theta = 0000$ and $\theta = 0111$ yield:

TABLE 5.4-5
THE SIXTEEN TWO-VALUED FUNCTIONS mgl(k,h) YIELDING STANDING WAVES FOR THE DYADIC SHIFT USING THE MINIMIZED CODE ACCORDING TO FIG.5.4-2. THE SIGNS + AND − ARE WRITTEN FOR +1 AND −1.

$h \rightarrow$	0	1	2	3	4	5	6	7	8	9	10	11	12	13	14	15
k $\downarrow$	A	B	C	D	E	F	G	H	I	J	K	L	M	N	O	P
0	+	+	+	+	+	+	+	+	+	+	+	+	+	+	+	+
1	+	+	+	+	−	−	−	−	−	−	−	−	+	+	+	+
2	+	+	+	−	+	+	+	−	−	−	−	+	−	−	−	+
3	+	+	+	−	−	−	−	+	+	+	+	−	−	−	−	+
4	+	+	−	+	+	+	−	+	−	−	+	−	−	−	+	−
5	+	+	−	+	−	−	+	−	+	+	−	+	−	−	+	−
6	+	+	−	−	+	+	−	−	+	+	−	−	+	+	−	−
7	+	+	−	−	−	−	+	+	−	−	+	+	+	+	−	−
8	+	−	+	+	+	−	+	+	−	+	−	−	−	+	−	−
9	+	−	+	+	−	+	−	−	+	−	+	+	−	+	−	−
10	+	−	+	−	+	−	+	−	+	−	+	−	+	−	+	−
11	+	−	+	−	−	+	−	+	−	+	−	+	+	−	+	−
12	+	−	−	+	+	−	−	+	+	−	−	+	+	−	−	+
13	+	−	−	+	−	+	+	−	−	+	+	−	+	−	−	+
14	+	−	−	−	+	−	−	−	−	+	+	+	−	+	+	+
15	+	−	−	−	−	+	+	+	+	−	−	−	−	+	+	+

$$D = \alpha_3 A, \; A = \alpha_3 D, \; \alpha_3 = \pm 1 \tag{15}$$

$$D = \alpha_3 A, \; H = \alpha_3 E \tag{16}$$

Finally, lines $\theta = 0000$ and $\theta = 1111$ yield:

$$E = \alpha_4 A, \; A = \alpha_4 E, \; \alpha_4 = \pm 1 \tag{17}$$

$$E = \alpha_4 A \tag{18}$$

Note that the number of relations in Eqs.(12), (14), (16), and (18) drops from 8 to 4, 2, and 1. The 2^4 possible combinations of the values $+1$ or -1 for α_1 to α_4 are listed in Table 5.4-4. The table also shows the samples A, B, ... , P expressed in terms of α_1 to α_4 according to Eqs.(12), (14), (16), and (18). The sixteen functions mgl(k,h) with sixteen sampled values A, B, ... , P each defined by Table 5.4-4 are listed in Table 5.4-5. The functions mgl$(k,h) =$ mgl$(k, \xi \oplus \theta)$ with ξ and θ taken from the minimized code with

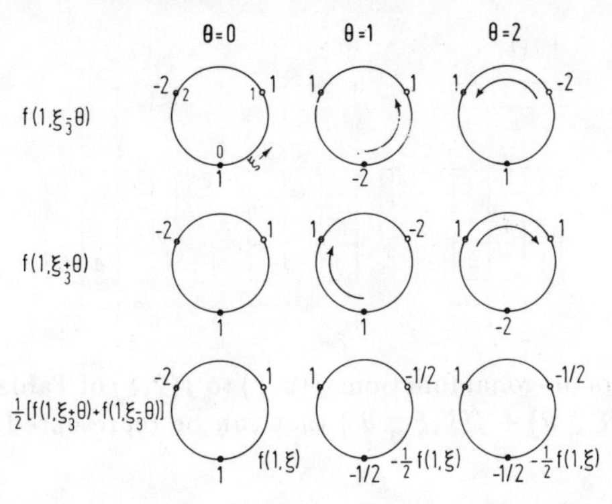

FIG.5.5-1. Ternary shifting by subtraction modulo 3 and addition modulo 3, and the sum of the two shifts.

16 elements in Table 5.2-2, are shown in Fig.5.4-7. It is quite evident from the symmetry of these patterns that the relation

$$\mathrm{mgl}(k,\xi)\,\mathrm{mgl}(k,\theta) = \mathrm{mgl}(k,\xi \oplus \theta) \tag{19}$$

is satisfied.

5.5 Nondyadic Coordinate Systems

Standing sinusoidal waves are characteristic for coordinate systems for the limit $2^N \gg 1$ in Fig.3.1-1, while the Walsh and other two-valued functions of Figs.5.3-6, 5.4-1, and 5.4-7 are characteristic for the other limit $2^N = 2$. Let us now consider the intermediate cases. We postulate that there are functions $f(k,\xi)$ that produce standing waves if shifted around a circle in Fig.3.1-1 in the mathematical positive direction, $f(\xi \underset{\mathrm{n}}{-} \theta)$, and in the mathematically negative direction, $f(\xi \underset{\mathrm{n}}{+} \theta)$, according to the equation:

$$f(k,\xi)g(k,\theta) = \frac{1}{2}\left[f(k,\xi \underset{\mathrm{n}}{+} \theta) + f(k,\xi \underset{\mathrm{n}}{-} \theta)\right] \tag{1}$$

The signs $\underset{\mathrm{n}}{-}$ and $\underset{\mathrm{n}}{+}$ indicate subtraction and addition modulo n. This equation is satisfied for $n \to \infty$ by Eq.(5.3-1), and for $n = 2$ by Eq.(5.3-3); for $n \to \infty$ we write $+$ and $-$, and for $n = 2$ we write $\oplus$ instead of $\underset{\mathrm{n}}{+}$ and $\underset{\mathrm{n}}{-}$.

For $n = 3$ the function $f(k,\xi)$ consists of three samples A, B, and C. The shifts $f(k,\xi \underset{3}{+} \theta)$ and $f(k,\xi \underset{3}{-} \theta)$ as well as the sum

$$\tfrac{1}{2}[f(k,\xi \underset{3}{+} \theta) + f(k,\xi \underset{3}{-} \theta)]$$

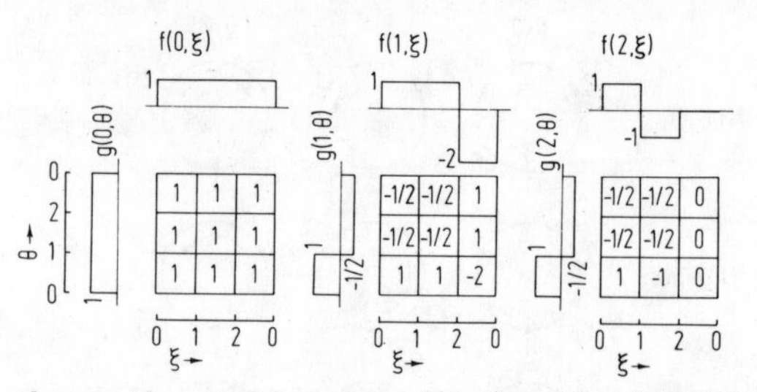

FIG.5.5-2. The three orthogonal functions $f(0,\xi)$ to $f(2,\xi)$ of Table 5.5-2 for $A = 1$ produce sums $\frac{1}{2}[f(k,\xi \underset{3}{+} \theta) + f(k,\xi \underset{3}{-} \theta)]$ that can be represented as products $f(k,\xi)g(k,\theta)$.

TABLE 5.5-1
THE TERNARY SHIFTS $f(k,\xi \underset{3}{+} \theta)$, $f(k,\xi \underset{3}{-} \theta)$ AND THEIR SUM.

	$f(k,\xi \underset{3}{+} \theta)$			$f(k,\xi \underset{3}{-} \theta)$			$f(k,\xi \underset{3}{+} \theta) + f(k,\xi \underset{3}{-} \theta)$		
	$\xi \rightarrow$			$\xi \rightarrow$			$\xi \rightarrow$		
	0	1	2	0	1	2	0	1	2
θ 0	A	B	C	A	B	C	$2A$	$2B$	$2C$
$\downarrow$ 1	B	C	A	C	A	B	$B+C$	$A+C$	$A+B$
2	C	B	A	B	C	A	$B+C$	$A+C$	$A+B$

are shown in Table 5.5-1, while Fig.5.5-1 shows the shifts and their sums for $A = 1$, $B = 1$, $C = -2$.

The lines $\theta = 1$ and $\theta = 2$ of the sum $f(k,\xi \underset{3}{+} \theta) + f(k,\xi \underset{3}{-} \theta)$ in Table 5.5-1 are equal. Hence, only the following three conditions must be satisfied to produce a standing wave according to Eq.(1):

$$B + C = 2\alpha A,\ A + C = 2\alpha B,\ A + B = 2\alpha C \tag{2}$$

For $2\alpha + 1 \neq 0$ one obtains readily

$$B = A,\ C = A,\ \alpha = +1, \tag{3}$$

while $2\alpha + 1 = 0$ yields

$$C = -(A + B),\ \alpha = -1/2 \tag{4}$$

where B can be chosen. For $B = A$ we obtain $C = -2A$, while $B = -A$ yields $C = 0$. The three solutions defined by Eqs.(3) and (4) form an orthogonal—not normalized—system of three functions $f(0,\xi)$ to $f(2,\xi)$ shown in Table 5.5-2.

TABLE 5.5-2

ORTHOGONAL SYSTEM OF THREE FUNCTIONS THAT PRODUCE STANDING WAVES IN A TERNARY COORDINATE SYSTEM.

			$\xi \rightarrow$		
			0	1	2
$\alpha = 1$		$f(0,\xi)$	A	A	A
$\alpha = -1/2$	$B = +A$	$f(1,\xi)$	A	A	$-2A$
$\alpha = -1/2$	$B = -A$	$f(2,\xi)$	A	$-A$	0

TABLE 5.5-3

THE QUATERNARY SHIFTS $f(k,\xi \mathop{+}_{4} \theta)$, $f(k,\xi \mathop{-}_{4} \theta)$ AND THEIR SUM.

$f(k,\xi \mathop{+}_{4} \theta)$					$f(k,\xi \mathop{-}_{4} \theta)$					$f(k,\xi \mathop{+}_{4} \theta) + f(k,\xi \mathop{-}_{4} \theta)$				
	$\xi \rightarrow$					$\xi \rightarrow$					$\xi \rightarrow$			
	0	1	2	3		0	1	2	3		0	1	2	3
θ 0	A	B	C	D	θ 0	A	B	C	D	θ 0	$2A$	$2B$	$2C$	$2D$
$\downarrow$ 1	B	C	D	A	$\downarrow$ 1	D	A	B	C	$\downarrow$ 1	$B+D$	$A+C$	$B+D$	$A+C$
2	C	D	A	B	2	C	D	A	B	2	$2C$	$2D$	$2A$	$2B$
3	D	A	B	C	3	B	C	D	A	3	$B+D$	$A+C$	$B+D$	$A+C$

The shifting of $f(1,\xi)$ of Table 5.5-2 according to Table 5.5-1 with $A = 1$ is shown in Fig.5.5-1 on a circle. The same shifting for all three functions $f(0,\xi)$ to $f(2,\xi)$ is shown in Fig.5.5-2, with the circle replaced by a straight line with the marks $\xi = 0, 1, 2, 0, \ldots$. It is evident from this illustration that the sum of shifts

$$\tfrac{1}{2}[f(k,\xi \mathop{+}_{3} \theta) + f(k,\xi \mathop{-}_{3} \theta)]$$

is equivalent to a multiplication $f(k,\xi)g(k,\theta)$ as demanded by Eq.(1). Contrary to our experience with sinusoidal or Walsh waves, the time function $g(k,\theta)$ and the space function $f(k,\xi)$ are not equal. The time function $g(k,\theta)$ for $\theta = 0, 1, 2$ is defined by $g(k,0) = 1, g(k,1) = g(k,2) = \alpha$.

Let us turn to the quaternary shifts $f(k,\xi \mathop{+}_{4} \theta)$ and $f(k,\xi \mathop{-}_{4} \theta)$ shown in Table 5.5-3. The rows $\theta = 1$ and $\theta = 3$ of the sum $f(k,\xi \mathop{+}_{4} \theta) + f(k,\xi \mathop{-}_{4} \theta)$ are equal. Only the two following sets of conditions must be satisfied to produce a standing wave:

$$(B+D)2\alpha_1 A,\ (A+C) = 2\alpha_1 B,\ (B+D) = 2\alpha_1 C,\ (A+C) = 2\alpha_1 D$$
$$C = \alpha_2 A,\ D = \alpha_2 B,\ A = \alpha_2 C,\ B = \alpha_2 D \qquad (5)$$

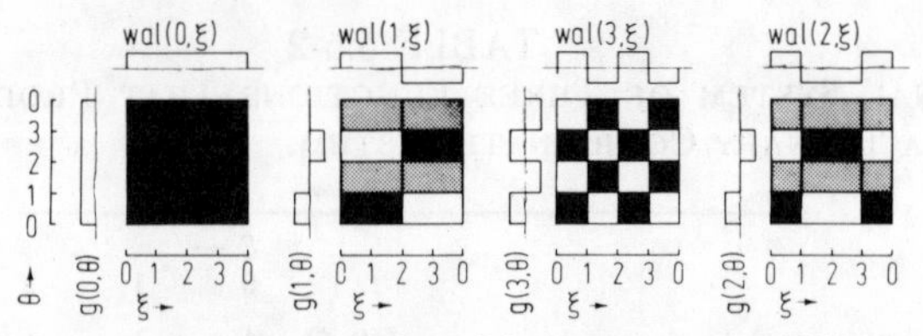

FIG.5.5-3. Standing wave patterns produced by quaternary shifting according to Tables 5.5-3 and 5.5-4. Black represents +1, grey 0, and white −1.

The resulting values for α_1, α_2, B , C, and D are listed in Table 5.5-4. We obtain a set of four orthogonal functions $f(k,\xi)$ and four time functions $g(k,\theta)$ defined by $g(k,0)=1$, $g(k,1)=g(k,3)=\alpha_1$, $g(k,2)=\alpha_2$. The functions $f(k,\xi)$ are the first four Walsh functions $\mathrm{wal}(0,\xi)$ to $\mathrm{wal}(3,\xi)$ but the patterns of the standing waves produced by them are not those of the dyadic shift. The standing wave patterns for the dyadic shift of $\mathrm{wal}(0,\xi)$ to $\mathrm{wal}(3,\xi)$ are shown by row 1, columns a to d of Fig.5.3-6, while the patterns for the quaternary shift are shown in Fig.5.5-3. Standing waves produced by dyadic shifting have the form

$$\mathrm{wal}(k,\xi)\,\mathrm{wal}(k,\theta)=\frac{1}{2}\left[\mathrm{wal}(k,\xi\oplus\theta)+\mathrm{wal}(k,\xi\ominus\theta)\right]$$

while standing waves produced by quaternary shifting have the form

$$\mathrm{wal}(k,\xi)g(k,\theta)=\frac{1}{2}[\mathrm{wal}(k,\xi\underset{4}{+}\theta)+\mathrm{wal}(k,\xi\underset{4}{-}\theta)],$$

where $g(k,\theta)$ is defined in Fig.5.5-3.

We turn next to the octonary shifts $f(k,\xi\underset{8}{+}\theta)$ and $f(k,\xi\underset{8}{-}\theta)$ on the ring 2^3 in Fig.3.1-1. These shifts and their sum are shown in Table 5.5-5. In analogy to the quaternary shift of Table 5.5-3 the rows $\theta=1$ and $\theta=7$, $\theta=2$ and $\theta=6$, as well as $\theta=3$ and $\theta=5$ of the sum $f(k,\xi\underset{8}{+}\theta)+f(k,\xi\underset{8}{-}\theta)$ are equal. Only the following four sets of eight conditions each must be satisfied:

$$\begin{aligned}
B+H&=2\alpha_1A, &&\ldots, & A+G&=2\alpha_1H\\
C+G&=2\alpha_2A, &&\ldots, & B+F&=2\alpha_2H\\
D+F&=2\alpha_3A, &&\ldots, & C+E&=2\alpha_3H\\
E&=\alpha_4A, &&\ldots, & D&=\alpha_4H
\end{aligned}$$

$$\alpha_5=\alpha_3,\ \alpha_6=\alpha_2,\ \alpha_7=\alpha_1 \tag{6}$$

TABLE 5.5-4
ORTHOGONAL SYSTEM OF FOUR FUNCTIONS THAT PRODUCE STANDING WAVES IN A QUATERNARY COORDINATE SYSTEM.

		$\xi \to$				
			0	1	2	3
$\alpha_1 = +1$	$\alpha_2 = +1$	$f(0,\xi)$	A	A	A	A
$\alpha_1 = 0$	$\alpha_2 = -1$	$f(1,\xi)$	A	A	$-A$	$-A$
$\alpha_1 = -1$	$\alpha_2 = +1$	$f(2,\xi)$	A	$-A$	A	$-A$
$\alpha_1 = 0$	$\alpha_2 = -1$	$f(3,\xi)$	A	$-A$	$-A$	A

TABLE 5.5-5
THE OCTONARY SHIFTS $f(k,\xi \underset{8}{+} \theta)$, $f(k,\xi \underset{8}{-} \theta)$ AND THEIR SUM.

		$\xi \to$ 0	1	2	3	4	5	6	7
$f(k,\xi \underset{8}{+} \theta) = a$									
θ	0	A	B	C	D	E	F	G	H
$\downarrow$	1	B	C	D	E	F	G	H	A
	2	C	D	E	F	G	H	A	B
	3	D	E	F	G	H	A	B	C
	4	E	F	G	H	A	B	C	D
	5	F	G	H	A	B	C	D	E
	6	G	H	A	B	C	D	E	F
	7	H	A	B	C	D	E	F	G
$f(k,\xi \underset{8}{-} \theta) = b$									
θ	0	A	B	C	D	E	F	G	H
$\downarrow$	1	H	A	B	C	D	E	F	G
	2	G	H	A	B	C	D	E	F
	3	F	G	H	A	B	C	D	E
	4	E	F	G	H	A	B	C	D
	5	D	E	F	G	H	A	B	C
	6	C	D	E	F	G	H	A	B
	7	B	C	D	E	F	G	H	A
SUM $a+b$									
θ	0	$2A$	$2B$	$2C$	$2D$	$2E$	$2F$	$2G$	$2H$
$\downarrow$	1	$B+H$	$A+C$	$B+D$	$C+E$	$D+F$	$E+G$	F+H	$A+G$
	2	$C+G$	$D+H$	$A+E$	$B+F$	$C+G$	$D+H$	$A+E$	$B+F$
	3	$D+F$	$E+G$	$F+H$	$A+G$	$B+H$	$A+C$	$B+D$	$C+E$
	4	$2E$	$2F$	$2G$	$2H$	$2A$	$2B$	$2C$	$2D$
	5	$D+F$	$E+G$	$F+H$	$A+G$	$B+H$	$A+C$	$B+D$	$C+E$
	6	$C+G$	$D+H$	$A+E$	$B+F$	$C+G$	$D+H$	$A+E$	$B+F$
	7	$B+H$	$A+C$	$B+D$	$C+E$	$D+F$	$E+G$	$F+H$	$A+G$

One obtains the eight sets of values for α_1, α_2, α_4 shown in Table 5.5-6 and $\alpha_3 = \alpha_4\alpha_1$. The constants B and C can be chosen. The particular choice shown in Table 5.5-6 yields the orthogonal set of eight functions $f(k,\xi)$ and the associated functions $g(k,\theta)$ that produce standing waves according to Eq.(1).

Although we have not proved that all coordinate systems based on rings have characteristic standing waves, the examples given clearly show that standing waves exist at least for some coordinate systems, and that the patterns of the complete sets of standing waves was different in all examples. One peculiarity occurred that has not yet been mentioned. For the dyadic coordinate systems in Section 5.3 and 5.4 we did not distinguish between one- and more-dimensional coordinate systems. For ternary, quaternary, and octonary systems discussed in this section, we restricted ourselves to only one dimension. This is in line with Eq.(5.3-1) for sinusoidal standing waves that also hold for one dimension x only. For two dimensions x, y we obtain a formula with four rather than two shifted waves:

$$\begin{aligned}\cos(2\pi kx/X)\cos(2\pi ky/X)\cos(2\pi kct/X)&\\ =\frac{1}{4}\Big\{\cos\big[2\pi k(x+y+ct)/X\big]&+\cos\big[2\pi k(x+y-ct)/X\big]\\ +\sin\big[2\pi k(x-y+ct)/X\big]&-\sin\big[2\pi k(x-y-ct)/X\big]\Big\}\end{aligned} \tag{7}$$

A possible generalization of Eq.(1) to two dimensions is thus:

$$\begin{aligned}f(k,\xi)f(k,\eta)g(k,\theta) = \frac{1}{4}\big[f(k,\xi \mathbin{\underset{\mathrm{n}}{+}} \eta \mathbin{\underset{\mathrm{n}}{+}} \theta)&\\ + f(k,\xi \mathbin{\underset{\mathrm{n}}{+}} \eta \mathbin{\underset{\mathrm{n}}{-}} \theta) + h(k,\xi \mathbin{\underset{\mathrm{n}}{-}} \eta \mathbin{\underset{\mathrm{n}}{+}} \theta) &- h(k,\xi \mathbin{\underset{\mathrm{n}}{-}} \eta \mathbin{\underset{\mathrm{n}}{-}} \theta)\big]\end{aligned} \tag{8}$$

For modulo 2 addition one obtains, however, from Eq.(5.3-2) a much simpler generalization:

$$f(k,\zeta)f(k,\eta)f(k,\theta) = f(k,\zeta\oplus\eta)f(k,\theta) = f(k,\zeta\oplus\eta\oplus\theta) \tag{9}$$

Hence, the two variables ζ and η can be treated as one variable $\zeta\oplus\eta=\xi$, which makes the one-dimensional case equal to any multidimensional case. We have used the one variable ξ to represent any number of two-valued variables x/X, y/Y, ... This mathematical formalism is seen now to have a physical base. Standing waves make no distinction between one- and

TABLE 5.5-6

ORTHOGONAL SYSTEM OF EIGHT FUNCTIONS $f(k,\xi)$ THAT PRODUCE STANDING WAVES IN AN OCTONARY COORDINATE SYSTEM, AND THE ASSOCIATED FUNCTIONS $g(k,\theta)$.

					$f(k,\xi)/A$ $\xi \rightarrow$								$g(k,\theta)$ $\theta \rightarrow$							
k	α_4	α_2	α_1		0	1	2	3	4	5	6	7	0	1	2	3	4	5	6	7
0	1	1	1	$B=+A$	1	1	1	1	1	1	1	1	1	1	1	1	1	1	1	1
1	1	1	1	$B=-A$	1	−1	1	−1	1	−1	1	−1	1	−1	1	−1	1	−1	1	−1
2	1	−1	−1	$B=+A$	1	1	−1	−1	1	1	−1	−1	1	0	−1	0	1	0	−1	0
3	1	−1	−1	$B=-A$	1	−1	−1	1	1	−1	−1	1	1	0	−1	0	1	0	−1	0
4	−1	0	$\frac{1}{\sqrt{2}}$	$C=+A$	1	$\sqrt{2}$	1	0	−1	$-\sqrt{2}$	−1	0	1	$\frac{1}{\sqrt{2}}$	0	$\frac{-1}{\sqrt{2}}$	−1	$\frac{-1}{\sqrt{2}}$	0	$\frac{1}{\sqrt{2}}$
5	−1	0	$\frac{1}{\sqrt{2}}$	$C=-A$	1	0	−1	$-\sqrt{2}$	−1	0	1	$\sqrt{2}$	1	$\frac{1}{\sqrt{2}}$	0	$\frac{-1}{\sqrt{2}}$	−1	$\frac{-1}{\sqrt{2}}$	0	$\frac{1}{\sqrt{2}}$
6	−1	0	$\frac{-1}{\sqrt{2}}$	$C=+A$	1	$-\sqrt{2}$	1	0	−1	$\sqrt{2}$	−1	0	1	$\frac{-1}{\sqrt{2}}$	0	$\frac{1}{\sqrt{2}}$	−1	$\frac{1}{\sqrt{2}}$	0	$\frac{-1}{\sqrt{2}}$
7	−1	0	$\frac{-1}{\sqrt{2}}$	$C=-A$	1	0	−1	$\sqrt{2}$	−1	0	1	$-\sqrt{2}$	1	$\frac{-1}{\sqrt{2}}$	0	$\frac{1}{\sqrt{2}}$	−1	$\frac{1}{\sqrt{2}}$	0	$\frac{-1}{\sqrt{2}}$

more-dimensional dyadic coordinate systems. Putting it differently, we do not need more than one coordinate variable—or "space variable"—if we use a dyadic coordinate system, just as we do not need more than one time variable.

5.6 Motion Based on Integer Number and Dyadic Topology

We have discounted in Chapter 2 the topology of the continuum, or the usual topology of the real numbers, as a mathematical abstraction that cannot be observed. The most obvious replacement is the integer number topology represented by the limit $2^N \to \infty$ of the rings in Fig.3.1-1, and the three-dimensional coordinate system in Fig.5.4-5. One might think that the difference between a discrete three-dimensional coordinate system according to Fig.5.4-5 and a three-dimensional continuum would vanish if the distances between adjacent points become sufficiently small, but this is not so. The continuum leads to differential equations, whereas the integer number topology leads to *difference* equations. The solutions of a difference equation do not necessarily converge to those of the differential equation when the differences Δx^i becomes very small, since an infinitesimal distance dx^i differs decisively from a small distance Δx^i. To show this, consider a distance of length X divided into intervals of length Δx. There shall be n such intervals, where n is an integer. When Δx is decreased, the number n will increase, and for $\Delta x \to 0$ the number of intervals will become denumerably infinite. On the other hand, we will need nondenumerably many infinitesimal intervals dx to cover the length X. Since the integer number topology thus does not lead automatically to the results of the real number topology, we must admit other topologies as physically possible. However, the real number topology has served us so well for so long that any other topology must by and large yield the same results within the realm of past observations to be acceptable. We will only accept deviations at the frontier of science, primarily at small distances and short times. We will specifically study motion in coordinate systems having integer number or dyadic topology. We will use the world line representation rather than the multidimensional structures preferred so far, since most readers are more familiar with it.

Consider a clock that advances the shown time at $t = 0$, $t = \Delta t = qT$, $t = 2\Delta t = 2qT$, ... Digital clocks work conspicuously in this way, while the advance in discrete steps is less obvious for other clocks. The time interval $nqT < t < (n+1)qT$ is called the time nqT, where $n = 0, 1, 2, \ldots$. Let a ruler for distance measurements according to Figs.1.3-1c or 2.2-1 have marks at $x = ncqT$. An observed point being closest to the mark n is said to have the distance $ncqT$ from the mark 0. The normalized

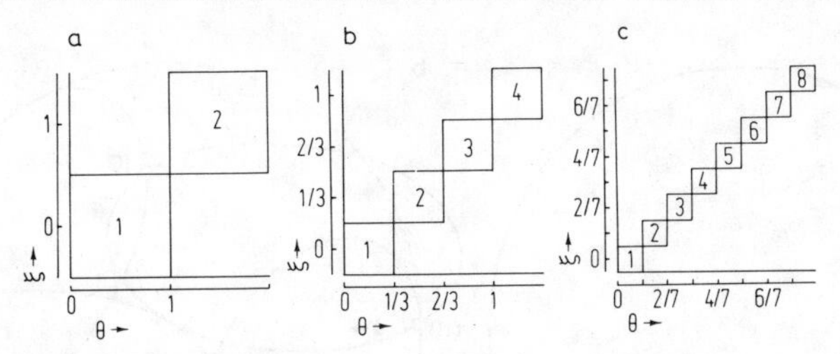

FIG.5.6-1. Integer number shifting $f(\xi-\theta)$. The size of the distance-time intervals decreases and the resolution increases from (a) to (c).

variables $\theta = t/T$ and $\xi = x/cT$ may be used instead of t and x. A measurement of the normalizing time T with our clock may be uncertain by an amount qT or $2qT$, but the relative uncertainty qT/T or $2qT/T$ can be made arbitrarily small by choosing qT sufficiently small compared with[1] T. Note that "arbitrarily small" does not imply that q becomes zero.

Let an observed point or a sample $f(0)$ be closest to the mark $n = 0$ or $\xi = nq = 0$ at the time $\theta = 0$, and let q have the value 1. The sample is then said to be in the interval[2] $0 < \theta < 1$, $-1/2 < \xi < 1/2$. The location of the sample is defined by square 1 in Fig.5.6-1a.

At the time $\theta = 1$ the sample must have propagated closest to the mark $n = \xi = 1$ in order for the relation $f(\xi - 1) = f(0)$ to be satisfied. The location of the sample is defined by square 2 in Fig.5.6-1a. This is the propagation according to the usual topology of integers.

Let q have the value 1/3. A sample $f(\xi - \theta)$ closest to the mark $\xi = 0$ at the time $\theta = 0$ means now the sample is in the interval $0 < \theta < 1/3$, $-1/6 < \xi < 1/6$. This is indicated by square 1 in Fig.5.6-1b. The location at the time $\theta = 1/3, 2/3, 1$ follows from $f(\xi - \theta) = f(0)$ to be $\xi = 1/3, 2/3, 1$. The propagation of the sample is thus represented by squares 2 to 4.

Figure 5.6-1c shows the propagation of the sample for $q = 1/7$. It is evident that for small values of q the sequence of squares approaches a line, and this is the reason why the term world line representation is used.

---

[1]The error of T will be zero if the clock used is by definition the time standard. Otherwise, the error T means the error of the clock used compared with a clock defining the time standard.

[2]The time variable θ and the coordinate variable ξ are treated differently here, because a high quality clock typically shows the time $t = nqT$ until it advances to $t = (n+1)qT$. When observing the location of a sample relative to a ruler, on the other hand, we observe it closest to the mark $x = ncqT$ when it is in the interval $(n-1/2)cqT < x < (n+1/2)cqT$, and not when it is in the interval $ncqT < x < (n+1)cqT$. The reason for this distinction is that it is easier to visualize an observed particle closest to a certain coordinate mark—which means that the mark and the particle are at the same location within the accuracy of observation—than to visualize an interval in coordinate systems like those in Figs.3.1-13 or 3.1-22.

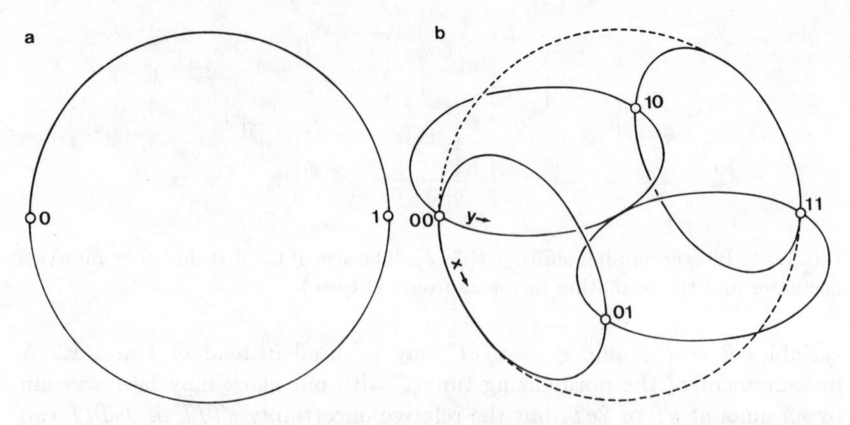

FIG.5.6-2. Generalization of Fig.2.2-1 for dyadic coordinate systems. (a) Two coordinate marks yield 1 bit of information about the location of a sample; (b) four marks yield 2 bit of information.

Let us consider next the propagation according to dyadic topology. This means that the propagation is defined by $f(\xi \oplus \theta)$ rather than by $f(\xi \ominus \theta)$. One could write $f(\xi \ominus \theta)$ instead of $f(\xi \oplus \theta)$, but we have pointed out before that modulo 2 addition and subtraction yield the same result.

First we have to investigate what an increased resolution or a decreased uncertainty about the location of a sample means in dyadic topology. In real-number-topology we used a ruler according to Fig.1.3-1c, and divided its scale into smaller intervals according to Fig.2.2-1 to obtain increased resolution. It is not readily apparent how this process can be applied to dyadic topology, since we have only discussed dyadic coordinate systems but not dyadic rulers. However, the ruler in Fig.1.3-1c is the x–axis of the coordinate system of Fig.1.3-1b. We can replace the two-dimensional coordinate system by one of its axes because the propagation $f(\xi-\theta)$ is along an Euclidean straight line. If one does not know that the propagation of a sample from point A to point B in Fig.1.3-1a is along such a straight line, one must subdivide the interval of the x– and the y–axes in Fig.1.3-1a—or generally of all the axes—according to Fig.2.2-1, if one wants to describe the propagation from point A to point B with increased resolution.

For the application of this line of thought to a dyadic coordinate system refer to Fig.5.6-2. On the left is shown the most primitive dyadic coordinate system with the marks $\xi = 0$ and $\xi = 1$ only. The observation of a sample closer to $\xi = 0$ than to $\xi = 1$ provides us with 1 bit of information, provided the probability for the sample being closer to 0 is the same as being closer to 1. The subdivision of the interval $0 < x/X < 1$ in Fig.2.2-1a into two

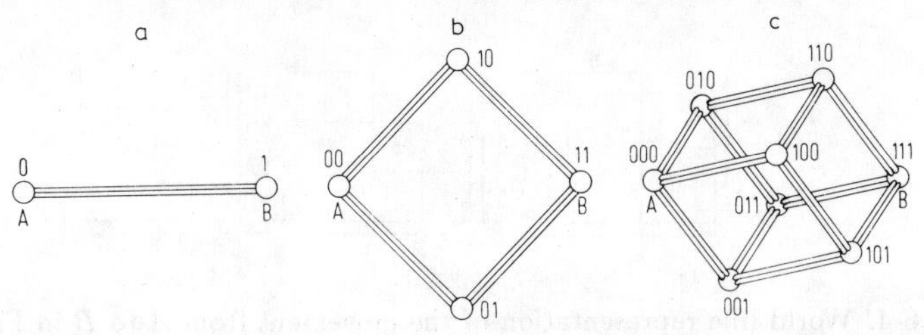

FIG.5.6-3. Simplified representation of the coordinate systems of Fig.5.6-2 (a, b); extension to eight coordinate marks yields 3 bit of information about the location of a sample (c).

intervals of half the length in Fig.2.1-1b requires now the replacement of the two marks 0 and 1 in Fig.5.6-2a by four marks 00, 01, 10, and 11, which define a two-dimensional dyadic coordinates system according to Fig.5.6-2b. The new marks 00 and 11 are the same as the marks 0 and 1 in Fig.5.6-2a. The additional marks 01 and 10 increase the information about the location of a sample closest to one of the four marks to 2 bit. Note that no assumption is introduced by the use of this dyadic coordinate system about how a sample propagates from the mark 00 to 11, just as no such assumption would be introduced by the use of a Cartesian coordinate system according to Fig.5.4-5. However, the Cartesian coordinate system is most convenient for propagation along an Euclidean straight line—in which case it can be replaced by one axis or ruler according to Fig.1.3-1—and the dyadic coordinate system will be most convenient for propagation along a geodesic described by the Gray code, minimized code, etc. as discussed in Section 5.2.

The extension of Fig.5.6-2b from four marks 00 ... 11 to eight marks 000 ... 111, and the resulting increase of the information about the location of a sample to 3 bit, is difficult to draw. We use again the simplified representation of Fig.3.1-20. Figures 5.6-3a and b show the simplified representation of Figs.5.6-2a and b, while Fig.5.6-3c shows the extension to eight marks 000 ... 111.

The propagation of a sample from A to B on a shortest way via the marks of the coordinate system is described by a minimized code, such as the one shown in Table 5.2-2. For instance, in Fig.5.6-3a it propagates from 0 to 1, in Fig.5.6-3b from 00 via 01 to 11, and in Fig.5.6-3c from 000 via 001 and 011 to 111.

Next we have to make some comments about the division of the clock scale. When the ruler of Fig.1.3-1c is used to observe successive locations of a sample propagating from A to B on a straight line, we will not hesitate to use the same division for the clock scale. If the marks 0, qc/T, $2qc/T$, ...

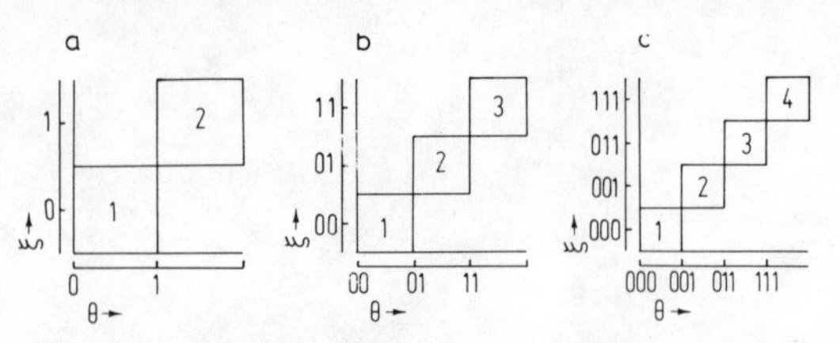

FIG.5.6-4. World line representation of the movement from A to B in Fig.5.6-3.

along the ruler are then reached by a sample at the times 0, q/T, $2q/T, \ldots$ we say that the sample moves with uniform velocity c. The integer numbers 0, 1, 2, ... of the usual scale of a ruler are replaced by the numbers 000, 001, 011, ... of the minimized code with 8 elements in Table 5.2-2, if the dyadic coordinate system or the "dyadic ruler" of Fig.5.6-3c is used, and if the observed sample propagates according to this code. Let the clock scale use the same division $t/T = 000, 001, 011, \ldots$. If the marks 000, 001, 011, ... of the coordinate system of Fig.5.6-3c are then reached by a sample at the times $t/T = 000, 001, 011, \ldots$ we will again say that the sample moves with uniform velocity c. Of course, the movement is now not along an Euclidean straight line but along a geodesic of a dyadic coordinate system defined by the particular minimized code used. Generally speaking, the successive marks of the scale of the clock should be numbered like the successive marks of the coordinate system in which the propagating sample is observed.

A sample $f(0)$ shall be closest to the mark $\xi = 0$ in Fig.5.6-3a at the time $\theta = 0$. Its location in Fig.5.6-4a is represented by square 1. In order to satisfy the relation $f(\xi \oplus 1) = f(0)$ at the time $\theta = 1$, the sample must have moved closest to the mark $\xi = 1$ in Fig.5.6-3a, and its location in Fig.5.6-4a is defined by square 2.

We consider next propagation in the coordinate system of Fig.5.6-3b. Again, a sample $f(0)$ shall be at $\xi = 00$ at the time $\theta = 00$. The equation $f(\xi \oplus \theta) = f(0)$ will be satisfied at the times $\theta = 01$ and 11 if ξ has the values $\xi = 01$ and 11. The motion is characterized in Fig.5.6-4b by the squares 1, 2, 3.

Figure 5.6-4c shows the world line representation of a sample propagating according to the equation $f(\xi \oplus \theta) = f(0)$ from $\xi = 000$ via 001 and 011 to $\xi = 111$ at the times $\theta = 000, 001, 011, 111$ in Fig.5.6-3c.

It is quite evident from Fig.5.6-1 and 5.6-4 that the equations of propagation $f(\xi - \theta) = f(0)$ and $f(\xi \oplus \theta) = f(0)$ yield the same world line or world graph representation[3]. It is also clear that the general equation

[3]Since the "world lines" in Fig.5.6-1 and 5.6-4 are actually sequences of small squares, it appears to be better to call them world graphs.

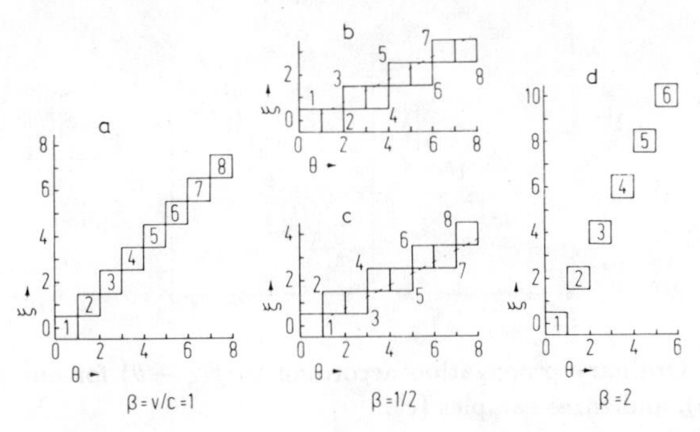

FIG.5.6-5. Propagation velocities $\beta = v/c = 1$ (a), $\beta = 1/2$ (b, c), and $\beta = 2$ (d).

$f(\xi \,\bar{_n}\, \theta) = f(0)$ must yield the same result, since all these equations are satisfied for $\xi = \theta$.

The definitions $\theta = t/T$ and $\xi = x/cT$ mean that the argument $\xi - \theta = (x - ct)/cT$ implies propagation with velocity c. The velocity v yields the equation $f(\xi - \beta\theta) = f(0)$ or—generally—$f(\xi \,\bar{_n}\, \beta\theta) = f(0)$, where $\beta = v/c$.

For the investigation of a velocity $\beta \neq 1$ we redraw first Fig.5.6-1c into Fig.5.6-5a with $q = 1/7$ replaced by $q = 1$ for simplification of writing. Let then $\beta = 1$ be replaced by $\beta = 1/2$. The relation $f(\xi - \theta/2) = f(0)$ yields $\xi = 0, 1, 2, \ldots$ for $\theta = 0, 2, 4, \ldots$. The respective squares of the world graph in Fig.5.6-5b are denoted 1, 3, 5, 7. For $\theta = 1, 3, 5, \ldots$ one obtains $\xi = 1/2, 3/2, \ldots$. Since only integer values of ξ are possible, we may choose either the sequence $\xi = 0, 1, 2, \ldots$ or the sequence $\xi = 1, 2, 3, \ldots$. The first choice is shown in Fig.5.6-5b, the second in Fig.5.6-5c. The squares 1, 3, 5, 7 of the world graph have the same location in both illustrations, but the squares 2, 4, 6, 8 are located differently. One may see that both sequences of the squares give an equally good approximation to the dashed lines that hold for a much better resolution of θ and ξ.

Consider now the velocity $\beta = v/c = 2$. The relation $f(\xi - 2\theta) = f(0)$ yields $\xi = 0, 2, 4, \ldots$ for $\theta = 0, 1, 2, \ldots$, but no time is assigned to the locations $\xi = 1, 3, 5, \ldots$. The resulting world graph is shown in Fig.5.6-5d. The squares 1, 2, 3, ... are no longer connected. If we observe a particle at successive times $\theta = 0, 1, 2, \ldots$ at locations $\xi = 0, 2, 4, \ldots$ but are not able to verify that the particle passed through the locations $\xi = 1, 3, 5, \ldots$, it becomes very difficult to claim that we observed the same particle; one may just as well claim that a particle disappeared at $\xi = 0$ and another, equal particle appeared at $\xi = 2$. If we do not assume a space-time continuum, there is always a velocity beyond which it becomes impossible to distinguish a very fast particle from a series of particles that appear and disappear. We

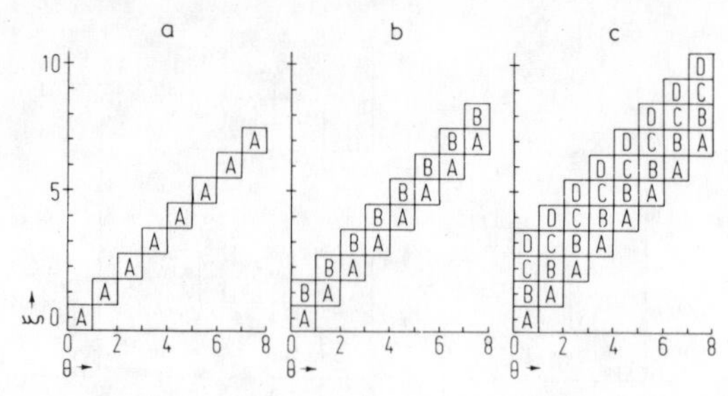

FIG.5.6-6. Ordinary propagation according to $f(\xi - \theta)$ for one sample (a), two samples (b), and three samples (c).

have denoted this velocity here with c, but we have not shown that c must be the velocity of light.

It is evident from Figs.5.6-5a – c that even for $\beta \leq 1$ it is difficult to state with certainty that one observes *one* moving particle, if the particle is so small that it fits into one resolution interval $n < \xi < n+1$ of the coordinate variable. Such a problem does not exist for an object that is so large that it covers many resolution intervals of ξ, or is observed at many coordinate points ξ.

The difficulty of distinguishing one moving particle from several appearing and disappearing particles is a hard-to-accept consequence of giving up the concept of a space-time continuum. A quote from Schrödinger may help:

> We must not admit the possibility of continuous observation .... The idea of a continuous range, so familiar to mathematicians in our days, is quite exorbitant, an enormous extrapolation of what is really accessible to us (Schrödinger, 1956).

If one replaces the relation $f(\xi - \beta\theta) = f(0)$ by $f(\xi \underset{n}{-} \beta\theta) = f(0)$ one obtains again the equation $\xi = \beta\theta$ that defines the world graphs in Fig.5.6-5. Hence, the results derived for integer number topology must hold generally for coordinate systems based on rings.

Let us turn to the propagation of more than one sample. Figure 5.6-6a shows once more the propagation of a sample denoted A according to the formula $f(\xi - \theta) = f(0)$. In addition to this sample A, a sample B propagates in Fig.5.6-6b according to the formula $f(\xi - \theta) = f(1)$, since B is located at $\xi = 1$ at the time $\theta = 0$. Figure 5.6-6c shows the additional samples C and D propagating to $f(\xi - \theta) = f(2)$ and $f(\xi - \theta) = f(3)$.

We investigate the propagation in dyadic topology according to the Gray code of Table 5.2-1. Figure 5.6-7a shows the propagation of one sample A, which follows from the relation $f(\xi \oplus \theta) = f(0)$ or $\xi = \theta$. An additional sample B located at $\xi = 001$ at the time $\theta = 000$ propagates in Fig.5.6-7b; its movement follows from $f(\xi \oplus \theta) = f(001)$ or $\xi = \theta \oplus 001$. Figure 5.6-7c

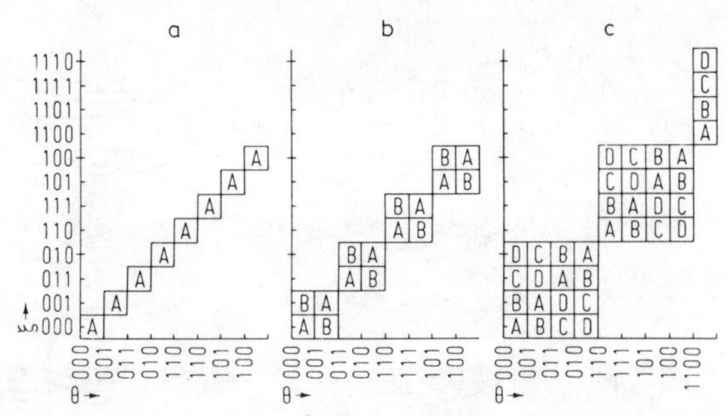

FIG.5.6-7. Dyadic propagation according to $f(\xi \oplus \theta)$ using the Gray code of Table 5.2-1 for one sample (a), two samples (b), and three samples (c).

shows the additional samples C and D at the locations $\xi = 011$ and $\xi = 010$ at the time $\theta = 0$; their propagation follows from $f(\xi \oplus \theta) = f(011)$ and $f(\xi \oplus \theta) = f(010)$, or $\xi = \theta \oplus 011$ and $\xi = \theta \oplus 010$.

There are two obvious similarities between Figs.5.6-6 and 5.6-7. First, sample A is shifted in the same way in both topologies, since it is shifted in its eigencoordinates. Second, the samples B, C, and D in Fig.5.6-7 propagate on the average like the samples B, C, and D in Fig.5.6-6.

The propagation in Fig.5.6-6 is so familiar that we do not need to give an example that it actually occurs outside abstract mathematics. It is not so evident that anything in the real world propagates according to Fig.5.6-7. For a possible example, let A stand for a proton and B for a neutron in Fig.5.6-7b. The pair AB may then represent a deuterium nucleus at the time $\theta = 000$. The dyadic shift causes an interchange of proton A and neutron B at the time $\theta = 001$. At the time $\theta = 011$ is again an interchange, but the nucleus has also moved from $\xi = 000$, 001 to $\xi = 011$, 010. At $\theta = 010$ is again an interchange of proton and neutron, etc. The exchange of proton and neutron by means of virtual pions is an accepted fact, and it is used to explain the attraction of proton and neutron by means of exchange forces. Dyadic topology yields this exchange automatically, whereas integer-number-topology according to Fig.5.6-6b shows no such exchange.

We note that the propagation of a deuterium nucleus according to Fig.5.6-7b corresponds to our usual concept of propagation if the samples A and B are not resolved but observed as only one sample. The particle exchange shows up only if the resolution is sufficiently good.

A similar interpretation applies to Fig.5.6-7c. Let A and C stand for protons, B and D for neutrons. The complex A, B, C, and D may then stand for a helium nucleus. On the average, the four particles move like in the usual propagation of Fig.5.6-6c. The exchange of the particles shows up only for sufficiently good resolution.

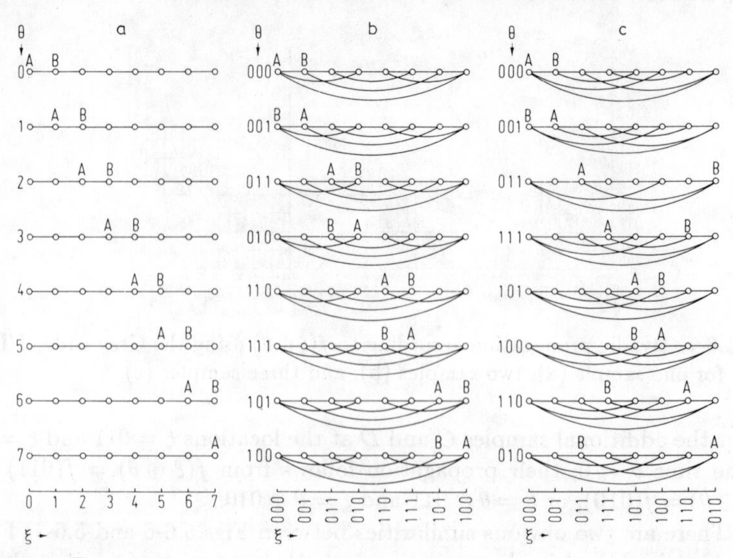

FIG.5.6-8. Propagation of two samples A and B along an integer numbers axis with simple or multiple connections; (a) usual numbers axis; (b) numbers axis with points connected according to the Gray code; (c) numbers axis with points connected according to a minimized code.

One may object to this interpretation that Fig.5.6-7 represents the propagation of particles from one mark to the next of the coordinate system of Fig.5.6-3c, while Fig.5.6-6 represents propagation along an Euclidean straight line. However, the important feature of the coordinate system of Fig.5.6-3c is the connection of the measuring marks 000 to 111, not their location "at the corners of a cube". We have not hesitated to replace the rings in Figs.3.1-11 and 3.1-19 by straight rods in Figs.3.1-12 and 3.1-20d. Furthermore, in Fig.5.4-6 we had left the coordinate marks at the location they had in the Cartesian coordinate system of Fig.5.4-5 but had put in connecting rods according to dyadic topology[4]. The topology of the coordinate system of Fig.5.6-3c is not changed if we treat the connecting rods as rubber strings that can be twisted and stretched, as long as the interconnections of the marks 000 to 111 remain unchanged.

We turn to Fig.5.6-8 for an explanation. The usual propagation of two samples A and B according to the world graph of Fig.5.6-6b is shown on the left (a). A ruler with marks $\xi = 0 \ldots 7$ is shown. At the time $\theta = 0$ the samples A and B are located at $\xi = 0$ and 1. With increasing time

[4]We use the term *dyadic topology* for "topology of the dyadic group with Hamming distance".

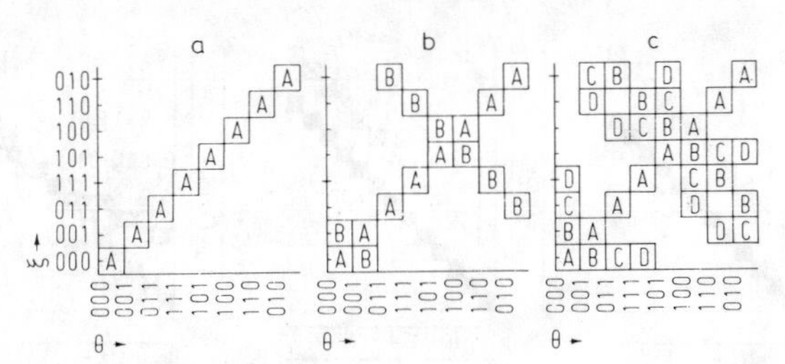

FIG.5.6-9. Dyadic propagation according to $f(\xi \oplus \theta)$ using the minimized code with eight elements of Table 5.2-2.

they propagate to larger values of ξ. The ruler is shown as an Euclidean straight line, but we automatically allow it to be some other geodesic, e.g. a meridian on the surface of the Earth.

The dyadic coordinate system of Fig.5.6-3c is shown in the center (b) of Fig.5.6-8 in this same "straight ruler representation". The marks $\xi = 000$ ... 100 are lined up as in Fig.5.6-8a, their ordering 000, 001, 011, 010, ... is that of the Gray code. The connections between the marks ξ are those shown by the coordinate system of Fig.5.6-3c. The propagation of the samples A and B for increasing times $\theta = 000, 001, \ldots$ is the same as in the world graph representation of Fig.5.6-7b. Both A and B move always from one mark ξ of the ruler to an adjacent mark, but each mark has now always three adjacent marks, while in Fig.5.6-8a there are only two or one ($\xi = 0, 7$) adjacent marks.

The usual ruler according to Fig.5.6-8a looks much more "natural" than the dyadic ruler of Fig.5.6-8b, but the flat Earth, the geocentric system, and Euclidean geometry also looked once very natural. There are many conceivable ways in which samples or particles can propagate in a discrete coordinate system, observed with a discrete clock. There is no *a priori* reason why they should propagate according to Fig.5.6-8a. Only observation can decide which of the thinkable modes of propagation actually occur.

To make the move of sample B from $\xi = 000$ at $\theta = 001$ to $\xi = 010$ at $\theta = 011$ in Fig.5.6-8b more acceptable, let us return to Fig.5.6-5. We had said there that one cannot actually say that a sample moves from square 1 to square 2, to square 3, etc. One can really only say that a sample disappears from square 1 and a sample appears in square 2. If the samples are individually marked like cars with a license plate, one can be sure they are the same sample or particle. For undistinguishable samples or particles, e.g. electrons, one cannot be sure. This is particularly evident for the velocity $\beta = 2$ in Fig.5.6-5d, where the squares do not touch each other, but it also holds in the other cases. If we assume the topology of the continuum for the coordinate system and the clock—or for space-time—

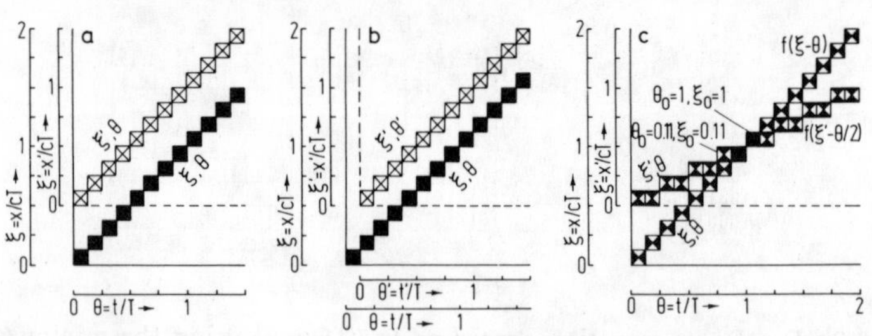

FIG.5.6-10. Dyadic shifting of two samples, using eigencoordinates for each sample. Trouble occurs only if the two samples cannot be distinguished and the connection between the samples and their eigencoordinates is lost.

as well as the possibility of continuous observation, we could observe how sample B in Fig.5.6-8b gets from $\xi = 000$ at $\theta = 001$ to $\xi = 010$ at $\theta = 011$. A coordinate system with finite resolution, a clock with finite resolution and—by implication—a finite number of observations, only permits us to say that an electron was observed at $\xi = 000$ at $\theta = 001$, and an electron was observed at $\xi = 010$ at $\theta = 011$, but one cannot say that it was the same electron or how it got from one location to the other. This will be particularly so if we make only two or a few observations. If we make many observations, plot the observed locations as function of the observation time θ, and the resulting points can readily be connected by a smooth curve, we will be convinced to have observed the same particle. However, in this case we observe the average propagation, and the average propagation is the same in Figs.5.6-6 and 5.6-7.

In Fig.5.6-4 we had plotted the propagation of one sample in dyadic coordinates according to the minimized code rather than the Gray code. Figure 5.6-9 shows the extension to two samples (b) and four samples (c). Figure 5.6-8c shows the propagation of two samples A, B along a ruler with multiple connected marks according to the minimized code.

In everyday life, there seems to be no question that a ruler according to Fig.5.6-8a rather than b should be used. However, sample A is propagating equally along both rulers since it is propagating in its eigencoordinates. Let us see how two samples can propagate both in their eigencoordinates.

Figure 5.6-10a shows two samples at the time $\theta = 0$ at the location $\xi = 0$ and $\xi' = 1/2$. Shifting the one sample in the coordinate system θ, ξ and the other in the coordinate system θ, ξ' makes both samples propagate as in integer-number-topology. Figure 5.6-10b shows that the same principle applies if the two samples start at different times as well as at different coordinate points, while Fig.5.6-10c shows the propagation of two samples with different velocity.

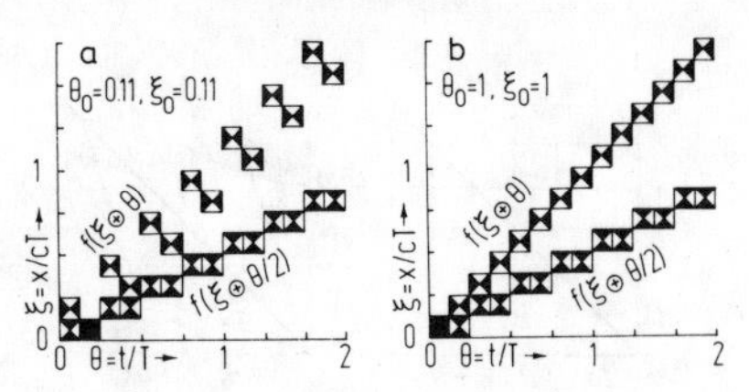

FIG.5.6-11. The two possible ways of dyadic shifting of $f(\xi \oplus \theta)$ and $f(\xi \oplus \theta/2)$, using the Gray code, after being in the same location-time interval in Fig.5.6-10c. The origin of the coordinate system influences the result, but no reason can be given why one choice of origin should be preferred over the other.

Let two cars leave from different places. As a rule, the drivers will keep track of their progress by measuring the elapsed time since their departure and the distance traveled from the point of departure. In other words, they use their eigencoordinates just as they are used for the two samples of Fig.5.6-10a or b. For the use of eigencoordinates it is necessary that a clock and a distance measuring device be carried by the samples—as in the case with cars—or that an observer can distinguish the samples and associate them with the proper time and location measurements. Let Fig.5.6-10 now represent the propagation of two equal particles like two electrons or two protons. They can be distinguished if they are so far apart in space-time[5] that an interaction is not possible. Once they get within interacting range, it becomes impossible to distinguish between them, and it is difficult to see how eigencoordinates can be defined. Let Fig.5.6-10c represent two electrons. Initially they are so far apart that their propagation can be represented in their eigencoordinates. However, the two electrons are in the same coordinate-time interval at the time $\theta_0 = 0.11$ or $\theta = 1$. Figure 5.6-11 shows their propagation in a common coordinate system from these two time points on. The average velocity is the same in Fig.5.6-11, but the faster electron ($\beta = 1$) has a jumpy movement superimposed on its average propagation in Fig.5.6-11a. This looks like a good argument against dyadic topology. However, an electron is supposed to have a jumpy movement—or a *Zitterbewegung*—in real-number-topology too, around the average path of propagation, which gives it the velocity of light even though the velocity in the direction of the average path is much slower (Schrödinger, 1930). Hence, dyadic topology yields qualitatively a result that would require an additional explanation in real-number-topology. We will return to this problem in Section 8.4, where we will be able to advance from this qualitative argument

[5] We use the short expression space-time to avoid cumbersome statements like "if their location in the coordinate system or the time of observation are sufficiently different . . . ".

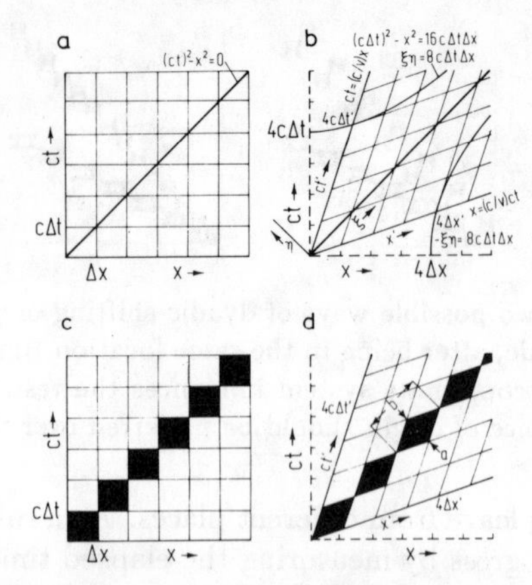

FIG.5.6-12. World line of a photon moving in the laboratory system (a) and in an inertial system having velocity v/c relative to the laboratory system (b), if real-number-topology is used. The corresponding world graphs for integer number or dyadic topology are shown in (c) and (d). Note that the *cells of uncertainty* have the area $c\Delta t\Delta x$, independent of the velocity v/c, due to the relation $ab/2 = |\xi\eta|/8 = c\Delta t\Delta x$.

to a quantitative result.

Let us close this section with a short discussion of moving coordinate systems. Figure 5.6-12 shows the generalization of our world graphs with finite resolution for moving coordinate systems. The plots in Figs.5.6-12a and b are standard illustrations of Lorentz geometry for the propagation of a photon. Figures 5.6-12c and d show the same, except that the continuous world line $(ct)^2 - x^2 = 0$ is replaced by world graphs consisting of black squares and rhombuses. The main lesson to be learned is that the area $c\Delta t\Delta x$ of the black squares in Fig.5.6-12c is not changed by the Lorentz transformation of Fig.5.6-12d. The resolution cells or cells of uncertainty have the same area $c\Delta t\Delta x$ for all inertial systems.

5.7 DYADIC CLOCKS

We have so far replaced the Cartesian coordinate system by dyadic coordinate systems, or other coordinate systems based on rings. We have also replaced the usual integer number ruler of Fig.5.6-8a by the dyadic rulers of Figs.5.6-b and c. The clock was never changed. We will suspect that the jumpy motion of some of the particles in Section 5.6 is caused by the use of an *usual-integer-number-clock*.

For an explanation of this last term refer to the usual-integer-number-ruler of Fig.5.6-8a. We bend it into a circle according to Fig.5.7-1a, add a

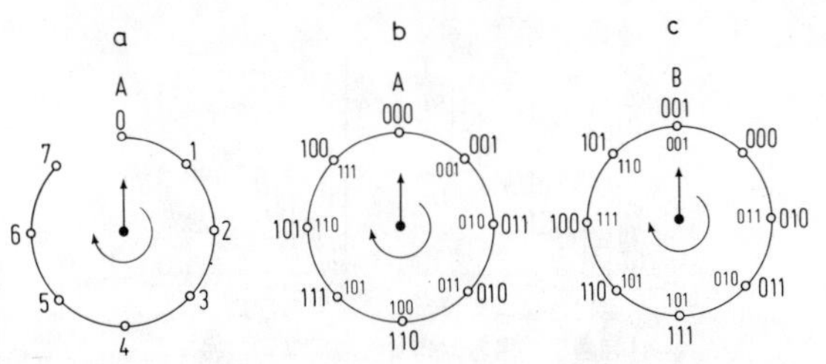

FIG.5.7-1. Ordinary clock based on the ruler and the movement of A in Fig.5.6-8a (a). Dyadic clock based on the ruler of Fig.5.6-8b and the movement of A and B (b, c). The large numbers on the outside of the circles give the time θ in Gray code, the small numbers inside the circles give θ in the usual binary code.

clock hand, and obtain a primitive version of our usual clock. Instead of this circular arrangement with a clock hand, we could use the straight ruler of Fig.5.6-8a and a particle A or B that move with constant velocity along this ruler[1].

Let us apply this principle to obtain a dyadic clock from Fig.5.6-8b. We bend the ruler of Fig.5.6-8b into a circle as shown in Fig.5.7-1b. The circle is closed since the points $\xi = 000$ and $\xi = 100$ of the ruler in Fig.5.6-8b are connected. The eight marks on the circle of Fig.5.7-1b are numbered in the sequence $\xi = 000, 001, \ldots, 100$ in which the particle A in Fig.5.6-8b passes these marks. By adding a clock hand in Fig.5.7-1b, that rotates with constant angular velocity, we obtain a simple dyadic clock. The numbers shown inside the circle are the usual binary code corresponding to the numbers of the Gray code according to Table 5.2-2. These numbers will be needed later to subdivide the time intervals for better time resolution.

Next we use the particle B in Fig.5.6-8 as indicator for our clock. We use in Fig.5.7-1c the same circle as in Fig.5.7-1b, but the eight marks are now numbered according to the sequence $\xi = 001, 000, 010, 011, 111, 110, 100, 101$ in which B moves in Fig.5.6-8b.

We use the dyadic clocks to represent the movement of A and B in Fig.5.6-8b in a world graph representation. With the usual clock we obtain Fig.5.6-7b. Using the dyadic clocks of Figs.5.7-1b and c we must plot the time θ according to the marks of these clocks; this is done in Fig.5.7-2, using the Gray code as well as the usual binary code. The coordinate ξ is marked as in Fig.5.6-7.

The jumpy movement of B in Fig.5.6-7b is removed in Fig.5.7-2b. This

[1]We have pointed out before that it is difficult to define a constant velocity before one has a clock. We typically avoid this problem by basing clocks on constant periods of oscillations, e.g. that of a pendulum, of quartz crystals, or of atoms. The constancy of the period of an oscillation can be checked without a clock by observing interference patterns and thus reducing the problem to one of distance measurements.

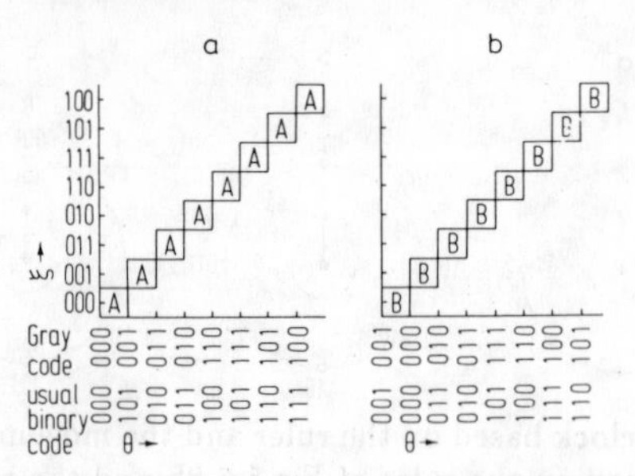

FIG.5.7-2. Dyadic propagation according to Fig.5.6-7b but with the usual time θ replaced by the times according to Fig.5.7-1b in (a) and according to Fig.5.7-1c in (b).

is no surprise from the way the time shown by the clocks was defined. We recall a quote used earlier:

Time is defined so that motion looks simple (Misner *et al.*, 1973, p. 23).

So far everything was so simple since we assumed the velocity $\beta = 1$. Let us turn to general values of β. The usual equation of propagation

$$f(\xi - \beta\theta) = f(\xi_0) \tag{1}$$

leads to the relations:

$$\xi - \beta\theta = \xi_0,\ \xi - \beta(\theta' + \theta_0) = \xi_0,\ \xi - \beta\theta' = \xi_0 - \beta\theta_0 = 0 \tag{2}$$

For $\xi_0 = 0$ a sample is shifted in its eigencoordinates. For $\xi \neq 0$ we can replace θ by the sum $\theta' + \theta_0$, and choose θ_0 so that $\xi_0 - \beta\theta_0$ equals zero. Hence, the definition of a new time $\theta' = \theta - \theta_0$ reduces the general case $\xi_0 \neq 0$ to the special case $\xi_0 = 0$ of shifting in eigencoordinates.

We apply Eqs.(1) and (2) to dyadic shifting and start with the following relation:

$$f(\xi \oplus \beta\theta) = f(\xi_0) \tag{3}$$

The next two steps according to Eq.(2) are still possible:

$$\xi \oplus \beta\theta = \xi_0,\ \xi \oplus \beta(\theta' \oplus \theta_0) = \xi_0 \tag{4}$$

The third step cannot be made since we cannot write $\xi \oplus \beta\theta' \oplus \beta\theta_0$ for $\xi \oplus \beta(\theta' \oplus \theta_0)$ in order to arrive at the equivalent of the third equation in Eq.(2). The problem is evidently that usual multiplication and addition are mixed with modulo 2 addition. We had pointed out in connection with Eq.(4.3-4) that an equation of the form $\xi \oplus \beta\theta = \xi_0$ implies a dyadic coordinate system but time added or subtracted as usual. If we use a dyadic clock we must use addition and multiplication modulo 2 as discussed in connection with Eq.(4.3-6). Instead of Eq.(3) we must write:

$$f(\xi \oplus \beta \otimes \theta) = f(\xi_0) \tag{5}$$

The dyadic relations equivalent to Eq.(2) follow:

$$\xi \oplus \beta \otimes \theta = \xi_0, \ \xi \oplus \beta \otimes (\theta' \oplus \theta_0) = \xi_0, \ \xi \oplus \beta \otimes \theta' \oplus \beta \otimes \theta_0 = \xi_0$$
$$\xi \oplus \beta \otimes \theta' = \xi_0 \oplus \beta \otimes \theta_0 = 0 \tag{6}$$

For $\beta = 1$ we obtain $\theta_0 = \xi_0$, which is the choice used in Fig.5.7-1c, but for $\beta = 1/2$ we obtain[2]:

$$\xi_0 \oplus \frac{1}{2} \otimes \theta_0 = 0, \ \theta_0 = 2\xi_0 \tag{7}$$

At successive times[3] $\theta' = 0, 1, 10, 11, \ldots$ one obtains the coordinates

$$\xi = \beta \otimes \theta' = \frac{1}{2} \otimes \theta' = \xi_0 \oplus \beta \otimes \theta \tag{8}$$

and the time:

$$\theta = \theta' \oplus \theta_0 = \theta' \oplus 2\xi_0 \tag{9}$$

The generalization of Figs.5.7-1 and 5.7-2 from $\beta = 1$ to $\beta = 1/2$ is shown by Figs.5.7-3 and 5.7-4. The clocks with eight time marks for the samples A and B in Fig.5.7-1 are replaced by clocks with 16 time marks in Fig.5.7-3, since twice as much time is required at half the velocity to cover the same distance. The clock for sample A shows the time increasing from $\theta = 0000$ in steps of 0001, both in Gray code (numbers outside the circle) and in the usual binary code (numbers inside the circle). The clock for sample B is more complicated. First the time θ_0 is determined from Eq.(7) for $\xi_0 = 0001$ to be 0010 (usual binary code) or 0011 (Gray code). The times $\theta' = 0, 1, 10, 11, 100, \ldots$ (usual binary code) or $\theta' = 0, 1, 11, 10, 110, \ldots$ (Gray code) are then added modulo 2 to θ_0 according to Eq.(9) to yield

[2] For $\beta = 2^n$, $n = 0, \pm 1, \pm 2, \ldots$, the modulo 2 multiplication yields the same result as the usual multiplication. See Box 4.3-1.

[3] The use of the usual binary code makes it easy to perform the multiplication $\beta \otimes \theta$ or $\beta \otimes \theta'$, while the use of the Gray code makes it easy to perform the addition $\xi_0 \oplus \beta \otimes \theta$ or $\theta' \oplus 2\xi_0$ since ξ and ξ_0 are given in Gray code in Figs.5.7-2 and 5.7-4. There is no good alternative to this switching back and forth between the two codes.

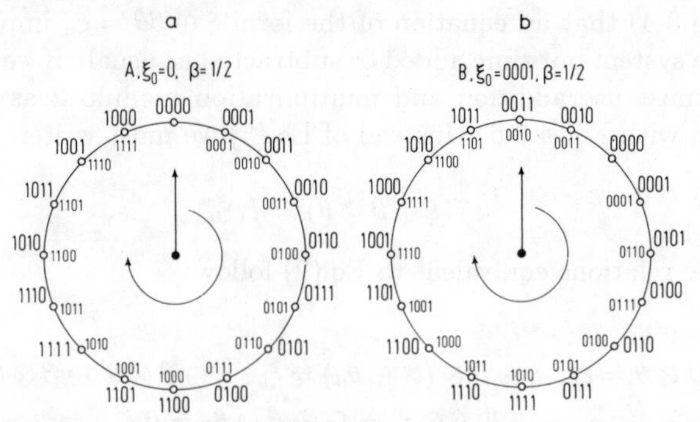

FIG.5.7-3. The clocks of Fig.5.7-1b and c but with 16 rather than 8 time marks θ in order to permit time observations for the propagation with velocity $\beta = 1/2$ from $\xi_0 = 000$ to $\xi = 100$ (a) or from $\xi_0 = 001$ to $\xi = 101$ (b). The time θ is given in Gray code (numbers outside the circles) and in usual binary code (numbers inside the circles).

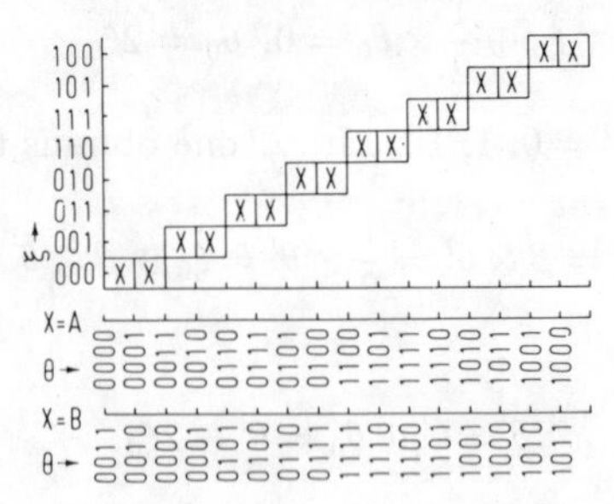

FIG.5.7-4. World graph of the propagation of samples A and B of Fig.5.7-2, but with the velocity $\beta = 1$ reduced to $\beta = 1/2$. The times θ are given in Gray code.

the other time marks. The propagation of the samples A and B according to Eqs.(5) and (6) is shown in Fig.5.7-4. Rather than drawing two equal illustrations with different time scales as in Fig.5.7-2, only one illustration with time scales either for sample A or for sample B is drawn.

We are well used to choosing coordinate systems that simplify calculations and help make results more understandable. We are also well used to carrying the underlying concepts further and assigning various metrics or topologies to what is usually called space. We do not treat time with the same pragmatism, even though we use the term space-time. As a result, a dyadic clock looks more artificial than a dyadic coordinate system. However, there does not seem to be any logical reason why one must use the usual type of clock, anymore than there is a logical reason why Euclidean geometry or the distance element $ds^2 = dx_i dx^i$ must be used.

6 Distinction of Sinusoidal Functions

6.1 Differential and Partial Differential Equations

We have shown in Sections 5.3 to 5.5 that standing waves characterize topology. It is generally agreed that the usual topology of the real numbers distinguishes sinusoidal functions, and that these functions should thus be distinguished if we live indeed in a space-time continuum having that topology. There is little doubt that sinusoidal functions are currently the most distinguished functions in science and engineering. However, this distinction may be based on reasons other than the topology of space-time.

A first reason is that the general solution of any ordinary differential equation with constant coefficients and the variable t consists of terms $e^{-t/\tau}\sin\omega t$ and $e^{-t/\tau}\cos\omega t$. Such differential equations are not only easy to solve but are also easy to implement practically, e.g. by time-invariant masses, springs, and dampers or inductors, capacitors, and resistors. Making these components time-variable is possible but difficult, and the resulting ordinary differential equations with variable coefficients usually have to be solved via series expansions at the singular points. Hence, we have little incentive to use time-variable differential equations and nonsinusoidal functions, but this has nothing to do with the topology of space-time.

Let us turn to partial differential equations with constant coefficients, and consider the example of the wave equation:

$$\frac{\partial^2 w}{\partial \theta^2} = \frac{\partial^2 w}{\partial \xi^2}, \quad \theta = \frac{t}{T}, \quad \xi = \frac{x}{cT} \tag{1}$$

Its general solution was found by d'Alembert in the eighteenth century:

$$w(\xi, \theta) = f(\xi - \theta) + g(\xi + \theta) \tag{2}$$

The wave equation distinguishes the time shifts $\xi - \theta$ and $\xi + \theta$. We have seen that these time shifts lead to sinusoidal functions for standing waves. This is a distinction, although a subtle one since the functions $f(\xi - \theta)$ and $g(\xi + \theta)$ do not have to be sinusoidal, nor do they have to be representable by a superposition of sinusoidal functions in the sense of uniform or mean-square convergence. Since the wave equation *assumes* a space-time continuum with the usual topology of real numbers, it is a logical consequence of the assumption that sinusoidal functions should be distinguished[1]. A less subtle distinction of sinusoidal functions is obtained—for a wrong reason—if Bernoulli's product method is used for the solution of partial differential equations. Let us apply this method to the wave equation:

$$w(\xi,\theta) = \varphi(\xi)\psi(\theta) \; . \tag{3}$$

Substitution of $\varphi(\xi)\psi(\theta)$ into Eq.(1) yields

$$\frac{1}{\psi(\theta)}\frac{\partial^2\psi(\theta)}{\partial\theta^2} = \frac{1}{\varphi(\xi)}\frac{\partial^2\varphi(\xi)}{\partial\xi^2} = \mu^2 \tag{4}$$

or

$$\frac{d^2\psi}{d\theta^2} - \mu^2\psi = 0 \tag{5}$$

$$\frac{d^2\varphi}{d\xi^2} - \mu^2\varphi = 0 \tag{6}$$

where μ^2 is a constant.

Since Eqs.(5) and (6) are ordinary differential equations with constant coefficients, their general solution consists of sine and cosine functions:

$$\begin{aligned} \psi(\theta) &= A\cos\mu\theta + B\sin\mu\theta \\ \varphi(\xi) &= C\cos\mu\xi + D\sin\mu\xi \end{aligned} \tag{7}$$

The distinction of the sine and cosine functions is strictly due to the use of Bernoulli's product method, which only yields particular solutions of the wave equation.

If we do not assume the topology of the continuum but use the topology of coordinate systems based on rings, the usual differential calculus is replaced by a more general one, and the role of exponential and sinusoidal

---

[1]In a previous book on radar engineering the author claimed that the wave equation and d'Alembert's solution of it do *not* distinguish sinusoidal functions (Harmuth, 1981). This referred to the fact that $f(\xi-\theta)$ and $g(\xi+\theta)$ in Eq.(2) do not have to be sinusoidal functions, and do not have to be representable by a Fourier series or transform, which yield mean-square convergence. It did not seem advisable to include a discussion of the topology of space-time in a book on radar engineering.

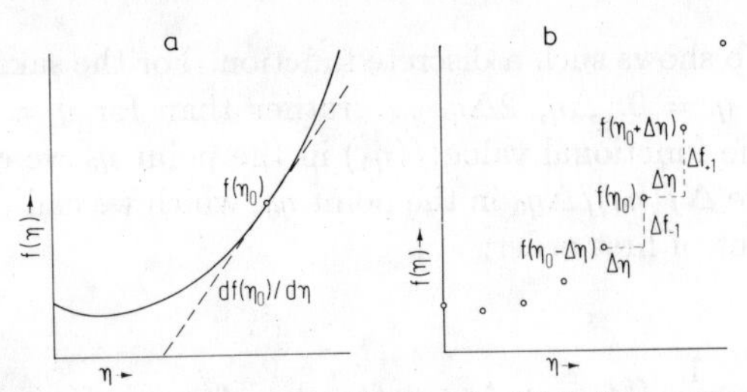

FIG.6.2-1. Continuous function with slope $df(\eta_0)/d\eta$ (a), and function defined for discrete numbers $\eta = 0, \Delta\eta, 2\Delta\eta, \ldots$ with average difference $\{[f(\eta_0 + \Delta\eta) - f(\eta_0)] + [f(\eta_0) - f(\eta_0 - \Delta\eta)]\}/2\Delta\eta$ in the point η_0.

functions by other functions. We will first study the limit ring $2^N \to \infty$, then the dyadic ring, and finally general rings.

6.2 Concepts of the Calculus of Finite Differences

Figure 6.2-1a shows a continuous curve. At the point η_0 we can define the functional value $f(\eta_0)$, its first derivative $df(\eta_0)/d\eta$, etc. The differential equation of first order,

$$\frac{df(\eta)}{d\eta} = qf(\eta) \tag{1}$$

is known to be satisfied by the exponential function

$$f(\eta) = Ae^{q\eta} \tag{2}$$

while the differential equation of second order,

$$\frac{d^2 f(\eta)}{d\eta^2} = -\omega^2 f(\eta) \tag{3}$$

is satisfied by the sine-cosine functions:

$$f(\eta) = A \sin \omega\eta + B \cos \omega\eta \tag{4}$$

It follows from Eq.(3) that

$$w(\xi, \theta) = A \sin \omega\theta \, \sin \omega\xi \tag{5}$$

is a particular solution of the wave equation of the continuum, Eq.(6.1-1).

Let us develop the corresponding concepts for a discrete function $f(\eta)$ that is defined for the values $\eta = 0, 1, 2, \ldots$ rather than for all real values of η. This is a function defined on the limit ring $2^N \to \infty$ in Fig.3.1-1.

Figure 6.2-1b shows such a discrete function. For the sake of generality it is defined for $\eta = 0, \Delta\eta, 2\Delta\eta, \ldots$ rather than for $\eta = 0, 1, 2, \ldots$. In addition to the functional value $f(\eta_0)$ in the point η_0 we can define the *average difference* $\Delta f(\eta_0)/\Delta\eta_0$ in the point η_0, which we call the symmetric difference quotient of first order:

$$\begin{aligned}\frac{\Delta f(\eta_0)}{\Delta\eta} &= \frac{1}{2\Delta\eta}\{[f(\eta_0+\Delta\eta)-f(\eta_0)]+[f(\eta_0)-f(\eta_0-\Delta\eta)]\}\\ &= \frac{1}{2\Delta\eta}[f(\eta_0+\Delta\eta)-f(\eta_0-\Delta\eta)] \qquad (6)\end{aligned}$$

The first differential quotient $df(\eta_0)/d\eta$ of Fig.6.2-1a is obtained as the limit $\Delta\eta \to d\eta$. However, we cannot make this transition without introducing the hypothesis that the finite or denumerable infinite points η in Fig.6.2-1b can be replaced by nondenumerable infinite points. As a result, the average difference of Eq.(6) cannot be interpreted as the slope of a tangent, since a tangent is a concept of continuum theory, just like the concepts of (continuous) line, surface, and space are. A proper interpretation of the average difference is to think of $f(\eta_0+\Delta\eta)$, $f(\eta_0)$, and $f(\eta_0-\Delta\eta)$ as temperatures at three adjacent points—measured at adjacent marks of either a ruler or a clock—and of $\Delta f(\eta_0)/\Delta\eta$ as the average temperature difference at the point η_0 relative to its two neighbors $\eta_0+\Delta\eta$ and $\eta_0-\Delta\eta$. From Eq.(6) we derive a difference equation that is the equivalent of the differential equation of first order, Eq.(1):

$$\begin{aligned}\frac{\Delta f(\eta)}{\Delta\eta} &= qf(\eta)\\ f(\eta+\Delta\eta)-2q\Delta\eta f(\eta)-f(\eta-\Delta\eta) &= 0 \qquad (7)\end{aligned}$$

This is a recursion formula or difference equation of second order, even though Eq.(1) is a differential equation of first order. The function that satisfied this equation will be the equivalent of the exponential function of Eq.(2).

For the solution of Eq.(7) we substitute $m = \eta/\Delta\eta = 0, 1, 2, \ldots$ and $p = q\Delta\eta$:

$$f(m+1)-2pf(m)-f(m-1) = 0 \qquad (8)$$

Such an ordinary difference equation with constant coefficients is solved by the substitution

TABLE 6.2-1
THE FIRST FEW FIBONACCI NUMBERS $f(m)$.

m	0	1	2	3	4	5	6	7	8	9	10
$f(m)$	0	1	1	2	3	5	8	13	21	34	55

$$f(m) = A\rho^m \tag{9}$$

just like differential equations with constant coefficients are solved by the substitution $f(m) = Ae^{qm}$ (Milne-Thomson, 1951; Nörlund, 1924). One obtains from Eq.(8):

$$\begin{gathered}
A\rho^{m-1}(\rho^2 - 2p\rho - 1) = 0 \\
\rho_1 = p + (1+p^2)^{1/2}, \ \rho_2 = p - (1+p^2)^{1/2} \\
f(m) = A\rho_1^m + B\rho_2^m
\end{gathered} \tag{10}$$

$$\begin{aligned}
f(\eta) = A\left\{q\Delta\eta + \left[1 + (q\Delta\eta)^2\right]^{1/2}\right\}^{\eta/\Delta\eta} \\
+ B\left\{q\Delta\eta - \left[1 - (q\Delta\eta)^2\right]^{1/2}\right\}^{\eta/\Delta\eta} \\
\approx A(1+q\Delta\eta)^{\eta/\Delta\eta} + B(-1+q\Delta\eta)^{\eta/\Delta\eta}, \ q\Delta\eta \ll 1
\end{aligned} \tag{11}$$

Let us note that Eq.(8) defines for the special case $2p = 1$ with the initial conditions $f(0) = 0$, $f(1) = 1$ the rapidly increasing Fibonacci numbers (Gelfond, 1958, p. 281). The first few are listed in Table 6.2-1. In the general solution of Eq.(11) one has to choose the constants $A = -B = 1/\sqrt{5}$ to satisfy the initial conditions $f(0) = 0$, $f(1) = 1$.

Let us turn to the difference quotient of second order. Instead of taking the average of the differences $f(\eta + \Delta\eta) - f(\eta)$ and $f(\eta) - f(\eta - \Delta\eta)$ we now take the difference of these two differences:

$$\begin{aligned}
\frac{\Delta^2 f(\eta_0)}{\Delta\eta^2} &= \frac{1}{(\Delta\eta)^2}\{[f(\eta_0 + \Delta\eta) - f(\eta_0)] - [f(\eta_0) - f(\eta_0 - \Delta\eta)]\} \\
&= \frac{1}{(\Delta\eta)^2}[f(\eta_0 + \Delta\eta) - 2f(\eta_0) + f(\eta_0 - \Delta\eta)]
\end{aligned} \tag{12}$$

A physical interpretation of this second order difference quotient is that it gives the difference between the temperature differences $f(\eta_0 + \Delta\eta) - f(\eta_0)$ and $f(\eta_0) - f(\eta_0 - \Delta\eta)$ at the marks $\eta_0 + \Delta\eta$, η_0, and $\eta_0 - \Delta\eta$.

The equivalent of the second order differential equation, Eq.(3), becomes:

$$\frac{\Delta^2 f(\eta)}{\Delta\eta^2} = -\omega^2 f(\eta)$$

$$f(\eta+\Delta\eta) - \left[2-(\omega\Delta\eta)^2\right] f(\eta) + f(\eta-\Delta\eta) = 0 \tag{13}$$

We substitute again $m = \eta/\Delta\eta$ as well as $2p^2 = (\omega\Delta\eta)^2$,

$$f(m+1) - 2(1-p^2)f(m) + f(m-1) = 0 \tag{14}$$

and obtain the characteristic equation

$$\rho^2 - 2(1-p^2)\rho + 1 = 0 \tag{15}$$

with the solutions:

$$\rho_1 = 1-p^2+ip(2-p^2)^{1/2},\ \rho_2 = 1-p^2-ip(2-p^2)^{1/2} \tag{16}$$

The general solution of Eq.(13) equals

$$f(m) = A\rho_1^m + B\rho_2^m$$

which assumes the following form for small values of $p = \omega\Delta\eta/\sqrt{2}$:

$$f(\eta) \approx A(1+i\omega\Delta\eta/\sqrt{2})^{\eta/\Delta\eta} + B(1-i\omega\Delta\eta/\sqrt{2})^{\eta/\Delta\eta} \tag{17}$$

The complex terms indicate that this function represents oscillations rather than a rapidly increasing function as $f(\eta)$ in Eq.(11). We will not discuss this solution any further at this time since we will have to come back to it later on. Here we note that it provides particular solutions of the discrete wave equation

$$\frac{\Delta^2 w(\xi,\theta)}{\Delta\xi^2} = \frac{\Delta^2 w(\xi,\theta)}{\Delta\theta^2} \tag{18}$$

just as the sinusoidal functions of Eqs.(4) and (5) do for the continuous wave equation, Eq.(6.1-1). To show this, let us form the second partial difference quotient of $w(\xi,\theta)$ if $w(\xi,\theta)$ can be represented by a product:

$$w(\xi,\theta) = f(\xi)g(\theta) \tag{19}$$

One obtains from Eq.(12):

$$\frac{\Delta^2 w(\xi,\theta)}{\Delta\xi^2} = \frac{\Delta^2 f(\xi)}{\Delta\xi^2} g(\theta)$$
$$\frac{\Delta^2 w(\xi,\theta)}{\Delta\theta^2} = \frac{\Delta^2 g(\theta)}{\Delta\theta^2} f(\xi)$$

Substitution into Eq.(18) yields:

$$\frac{\Delta^2 f(\xi)}{\Delta\xi^2}\frac{1}{f(\xi)} = \frac{\Delta^2 g(\theta)}{\Delta\theta^2}\frac{1}{g(\theta)} = -\omega^2 \tag{20}$$

A comparison with Eq.(13) shows that Eq.(19) is indeed a particular solution of the discrete wave equation (18), if $f(\xi)$ and $g(\theta)$ satisfy Eq.(20).

It seems fairly clear from these examples that we do not need the concept of a continuum to obtain a wave equation. For the limit ring $2^N \to \infty$ the exponential function is replaced by the function $f(\eta)$ defined in Eq.(11), while sine-cosine functions are replaced by the function defined by Eq.(17).

We will next turn to the other extreme of Fig.3.1-1 and explore differentiation, differential equations, and the wave equation for the dyadic ring $2^N = 2$.

6.3 Concepts of the Dyadic Calculus

For the generalization of differential calculus for a finite ring we turn to the dyadic coordinate system of Fig.3.1-19. For simplification we use its representation by a cubical structure as developed in Fig.3.1-20. Such a simplified dyadic coordinate system is shown in Fig.6.3-1a. Let us consider a particular point m of the eight points 000 ... 111 of this coordinate system. It has the three neighbors $m \oplus 001$, $m \oplus 010$, $m \oplus 100$, all with the same distance 1 or—generally—$\Delta\eta$ from m as shown in Fig.6.3-1b. In analogy to Eq.(6.2-6) we define the *average dyadic difference* $Df(m)/Dm$ in the point m, where $f(m)$ is a scalar value assigned to the point m, just as $f(\eta)$ or $f(\eta_0)$ was assigned to the points η or η_0 in[1] Fig.6.2-1:

$$\begin{aligned}\frac{Df(m)}{Dm} &= \frac{1}{3\Delta\eta}\{\,[f(m)-f(m\oplus 001)] + [f(m)-f(m\oplus 010)] \\ &\qquad\qquad + [f(m)-f(m\oplus 100)]\,\} \\ &= \frac{1}{\Delta\eta}\{f(m) - \frac{1}{3}[f(m\oplus 001) + f(m\oplus 010) + f(m\oplus 100)]\,\}\end{aligned} \tag{1}$$

[1]Note that we use the normalized variable $m = 0, 1, 2, \ldots$ while in Section 6.2 we started out with the normalized variable η and switched to the normalized variable $m = \eta/\Delta\eta$ at Eq.(6.2-8).

It turns out that a very similar operation was introduced by Gibbs[2]. Instead of the coordinate system of Fig.6.3-1 with equally long rods of length $\Delta\eta$ one must use the coordinate system with a rod of length $\Delta\eta$ between the points m and $m \oplus 001$, a rod of length $2\Delta\eta$ between m and $m \oplus 010$, a rod of length $4\Delta\eta$ between m and $m \oplus 100$, etc. The point m and its three neighbors in a three-dimensional coordinate system with rods of length $\Delta\eta$, $2\Delta\eta$, and $4\Delta\eta$ is shown in Fig.6.3-1c. Equation (1) modified for this coordinate system becomes:

$$\frac{Gf(m)}{Gm} = \frac{1}{3\Delta\eta}\{[f(m) - f(m \oplus 001)] + \frac{1}{2}[f(m) - f(m \oplus 010)] + \frac{1}{4}[f(m) - f(m \oplus 100)]\} \quad (2)$$

The dyadic differentiation by Gibbs uses the factor $2^{3-2} = 2$ instead of $1/3\Delta\eta$ for a reason to be explained later. We generalize from $r = 3$ to r dimensions and obtain the dyadic differentiation according to Gibbs:

$$\frac{gf(m)}{gm} = 2^{r-2}\{2^0[f(m) - f(m \oplus 1)] + 2^{-1}[f(m) - f(m \oplus 10)] + 2^{-2}[f(m) - f(m \oplus 100)] + \dots + 2^{-(r-1)}[f(m) - f(m \oplus 10\dots0)]\} \quad (3)$$

For an interpretation of the physical meaning of dyadic differentiation let $f(m)$, $f(m \oplus 001)$, $f(m \oplus 010)$, and $f(m \oplus 100)$ denote four temperatures recorded at the points m, $m \oplus 001$, $m \oplus 010$, and $m \oplus 100$ of a coordinate system according to Fig.6.3-1c. The quantities $(\Delta\eta)^{-1}[f(m) - f(m \oplus 001)]$, $(2\Delta\eta)^{-1}[f(m) - f(m \oplus 010)]$, $(4\Delta\eta)^{-1}[f(m) - f(m \oplus 100)]$ are then the temperature differences per unit distance $\Delta\eta$ between the point m and its three neighbors. The average temperature difference, $\Delta T_K/\Delta\eta$, becomes:

$$\frac{\Delta T_K}{\Delta\eta} = \frac{1}{3\Delta\eta}\{[f(m) - f(m \oplus 001)] + \frac{1}{2}[f(m) - f(m \oplus 010] + \frac{1}{4}[f(m) - f(m \oplus 100]\} \quad (4)$$

[2] See, e.g. Gibbs and Ireland (1974). We use the symbol $gf(x)/gx$ in analogy to $df(x)/dx$ in honor of Gibbs.

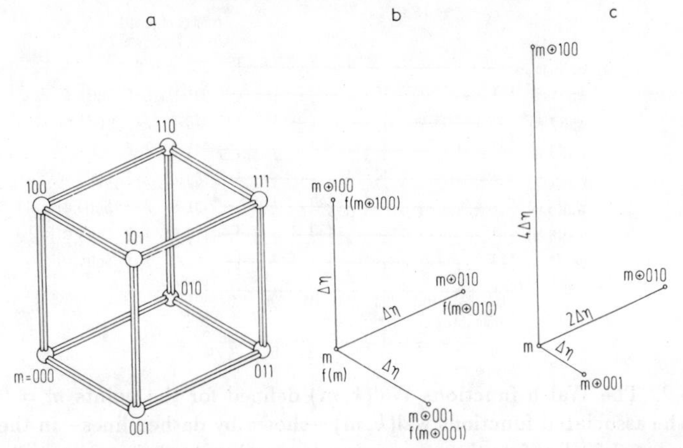

FIG.6.3-1. Three-dimensional dyadic coordinate system (a). A point m of this coordinate system with three neighbors $m \oplus 001$, $m \oplus 010$, $m \oplus 100$, all having the same distance $\Delta\eta$ from m (b). Modification of (b) so that $m \oplus 001$ has the distance $\Delta\eta$, $m \oplus 010$ the distance $2\Delta\eta$, and $m \oplus 100$ the distance $4\Delta\eta$ from m.

This equation equals Eq.(3) for $r = 3$ except for a constant factor. Furthermore, its comparison with Eq.(1) shows that the interpretation of the first dyadic difference quotient as an average difference at point m also carries over to Eq.(1), if the coordinate system of Fig.6.3-1b rather than that of Fig.6.3-1c is used.

Let us see why Gibbs chose the particular definition of Eq.(3) for the dyadic difference quotient $gf(m)/gm$. We *difference* the eight Walsh functions $\mathrm{Wal}(k, m)$ in $r = 3$ dimensions with $k = 0 \ldots 7$, $m = 0 \ldots 7$:

$$\begin{aligned}\frac{g\,\mathrm{Wal}(k,m)}{gm} &= 2\{\,[\mathrm{Wal}(k,m) - \mathrm{Wal}(\mathrm{k}, \mathrm{m} \oplus 001)] \\ &+ \frac{1}{2}[\mathrm{Wal}(k,m) - \mathrm{Wal}(k, m \oplus 010] + \frac{1}{4}[\mathrm{Wal}(k,m) - \mathrm{Wal}(k, m \oplus 100)]\,\} \\ &= \mathrm{Wal}(k,m)\{2\,[1 - \mathrm{Wal}(k,001)] + [1 - \mathrm{Wal}(k,010)] \\ &\qquad + \frac{1}{2}[1 - \mathrm{Wal}(k,100)]\,\}\end{aligned} \tag{5}$$

For the calculation of the terms in braces refer to Fig.6.3-2 and Table 6.3-1. Figure 6.3-2 shows the Walsh functions in the interval $-0.5 \leq m \leq 7.5$. These functions are equal to the functions $\mathrm{Wal}(k, \theta)$ in Fig.5.3-1 but the interval has been changed, and the variable m is listed in Gray code while θ in Fig.5.3-1 is listed in decimal numbers. From Fig.6.3-2 we take the values of $\mathrm{Wal}(k, 001)$, $\mathrm{Wal}(k, 010)$, and $\mathrm{Wal}(k, 100)$ for Table 6.3-1. One readily

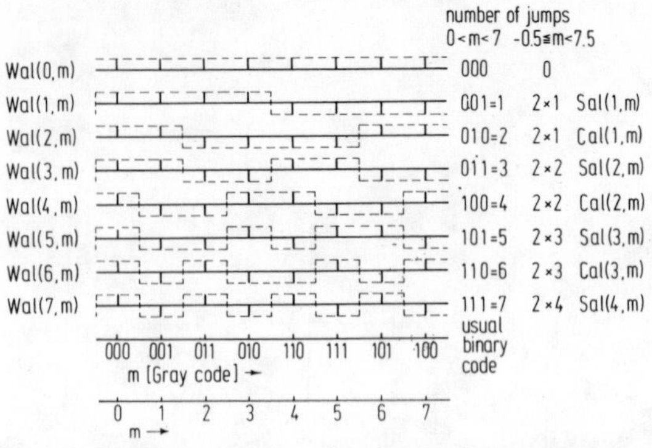

FIG.6.3-2. The Walsh functions $\mathrm{Wal}(k,m)$ defined for the points $m = 0, 1, \ldots,$ 7, and the associated functions $\mathrm{wal}(k,m)$—shown by dashed lines—in the interval $-0.5 \leq m \leq 7.5$. The functions are ordered according to their sequency k, which is the number of sign changes or zero crossings of the functions $\mathrm{wal}(k,m)$ in the open interval $-0.5 < m < 7.5$. The term sequency was originally introduced for one half the number of zero crossings in the half open interval $-0.5 \leq m < 7.5$ listed on the right of the illustration. This leads to the notation $\mathrm{Sal}(i,m)$, $\mathrm{Cal}(i,m)$ for the Walsh functions, which stresses the similarity of the odd functions $\mathrm{Sal}(i,m)$ with sinusoidal functions, and of the even functions $\mathrm{Cal}(i,m)$ with cosinusoidal functions (Pichler, 1967; Harmuth, 1977).

TABLE 6.3-1

EVALUATION OF EQ.(5) WITH THE HELP OF THE FUNCTIONS $\mathrm{Wal}(k,m)$ IN FIG.6.3-2.

k	0	1	2	3	4	5	6	7
$\mathrm{Wal}(k,001)$	1	1	1	1	−1	−1	−1	−1
$\mathrm{Wal}(k,010)$	1	1	−1	−1	1	1	−1	−1
$\mathrm{Wal}(k,100)$	1	−1	1	−1	1	−1	1	−1
$2[1-\mathrm{Wal}(k,001)]$	0	0	0	0	4	4	4	4
$1-\mathrm{Wal}(k,010)$	0	0	2	2	0	0	2	2
$0.5[1-\mathrm{Wal}(k,100)]$	0	1	0	1	0	1	0	1
sum	0	1	2	3	4	5	6	7

derives from them $2[1-\mathrm{Wal}(k,001)]$, $1-\mathrm{Wal}(k,010)$, and $0.5[1-\mathrm{Wal}(k,100)]$ as well as the sum of these three terms. This sum equals k. Hence, we may rewrite Eq.(5) as follows:

$$\frac{g\,\mathrm{Wal}(k,m)}{gm} = k\,\mathrm{Wal}(k,m) \tag{6}$$

Hence, the Walsh functions play for the dyadic difference quotient of first

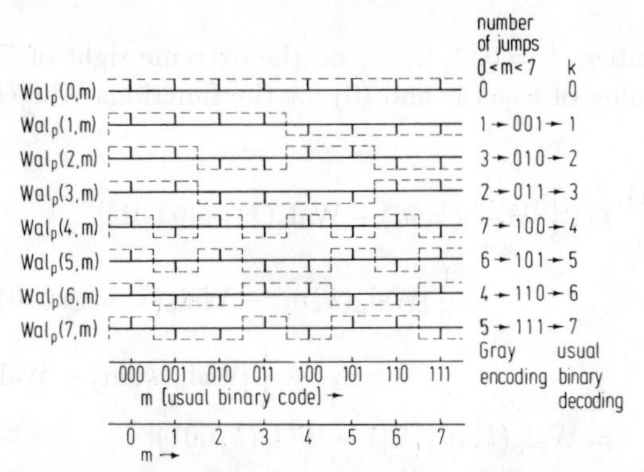

FIG.6.3-3. The Walsh functions $\mathrm{Wal_p}(k,m)$ in Paley ordering or Gray-code-of-sequency ordering.

TABLE 6.3-2

EVALUATION OF EQ.(7) WITH THE HELP OF THE FUNCTIONS $\mathrm{Wal_p}(k,m)$ IN FIG.6.3-3.

k	0	1	2	3	4	5	6	7
$\mathrm{Wal_p}(k,001)$	1	1	1	1	−1	−1	−1	−1
$\mathrm{Wal_p}(k,010)$	1	1	−1	−1	1	1	−1	−1
$\mathrm{Wal_p}(k,100)$	1	−1	1	−1	1	−1	1	−1
$2[1-\mathrm{Wal_p}(k,001)]$	0	0	0	0	4	4	4	4
$1-\mathrm{Wal_p}(k,010)$	0	0	2	2	0	0	2	2
$0.5[1-\mathrm{Wal_p}(k,100)]$	0	1	0	1	0	1	0	1
sum	0	1	2	3	4	5	6	7

order according to Gibbs exactly the same role as the exponential function for the differential quotient of first order according to Eqs.(6.2-1) and (6.2-2) if we substitute $m = \eta/\Delta\eta$ and $k = q\Delta\eta$.

We had to list m in the Gray code and k according to sequency to obtain Eq.(6). For completeness sake we also point out that one can list k according to the Gray code of sequency and m in the usual binary code. The Walsh functions in this ordering are denoted $\mathrm{Wal_p}(k,m)$—following a recommendation by Ahmed *et al.* (1973)—where the subscript p refers to Paley, who introduced this ordering (Paley, 1932). Figure 6.3-3 shows the first eight functions so ordered. The number of jumps of the function $\mathrm{wal_p}(k,m)$—shown by the dashed lines—are listed on the right of Fig.6.3-3. With the help of Table 5.2-2 we replace the numbers 0, 1, 3, 2, 7, ... by their Gray code. The resulting binary numbers 000, 001, 010, 011, ... increase like the numbers of the usual binary code, and they are decoded accordingly into the numbers $k = 0, 1, 2, \ldots$ on the extreme right of Fig.6.3-2.

The analog of Eqs.(5) and (6) for the functions $\mathrm{Wal_p}(k, m)$ becomes:

$$\begin{aligned}\frac{g\,\mathrm{Wal_p}(k,m)}{gm} &= 2\{\,[\mathrm{Wal_p}(k,m) - \mathrm{Wal_p}(k, m \oplus 001)] \\ &\qquad + \frac{1}{2}\,[\mathrm{Wal_p}(k,m) - \mathrm{Wal_p}(k, m \oplus 010)] \\ &\qquad\qquad + \frac{1}{4}\,[\mathrm{Wal_p}(k,m) - \mathrm{Wal_p}(k, m \oplus 100)]\,\} \\ &= \mathrm{Wal_p}(k,m)\{2\,[1 - \mathrm{Wal_p}(k,001)] \\ &\qquad + [1 - \mathrm{Wal_p}(k,010)] + \frac{1}{2}\,[1 - \mathrm{Wal_p}(k,100)]\,\} \\ &= k\,\mathrm{Wal_p}(k,m) \end{aligned} \tag{7}$$

Table 6.3-2 shows how the result $k\,\mathrm{Wal_p}(k,m)$ is obtained, using the functions $\mathrm{Wal_p}(k,001)$, $\mathrm{Wal_p}(k,010)$, and $\mathrm{Wal_p}(k,100)$ of Fig.6.3-3.

Consider the first dyadic derivative of a constant $f(m) = C$. Both Eqs.(2) and (3) yield zero:

$$\frac{DC}{Dm} = \frac{gC}{gm} = 0 \tag{8}$$

Next let us calculate the first dyadic derivative of the function $f(m) = m$ according to Eq.(3) for $r = 3$:

$$\frac{gm}{gm} = 2\{\,[m - (m \oplus 001)] + \frac{1}{2}\,[m - (m \oplus 010)] + \frac{1}{4}\,[m - (m \oplus 100)]\,\} \tag{9}$$

Table 6.3-3 shows the evaluation of this equation. The function $f(m) = m = 000 \ldots 100$ is given as binary number in Gray code in accordance with Fig.6.3-2, as well as in the decimal equivalents. Using the binary numbers, one may readily obtain $m \oplus 001$, $m \oplus 010$, and $m \oplus 100$ as well as their decimal equivalents. The ordinary subtractions and additions are done with these decimal numbers more readily than with the binary numbers. The sum in the last row of Table 6.3-3 represents gm/gm.

For a physical interpretation of this derivative we will consider the three-dimensional coordinate system of Fig.6.3-4a which has rods of length $\Delta\eta$ between any two points m and $m \oplus 001$, $2\Delta\eta$ between any two points m and $m \oplus 010$, and $4\Delta\eta$ between any two points m and $m \oplus 100$ in accordance with Fig.6.3-1c. The 8 marks of this coordinate system are numbered 000 ... 100 like m in Table 6.3-3; in addition, the decimal numbers of m are

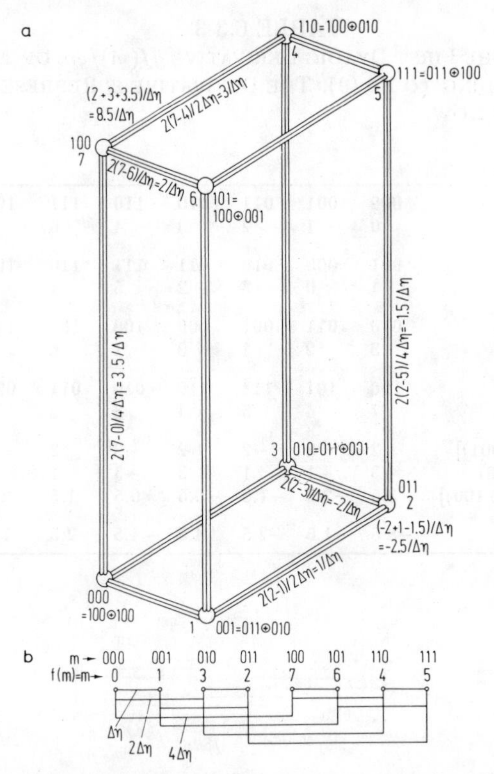

FIG.6.3-4. A three-dimensional coordinate system showing the functional values $f(m) = m$ at its 8 coordinate marks. The derivation of the dyadic derivative according to Eq.(9) is shown for the points $m = 011$ and 100 (a). Representation of the same coordinate system by a topologically equivalent, multiple connected ruler (b). Note that m is given in Gray code.

shown. Let these numbers represent temperatures measured at the marks of the coordinate system. The temperature difference per unit distance—multiplied by the weighting factors 2, 1, 1/2 of Eq.(9)—between the point $m = 011 = 2$ and its three neighbors are shown ($1/\Delta\eta$, $-2/\Delta\eta$, $-1.5\Delta\eta$). Their sum yields $-2.5/\Delta\eta$, which is the average temperature difference per unit distance in the point $m = 011$; this is the value shown in the last row of Table 6.3-3.

Figure 6.3-4b shows a ruler with multiple connected marks that has the same topological structure as the three-dimensional coordinate system of Fig.6.3-4a. One may readily verify that the mark $m = 011$ has the distance $\Delta\eta$ from $m = 010$, $2\Delta\eta$ from $m = 001$, and $4\Delta\eta$ from $m =$

TABLE 6.3-3

EVALUATION OF THE FIRST DYADIC DERIVATIVE $gf(m)/gm$ OF THE FUNCTION $f(m) = m$ ACCORDING TO EQ.(9). THE DERIVATIVE IS REPRESENTED BY THE SUM IN THE LAST ROW.

m	000	001	011	010	110	111	101	100
decimal	0	1	2	3	4	5	6	7
$m \oplus 001$	001	000	010	011	111	110	100	101
decimal	1	0	3	2	5	4	7	6
$m \oplus 010$	010	011	001	000	100	101	111	110
decimal	3	2	1	0	7	6	5	4
$m \oplus 100$	100	101	111	110	010	011	001	000
decimal	7	6	5	4	3	2	1	0
$2[m - (m \oplus 001)]$	−2	2	−2	2	−2	2	−2	2
$m - (m \oplus 010)$	−3	−1	1	3	−3	−1	1	3
$0.5[m - (m \oplus 100)]$	−3.5	−2.5	−1.5	−0.5	0.5	1.5	2.5	3.5
sum	−8.5	−1.5	−2.5	4.5	−4.5	2.5	1.5	8.5

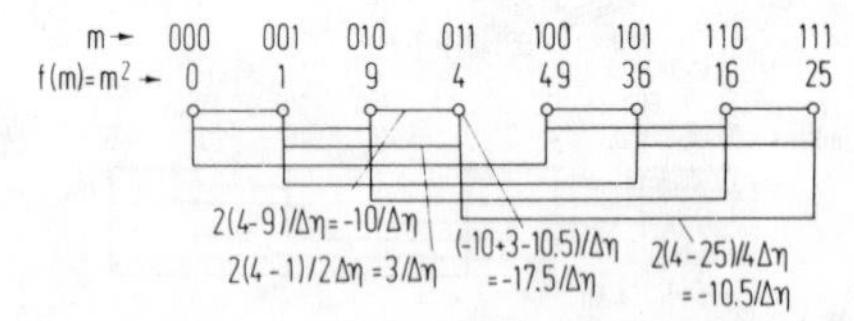

FIG.6.3-5. Multiple connected ruler according to Fig.6.3-4b, and the derivation of the average difference per unit distance for the function $f(m) = m^2$ at the point $m = 011 = 2$. Note that m is given in Gray code.

111, just like in Fig.6.3-4a. This possibility of using a ruler with multiple connected marks for the topologically correct representation—in addition to the analogy of Eq.(6) with Eqs.(6.2-1) and (6.2-2)—is the advantage of the dyadic difference quotient according to Gibbs[3].

Let us advance to the dyadic derivative of the function $f(m) = m^2$:

$$\frac{gm^2}{gm} = 2\{ [m^2 - (m \oplus 001)^2] + \frac{1}{2}[m^2 - (m \oplus 010)^2] + \frac{1}{4}[m^2 - (m \oplus 100)^2] \} \quad (10)$$

[3]No satisfactory equivalent to Eq.(6) exists for the derivative $Df(m)/Dm$. This is discussed in Section 12.11.

Table 6.3-4 shows the evaluation of gm^2/gm in analogy to the evaluation of gm/gm in Table 6.3-3. Figure 6.3-5 shows the ruler of Fig.6.3-4b with the functional values $f(m) = m^2$, and the derivation of the average difference per unit distance for the mark $m = 011$.

The Fourier expansion plays a significant role in differential calculus. A similar role is played by the Walsh expansion in dyadic calculus. Consider the expansion $u(k)$ of a function $f(m)$ in a system of Walsh functions:

$$u(k) = 2^{-r} \sum_{m=0}^{2^r-1} f(m)\,\mathrm{Wal}(k,m) \tag{11}$$

$$f(m) = 2^{-r} \sum_{k=0}^{2^r-1} u(k)\,\mathrm{Wal}(k,m) \tag{12}$$

Since we have only 2^r points for which $u(k)$ or $f(m)$ are defined, we do not have to be concerned about the convergence of the sums. Furthermore, the series of Eq.(12) represents $f(m)$ exactly, not only with a vanishing mean square error.

As an example consider the Walsh expansion of the function $f(m) = m$ for $2^r = 8$. Table 6.3-5 shows m and $f(m)$ on top and the functions $\mathrm{Wal}(0,m)$ to $\mathrm{Wal}(7,m)$ below. The function $u(k)$ according to Eq.(11) is shown on the right. The Walsh expansion of $f(m) = m$ has thus the following form:

$$f(m) = m = \frac{7}{2}\,\mathrm{Wal}(0,m) - 2\,\mathrm{Wal}(1,m) - \mathrm{Wal}(3,m) - \frac{1}{2}\,\mathrm{Wal}(7,m) \tag{13}$$

Since there is no problem of convergence, we can obtain the dyadic derivative of a function $f(m)$ from its Walsh expansion and Eq.(6):

$$\frac{gf(m)}{gm} = \frac{g}{gm}\left(2^{-r} \sum_{k=0}^{2^r-1} u(k)\,\mathrm{Wal}(k,m)\right)$$
$$= 2^{-r} \sum_{k=0}^{2^r-1} k u(k)\,\mathrm{Wal}(k,m) \tag{14}$$

As an example, let us derive gm/gm from Eq.(13), and show that the result is the same as obtained from Eq.(9) and Table 6.3-3. We obtain from Eq.(13) according to Eq.(14):

TABLE 6.3-4
EVALUATION OF THE FIRST DERIVATIVE $gf(m)/gm$ OF THE FUNCTION $f(m) = m^2$ ACCORDING TO EQ.(10).

m	000	001	011	010	110	111	101	100
decimal	0	1	2	3	4	5	6	7
$m \oplus 001$	001	000	010	011	111	110	100	101
decimal	1	0	3	2	5	4	7	6
$m \oplus 010$	010	011	001	000	100	101	111	110
decimal	3	2	1	0	7	6	5	4
$m \oplus 100$	100	101	111	110	010	011	001	000
decimal	7	6	5	4	3	2	1	0
m^2	0	1	4	9	16	25	36	49
$(m \oplus 001)^2$	1	0	9	4	25	16	49	36
$(m \oplus 010)^2$	9	4	1	0	49	36	25	16
$(m \oplus 100)^2$	49	36	25	16	9	4	1	0
gm^2/gm	−35.5	−18.5	−17.5	15.5	−47.5	17.5	2.5	83.5

TABLE 6.3-5
WALSH EXPANSION OF THE FUNCTION $f(m) = m$ FOR $2^r = 8$.

		$m = f(m) \rightarrow$								
k	$\mathrm{Wal}(k, m)$	0	1	2	3	4	5	6	7	$\frac{1}{8}\sum_{m=0}^{7} f(m)\,\mathrm{Wal}(k, m)$
0	$\mathrm{Wal}(0, m)$	1	1	1	1	1	1	1	1	7/2
1	$\mathrm{Wal}(1, m)$	1	1	1	1	−1	−1	−1	−1	−2
2	$\mathrm{Wal}(2, m)$	1	1	−1	−1	−1	−1	1	1	0
3	$\mathrm{Wal}(3, m)$	1	1	−1	−1	1	1	−1	−1	−1
4	$\mathrm{Wal}(4, m)$	1	−1	−1	1	1	−1	−1	1	0
5	$\mathrm{Wal}(5, m)$	1	−1	−1	1	−1	1	1	−1	0
6	$\mathrm{Wal}(6, m)$	1	−1	1	−1	−1	1	−1	1	0
7	$\mathrm{Wal}(7, m)$	1	−1	1	−1	1	−1	1	−1	−1/2

$$\frac{gm}{gm} = -2\,\mathrm{Wal}(1, m) - 3\,\mathrm{Wal}(3, m) - \frac{7}{2}\,\mathrm{Wal}(7, m) \tag{15}$$

Table 6.3-6 shows the functions $\mathrm{Wal}(1, m)$, $\mathrm{Wal}(3, m)$. $\mathrm{Wal}(7, m)$, and the weighted sum according to Eq.(15). The result gm/gm is the same as in the last row of Table 6.3-3.

Dyadic integration is defined as the reversal of dyadic differentiation.

TABLE 6.3-6

DIFFERENTIATION OF THE FUNCTION $f(m) = m$ VIA A WALSH EXPANSION ACCORDING TO EQ.(15).

m	0	1	2	3	4	5	6	7
$\mathrm{Wal}(1,m)$	1	1	1	1	−1	−1	−1	−1
$\mathrm{Wal}(3,m)$	1	1	−1	−1	1	1	−1	−1
$\mathrm{Wal}(7,m)$	1	−1	1	−1	1	−1	1	−1
gm/gm	-8.5	-1.5	-2.5	4.5	-4.5	2.5	1.5	8.5

We may define $Df(m)/Dm$ in Eq.(1) or $gf(m)/gm$ in Eq.(3) for all values of m, and solve the resulting systems of linear equations for $f(m)$. The Walsh expansion of a function permits us to obtain the dyadic integral in a different way in the case of the reversal of Eq.(3). From Eq.(6) follows immediately the dyadic integral of $\mathrm{Wal}(k,m)$:

$$\oint \mathrm{Wal}(k,m)\, gm = \frac{1}{k} \mathrm{Wal}(k,m) \tag{16}$$

The value $k = 0$ must be excluded.

With the help of Eqs.(11) and (12) one may then define the integration for any function $f(m)$:

$$\begin{aligned} \oint f(m)\, gm &= \oint 2^{-r} \sum_{k=0}^{2^r+1} u(k) \mathrm{Wal}(k,m)\, gm \\ &= 2^{-r} \sum_{k=0}^{2^r-1} \frac{1}{k} u(k) \mathrm{Wal}(k,m) \end{aligned} \tag{17}$$

As an example, consider the integration of $f(m) = m$ in Eq.(13). Since we cannot integrate $\mathrm{Wal}(0,m)$, we integrate $m - 3.5\,\mathrm{Wal}(0,m) = m - 7/2$:

$$\oint \left(m - \frac{7}{2}\right) gm = -2\,\mathrm{Wal}(1,m) - \frac{1}{3} \mathrm{Wal}(3,m) - \frac{1}{14} \mathrm{Wal}(7,m) \tag{18}$$

Table 6.3-7 shows the function $f(m) = m - 7/2$ and its dyadic integral obtained according to Eq.(18).

Let us turn to the physical interpretation of the dyadic integration. Figure 6.3-6 shows again the coordinate system of Fig.6.3-4a with the marks $m = 000, 001, \ldots$. We assign the functional values $f(m) = -2.40, -2.26, \ldots$

TABLE 6.3-7

INTEGRATION OF A FUNCTION $f(m) = m - 7/2$ VIA A WALSH EXPANSION ACCORDING TO EQ.(18).

m	000	001	011	010	110	111	101	100
decimal	0	1	2	3	4	5	6	7
$m - 7/2$	−3.5	−2.5	−1.5	−0.5	0.5	1.5	2.5	3.5
$\mathrm{Wal}(1, m)$	1	1	1	1	−1	−1	−1	−1
$\mathrm{Wal}(3, m)$	1	1	−1	−1	1	1	−1	−1
$\mathrm{Wal}(7, m)$	1	−1	1	−1	1	−1	1	−1
$\mathcal{S}(m - 7/2)\, gm$	−2.40	−2.26	−1.74	−1.60	1.60	1.74	2.26	2.40

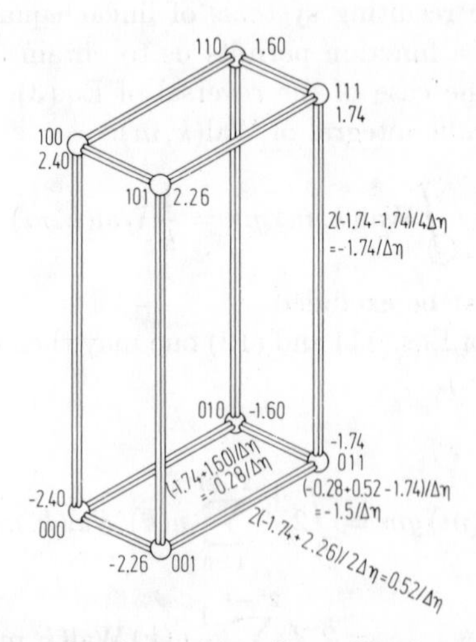

FIG.6.3-6. Dyadic integration in a three-dimensional coordinate system. The values $m - 7/2$ in Table 6.3-7 are obtained for the average difference per unit distance if the values $\mathcal{S}(m - 7/2)\, gm$ are assigned to the marks of the coordinate system.

according to Eq.(18) and Table 6.3-7 to these marks, and interpret $f(m)$ as the temperature measured at the mark m. We then produce the average temperature difference per unit distance for each point m; this is shown in detail for the point $m = 011$ in Fig.6.3-6. We obtain $-1.5/\Delta\eta$ or -1.5 for $\Delta\eta = 1$ in the point $m = 011$, and generally the values $m - 7/2$ shown in Table 6.3-7. Hence, dyadic integration solves the problem of finding which temperatures at the points m will produce a given average temperature difference per unit distance at the points m.

Let us turn to the dyadic difference quotient of second order. First, we try to define it in analogy to Eq.(6.2-12) as the difference of differences. For instant, the equivalent of Eq.(6.2-12) would contain terms of the form

$$[f(m\oplus 1)-f(m)]-[f(m)-f(m\ominus 1)]=2\,[f(m\oplus 1)-f(m)] \tag{19}$$

The equality of addition and subtraction modulo 2 reduces the difference of differences to the original difference, and this approach leads back to the first order difference quotient of Eq.(1).

A second approach is to define the second order difference quotient by the repeated application of the first order difference quotient. This is what one does to derive the differential quotient $d^2f(x)/dx^2$ from $df(x)/dx$, but it is not the way Eq.(6.2-12) was derived from Eq.(6.2-6). We first generalize Eq.(1) from 3 to r dimensions:

$$\frac{Df(m)}{Dm}=\frac{1}{\Delta\eta}\left[f(m)-\frac{1}{r}\sum_{i=0}^{r-1}f(m\oplus 2^i)\right] \tag{20}$$

$$\begin{aligned}
\frac{D^2f(m)}{Dm^2}&=\frac{1}{\Delta\eta}\left[\frac{Df(m)}{Dm}-\frac{1}{r}\sum_{k=0}^{r-1}\frac{Df(m\oplus 2^k)}{Dm}\right]\\
&=\frac{1}{\Delta\eta}\left\{\frac{1}{\Delta\eta}\left[f(m)-\frac{1}{r}\sum_{i=0}^{r-1}f(m\oplus 2^i)\right]\right.\\
&\qquad\left.-\frac{1}{r\Delta\eta}\sum_{k=0}^{r-1}\left[f(m\oplus 2^k)-\frac{1}{r}\sum_{i=0}^{r-1}f(m\oplus 2^k\oplus 2^i)\right]\right\}\\
&=\frac{1}{(\Delta\eta)^2}\left[\frac{r+1}{r}f(m)-\frac{2}{r}\sum_{i=0}^{r-1}f(m\oplus 2^i)\right.\\
&\qquad\left.+\frac{1}{r^2}\sum_{\substack{i=0\\ i\neq k}}^{r-1}\sum_{k=0}^{r-1}f(m\oplus 2^i\oplus 2^k)\right]
\end{aligned} \tag{21}$$

The first derivative in Eq.(20) contains the function $f(m)$ at the point m and its *first neighbors* $m\oplus 1$, $m\oplus 10$, $m\oplus 100$, ..., having the absolute distance 1 or $\Delta\eta$ from m. The second difference contains—in the double sum of Eq.(21)—also the function $f(m)$ at the *second neighbors* $m\oplus 11$, $m\oplus 101$, $m\oplus 110$, ..., having the absolute distance 2 or $2\Delta\eta$ from m. Mathematically, there is no objection to this, but from the standpoint of physics one introduces in essence an "action at a distance" if one permits an effect at point m that comes from a point with more than the smallest possible distance from m. We will discuss this in Section 8.3. Let us observe here that the second order difference quotient $\Delta^2 f(\eta_0)/\Delta\eta^2$ on the limit ring $2^N \to \infty$ according to Eq.(6.2-12) in the point η_0 contains only the function in the point η_0 and in the two first neighbor points $\eta_0 + \Delta\eta$ and $\eta_0 - \Delta\eta$. If we had defined there the second order difference quotient as a repetition of the first order difference quotient according to Eq.(6.2-6) we would have introduced functional values $f(\eta_0 + 2\Delta\eta)$ and $f(\eta_0 - 2\Delta\eta)$ at second neighbor points.

A number of variations of Eq.(19) have been tried, but no way was found to define a second order difference quotient that treats all first neighbor points $m \oplus 2^i$ equally and did not require second neighbor points $m \oplus 2^i \oplus 2^k$, $i \neq k$.

Consider next the dyadic wave equation obtained from Eq.(6.1-1) by the substitution of dyadic operators for the differential operators[4]:

$$\frac{D^2 w}{Dm^2} = \frac{D^2 w}{Ds^2} \qquad \begin{aligned} m &= \theta/\Delta\theta = (t/T)/(\Delta T/T) = 0,\,1,\,2,\ldots \\ s &= \xi/\Delta\xi = (x/cT)/(\Delta X/cT) = 0,\,1,\,2,\ldots \end{aligned} \tag{22}$$

Instead of the operator D/Dm we may also use the Gibbs operator g/gm of Eq.(3), which leads more readily to a solution:

$$\frac{g^2 w}{gm^2} = \frac{g^2 w}{gs^2} \tag{23}$$

Consider the function $f(m \oplus s)$ as a possible solution of Eq.(22), in analogy to d'Alembert's solution, Eq.(6.1-2), of the wave equation. We expand $f(m \oplus s)$ in a Walsh series:

$$\begin{aligned} f(m \oplus s) &= 2^{-r} \sum_{k=0}^{2^r-1} u(k)\,\mathrm{Wal}(k, m \oplus s) \\ &= 2^{-r} \sum_{k=0}^{2^r-1} u(k)\,\mathrm{Wal}(k, m)\,\mathrm{Wal}(k, s) \end{aligned} \tag{24}$$

With the help of Eq.(6) one obtains

[4]We use the same symbols D and g for difference and partial difference operators.

TABLE 6.3-8

EVALUATION OF THE FIRST DERIVATIVE $gf(m)/gm$ OF THE FUNCTION $f(m) = m$ WITH THE HELP OF EQ.(9) GENERALIZED FROM $r = 3$ TO $r = 4$.

m	0000	0001	0011	0010	0110	0111	0101	0100	1100	1101	1111	1110	1010	1011	1001	1000
decimal	0	1	2	3	4	5	6	7	8	9	10	11	12	13	14	15
$m \oplus 0001$	0001	0000	0010	0011	0111	0110	0100	0101	1101	1100	1110	1111	1011	1010	1000	1001
decimal	1	0	3	2	5	4	7	6	9	8	11	10	13	12	15	14
$m \oplus 0010$	0010	0011	0001	0000	0100	0101	0111	0110	1110	1111	1101	1100	1000	1001	1011	1010
decimal	3	2	1	0	7	6	5	4	11	10	9	8	15	14	13	12
$m \oplus 0100$	0100	0101	0111	0110	0010	0011	0001	0000	1000	1001	1011	1010	1110	1111	1101	1100
decimal	7	6	5	4	3	2	1	0	15	14	13	12	11	10	9	8
$m \oplus 1000$	1000	1001	1011	1010	1110	1111	1101	1100	0100	0101	0111	0110	0010	0011	0001	0000
decimal	15	14	13	12	11	10	9	8	7	6	5	4	3	2	1	0
$4[m - (m \oplus 0001)]$	-4	4	-4	4	-4	4	-4	4	-4	4	-4	4	-4	4	-4	4
$2[m - (m \oplus 0010)]$	-6	-2	2	6	-6	-2	2	6	-6	-2	2	6	-6	-2	2	6
$m - (m \oplus 0100)$	-7	-5	-3	-1	1	3	5	7	-7	-5	-3	-1	1	3	5	7
$\frac{1}{2}[m - (m \oplus 1000)]$	-7.5	-6.5	-5.5	-4.5	-3.5	-2.5	-1.5	-0.5	0.5	1.5	2.5	3.5	4.5	5.5	6.5	7.5
sum	-24.5	-9.5	-10.5	4.5	-12.5	2.5	1.5	16.5	-16.5	-1.5	-2.5	12.5	-4.5	10.5	9.5	24.5

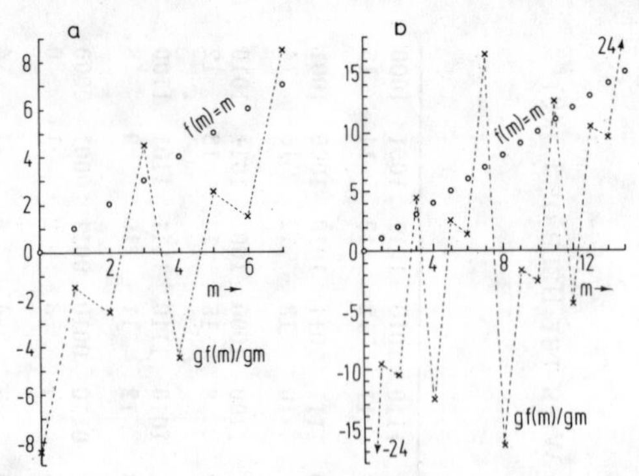

FIG.6.3-7. The function $f(m)$ and its first derivative $gf(m)/gm$ plotted in a discrete Cartesian coordinate system with one independent variable for $2^r = 8$ (a) and $2^r = 16$ (b) points.

$$\frac{g^2 f(m \oplus s)}{gm^2} = 2^{-r} \sum_{k=0}^{2^r+1} k^2 u(k) \operatorname{Wal}(k,m) \operatorname{Wal} k,s) \tag{25}$$

and an equal equation for $g^2 f(m \oplus s)/gs^2$. Hence, $\operatorname{Wal}(k, m \oplus s)$ is a particular solution of the dyadic wave equation, and

$$w(m,s) = f(m \oplus s) \tag{26}$$

is the general solution. It is known as the Butzer-Wagner solution (Butzer and Wagner, 1973).

Gibbs introduced his dyadic difference operator in 1974, but attempts to give it a physical meaning failed until dyadic coordinate systems were introduced, operations in such systems were studied, and the similarity between Eqs.(1) and (3) was recognized. For an example how confusing dyadic differentiation looks in the usual Cartesian coordinate systems, consider the function $f(m) = m$ and its derivative $gf(m)/gm$ of Eq.(9) and Table 6.3-3. Both functions are plotted in Fig.6.3-7a. The function $f(m) = m$ looks most reasonable in this representation, which misleads one into believing that the representation should also be meaningful for $gf(m)/gm$. Since no physical relevance of $gf(m)/gm$ can be seen in Fig.6.3-7a, one may try to increase the resolution from 2^3 to 2^4 points m. The derivative $gf(m)/gm$ is defined for this case by Eq.(3) with $r = 4$, and computed in Table 6.3-8 with the help of Fig.6.3-8; it is plotted, together with $f(m) = m$, in Fig.6.3-7b. No insight is gained by the increase of resolution[5].

[5]See also Harmuth (1972), p. 66, for a discussion of dyadic differentiation and integration before its connection with dyadic coordinate systems was recognized.

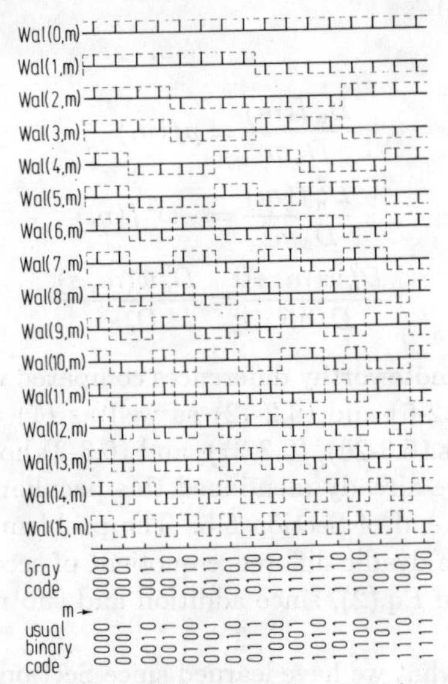

FIG.6.3-8. The 16 sampled Walsh functions Wal(k, m), $k = 0 \ldots 15$.

6.4 Difference Calculus on General Rings

Consider a function $f(m)$ defined for each point of a ring with n points, where n is between the limits $2^N = 2$ and $2^N \to \infty$ considered so far. We use the difference symbol D_n instead of the symbols D and Δ used in Eqs.(6.3-1) and (6.2-6). The first order difference quotient is readily defined in analogy to Eq.(6.2-6):

$$\begin{aligned}\frac{D_\mathrm{n} f(m)}{D_\mathrm{n} m} &= \frac{1}{2}\left\{\left[f(m \underset{\mathrm{n}}{+} 1) - f(m)\right] + \left[f(m) - f(m \underset{\mathrm{n}}{-} 1)\right]\right\} \\ &= \frac{1}{2}\left[f(m \underset{\mathrm{n}}{+} 1) - f(m \underset{\mathrm{n}}{-} 1)\right] \qquad m = 0, 1, 2, \ldots \qquad (1)\end{aligned}$$

A second order difference quotient may be defined in analogy to Eq.(6.2-12):

$$\begin{aligned}\frac{D_\mathrm{n}^2 f(m)}{D_\mathrm{n} m^2} &= \left\{\left[f(m \underset{\mathrm{n}}{+} 1) - f(m)\right] - \left[f(m) - f(m \underset{\mathrm{n}}{-} 1)\right]\right\} \\ &= f(m \underset{\mathrm{n}}{+} 1) - 2f(m) + f(m \underset{\mathrm{n}}{-} 1) \qquad (2)\end{aligned}$$

Difference and partial difference equations, in particular the wave equation, can be defined as before:

$$\frac{D_{\mathrm{n}} f(m)}{D_{\mathrm{n}} m} = p f(m) \tag{3}$$

$$\frac{D_{\mathrm{n}}^2 f(m)}{D_{\mathrm{n}} m^2} = -\omega^2 f(m) \tag{4}$$

$$\frac{D_{\mathrm{n}}^2 w(m,s)}{D_{\mathrm{n}} m^2} = \frac{D_{\mathrm{n}}^2 w(m,s)}{D_{\mathrm{n}} s^2} \tag{5}$$

There are a few noteworthy differences compared with the dyadic equations. Equations (6.2-6) and (6.2-12) as well as (1) and (2) hold for one dimension, while Eqs.(6.3-20), (6.3-21), and (6.3-3) hold for any number of dimensions. We have already mentioned this peculiarity of dyadic coordinate systems at the end of Section 5.5. The problem of second neighbors encountered with the dyadic difference quotient of second order of Eq.(6.3-21) does not occur in Eq.(2), since addition and subtraction modulo n are not equal for $n > 2$.

Let us sum up what we have learned since Section 6.1. The distinction of sinusoidal and exponential functions is clearly due to the assumption of a continuum with the usual topology of the real numbers. The replacement of the real numbers axis by rings with discrete points leads to a general calculus of finite differences, that distinguishes functions other than the sinusoidal and exponential functions, depending on the number of points on the particular ring chosen. In the limit $2^N \to \infty$ of denumerably many points one obtains the usual calculus of finite differences, which has been known for a long time even though its state of development does not come close to that of differential calculus. For the other limit, $2^N = 2$, one obtains a difference calculus that can be made equal, by small modifications, to the one introduced by Gibbs some ten years ago under the name logic calculus into abstract mathematics. This calculus is very little developed, but it permitted us to show that the Walsh functions play the role of exponential and sinusoidal functions for the topology of the dyadic group with Hamming distance, modified by the weighting factors used in Gibbs' version of the dyadic difference quotient. For rings between the limits $2^N \to \infty$ and $2^N = 2$ we do not seem to have yet an abstract mathematical theory, but it clearly can be developed. This shows that the mathematical distinction of sinusoidal functions proves about as much as the Euclidean parallels once proved. Since there are so many possible assumptions that lead to different results one can only decide by observation which metric or which topology should be used.

We will now turn from mathematics to physical observations that seem to distinguish sinusoidal functions, and thus support the assumption that there is something with the topology of the continuum which can be called space or space-time.

6.5 Spectral Decomposition of Light

We have seen that sinusoidal functions are distinguished by ordinary differential equations with constant coefficients, and by partial differential equations whose solution by means of Bernoulli's product method leads to ordinary differential equations with constant coefficients. Let us apply this knowledge to the spectral decomposition of light.

Light is always decomposed into sinusoidal waves with various wavelengths. Such a decomposition does not prove that light consists of sinusoidal waves, but only that prisms, diffraction gratings, etc. decompose light according to the system of sine-cosine functions. In the case of diffraction gratings and other devices based on interference, one immediately suspects that the time invariance of these devices is the reason why they decompose light into sinusoidal waves. The wave theory of light was developed by Huygens (1690) without assuming a sinusoidal time variation of the waves. The development of a light source for sinusoidal waves in the form of the laser was a major technical feat. There are many references which show that physicists of the nineteenth century were well aware that the decomposition of light into sinusoidal waves was only one out of many possible ones:

> ... we are at liberty, whenever it is convenient, to present white light by superposing a number of homogeneous vibrations having periods which lie very close together. But we are equally at liberty to assume any other representation so long as its resolution by Fourier's theorem gives us a distribution of intensity equal to that of the observed one[1] (Schuster, 1904, p. 321).

> ... one can consider white light to be composed of a sequence of random pulses, or of constantly distorted vibrations like the trembling movement which, for some physicists, constitute the heat movement[2] Gouy, 1886, p. 362).

---

[1] Rayleigh points out the possibility of non-Fourier decompositions for sound (1894, vol. 1, pp. 24, 25) but writes as follows about light (1889): "There is nothing arbitrary in the use of circular functions to represent the waves. As a general rule this is the only kind of wave which can be propagated without a change of form; and, even in the exceptional cases where the velocity is independent of wave-length, no generality is really lost by this process, because in accordance with Fourier's theorem any kind of periodic wave may be regarded as composed of a series of such [sinusoidal waves], with wave-lengths in harmonic progression." The interesting point is the unquestioned assumption of periodic waves. Since any observed wave occupies a finite time interval while a periodic wave must occupy an infinite time interval, there is obviously a question when the mathematical model of the periodic wave can be used for the analysis of observed waves.

[2] Translation by the author of the following statement: "... on peut regarder la

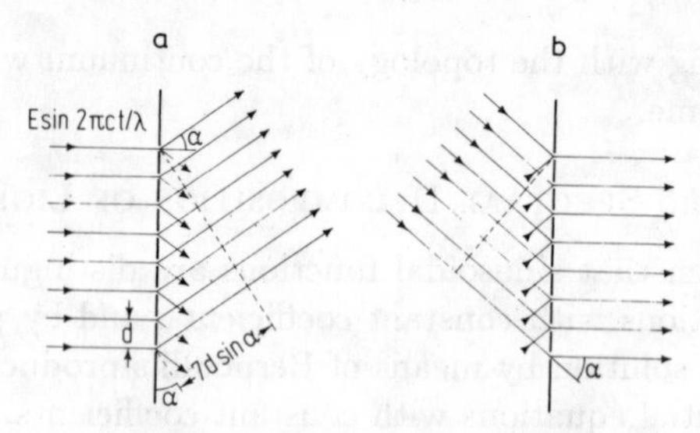

FIG.6.5-1. Linear, time-invariant diffraction grating with fixed light source and movable detector (a), as well as with fixed detector and movable light source (b). Maxima occur for $\sin\alpha = k\lambda/d$, minima for $\sin\alpha = (2k+1)\lambda/2d$ where $k = 0, 1, 2, \ldots$.

What is needed is an instruction how to build a spectrometer that decomposes light into other than sinusoidal waves. To find such an instruction, let us investigate the operation of a diffraction grating. Figure 6.5-1a shows a diffraction grating with eight transparent slots at a distance d. A sinusoidal wave with wavelength λ coming from the left produces eight spherical waves on the right side of the grating. The waves add up for the angle $\alpha = 0$, they cancel to yield the first minimum for $\sin\alpha = \lambda/2d$, they sum again to yield the second maximum for $\sin\alpha = \lambda/d$, and so forth.

If a sum of sinusoidal waves is received from the left, one obtains minima and maxima for each wave. This means that the incident light is decomposed into sinusoidal functions. Since we know from Fourier analysis that any signal with finite energy can be decomposed with a vanishing mean-square-error into sinusoidal functions, the pattern of minima and maxima produced by the diffraction grating will prove to us that this device has the necessary features actually to perform such a decomposition. In other words, the diffraction pattern proves, among other things, that the diffraction grating is a linear, time-invariant device. A linear but time-variable diffraction grating will decompose light into some other system of functions.

Figure 6.5-1b shows a modification of the diffraction grating of Fig.6.5-1a. The detector observes the light emitted vertically to the grating while the light source is moved to provide angles of incidence from $\alpha = 0°$ to $\alpha = 90°$. In Fig.6.5-1a, on the other hand, the detector has to be moved while the light source is fixed.

Figure 6.5-2 shows how the diffraction grating of Fig.6.5-1b has to be modified to decompose light into Walsh waves. The Walsh functions

---

lumière blanche comme formée par une suit d'impulsions tout à fait irrégulières, ou de vibrations sans cesse troublées, analogues au movement de trépidation qui, pour quelques physiciens, constitue le mouvement calorifique".

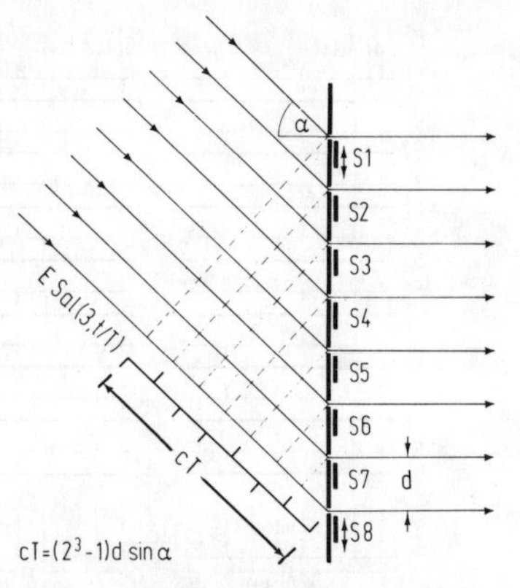

FIG.6.5-2. Linear, time-variable diffraction grating with fixed detector. The time variation is provided by the shutters S1 to S8.

$\mathrm{Cal}(l, t/T) = \mathrm{Cal}(l, \theta) = \mathrm{Wal}(2l, \theta)$ and $\mathrm{Sal}(l, \theta) = \mathrm{Wal}(2l-1, \theta)$ have the two parameters *time base* T and *sequency* $\varphi = l/T$, which are significant for our purpose, while only the frequency or wavelength was important for sinusoidal functions[3]. The time base is determined by the choice of angle of incidence α. If the diffraction grating has eight slots with distance d one has to choose α according to the formula

$$cT = 7d \sin \alpha \tag{1}$$

where c is the velocity of light. In the general case of 2^n slots, one has to substitute $2^n - 1$ for 7.

We had introduced in Fig.5.3-1 the sampled Walsh functions $\mathrm{Wal}(k, \theta)$ and the continuous Walsh functions $\mathrm{wal}(k, \theta)$. The sampled functions should be visualized as voltages or field strengths that are zero most of the time and have short pulses with amplitude $+E$ or $-E$. We divide these functions into even functions $\mathrm{Cal}(i, \theta) = \mathrm{Wal}(2i, \theta)$ and odd functions $\mathrm{Sal}(i, \theta) = \mathrm{Wal}(2i-1, \theta)$ having the same sequency $\varphi = i/T$, just as we divide sinusoidal functions into even and odd functions, $\cos 2\pi i t/T$ and $\sin 2\pi i t/T$, having the same frequency $f = i/T$. The Walsh functions $\mathrm{Sal}(i, t/T)$ and $\mathrm{Cal}(i, t/T)$

[3]The sequency of a function is defined as one half the number of zero crossings per unit time. For instance, the functions $\mathrm{wal}(6, \theta) = \mathrm{cal}(3, \theta)$ and $\mathrm{wal}(5, \theta) = \mathrm{sal}(3, \theta)$ in Fig.5.3-1 have 6 zero crossings, and thus have the sequency $\varphi = 3/T$. A short reflection shows that the sampled functions $\mathrm{Wal}(6, \theta) = \mathrm{Cal}(3, \theta)$ and $\mathrm{Wal}(5, \theta) = \mathrm{Sal}(3, \theta)$—periodically continued—also have 6 zero crossings per time interval T, where $T = t/\theta$. There are other complete systems of orthogonal functions whose individual functions are characterized by the number of zero crossings, e.g. Legendre polynomials and parabolic cylinder functions, which are Hermite polynomials multiplied by $\exp(-x^2)$. Sequency and frequency are identical for sinusoidal functions (Harmuth, 1972, pp. 4, 14, 22).

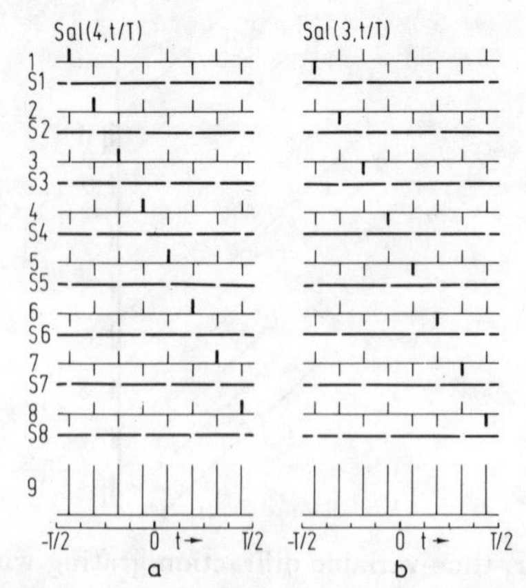

FIG.6.5-3. Operation of shutters S1 to S8 for passage of the wave sal$(4, t/T)$ and sal$(3, t/T)$; black: shutter closed; white: shutter open. The amplitude samples shown by a heavy line illustrate the delay between the eight waves.

with time base T but various *normalized sequencies* $i = \varphi T$ are separated by making the transparent slots of the diffraction grating time variable. In the case of the Walsh functions this time variation is provided by shutters S1 to S8 in Fig.6.5-2 that are either open or closed. For other systems of functions one would generally need a more complicated time variation. Let us ignore for the moment how such shutters could be implemented and let us find the rule for their operation.

Figure 6.5-3a shows the operation of shutters S1 to S8 to permit a wave with the time variation of Sal$(4, t/T)$ to pass. Lines 1 to 8 show this function, periodically repeated sufficiently often outside the interval $-T/2 \leq t \leq T/2$, with time shifts $0, T/8, 2T/8, \dots, 7T/8$. These time shifts correspond to the arrival of a wavefront at the eight transparent slots in Fig.6.5-2. Shutters S1 to S8 are open (white) or closed (black) as shown. There are always four positive samples of the Walsh wave that are shown superimposed in line 9, due to the properly opened shutters, while the negative samples are blocked by closed shutters. One can readily see that the opening of shutter S1 coincides with the four positive samples of Sal$(4, t/T)$ in Fig.6.5-3a, and with those of Sal$(3, t/T)$ in Fig.6.5-3b. The opening times of shutters S2 to S8 are obtained from a cylical shift of the opening times of shutter S1.

Figure 6.5-4a shows the shutters operated for passage of Sal$(3, t/T)$ but the function Sal$(4, t/T)$ being applied. Two positive and two negative samples are passed through the diffraction grating at any time, and their sum yields zero. Figure 6.5-4b shows the shutters operated for passage of Sal$(4, t/T)$ but Sal$(3, t/T)$ being applied. Again the positive and negative

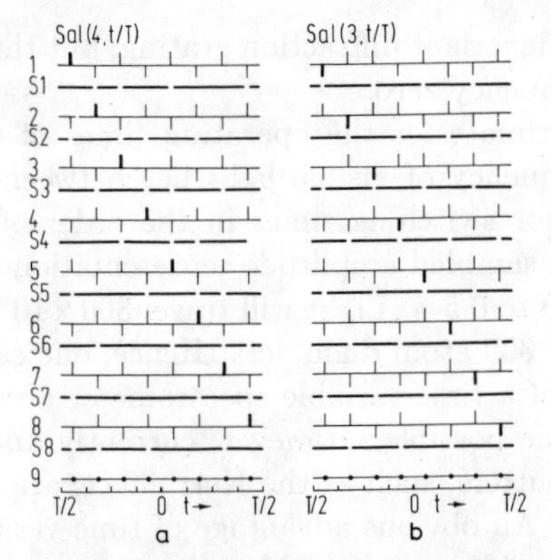

FIG.6.5-4. Shutters S1 to S8 operating to pass $\mathrm{sal}(3, t/T)$ and block $\mathrm{sal}(4, t/T)$ (a). Shutters operating to pass $\mathrm{sal}(4, t/T)$ and block $\mathrm{sal}(3, t/T)$ (b). The amplitude samples shown by a heavy line illustrate the delay between the eight waves.

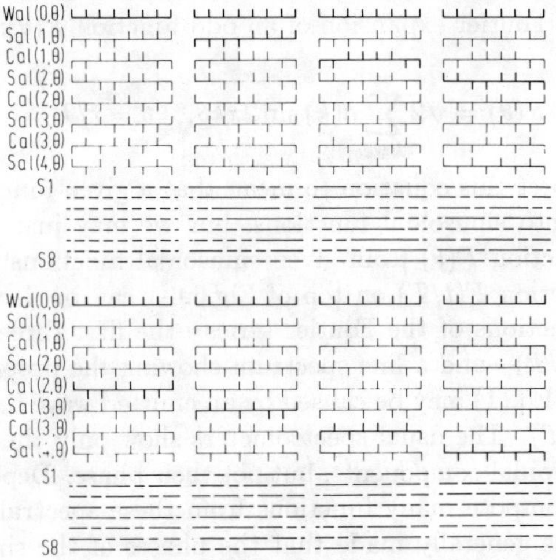

FIG.6.5-5. Operation of shutters S1 to S8 for separation of the first eight Walsh functions.

samples cancel at all times.

Figure 6.5-5 shows the operation of shutters S1 to S8 for separation of the first eight Walsh functions. The functions shown in heavy lines pass the shutters operated according to the pattern shown below each set of functions. All the other functions cancel by interference. Note that the constant function $\mathrm{Wal}(0, \theta)$ is always passed in addition to the desired function $\mathrm{Sal}(i, \theta)$ or $\mathrm{Cal}(i, \theta)$. A short reflection of Fig.6.5-1 shows that the same holds true for a time-invariant diffraction grating, but there is no light wave with frequency or sequency zero.

To obtain an estimate for the operating times of the shutters let us observe that the frequency of visible light lies between 4×10^{14} and 8×10^{14} Hz. This calls for switching times in the order of 10^{-16} s or less to permit the use of the sampled amplitude representation of Walsh functions as shown in Figs.6.5-3 to 6.5-5. Light will travel 300×10^{-8} cm in 10^{-16} s, or about the distance of 300 atom diameters. Hence, one cannot rule out that the implementation of a time-variable spectrometer according to Fig.6.5-2 will eventually become possible. However, currently known effects for the implementation of shutters, such as the Kerr effect, are some 5 to 6 orders of magnitude slower. An obvious advantage of time-variable spectrometers would be that they would not only yield a power spectrum but also uncover any relationship between times of emission.

6.6 Laser for Nonsinusoidal Waves

Consider the Fourier expansion of an odd function $F(\theta)$:

$$F(\theta) = \sqrt{2} \sum_{k=1}^{\infty} a(k) \sin 2\pi k\theta, \quad \theta = t/T \tag{1}$$

We usually interpret this equation to mean that a given function $F(\theta)$ can be decomposed into sinusoidal functions, but we may just as well use it to compose a function $F(\theta)$ from given sinusoidal functions $\sin 2\pi k\theta$. For instance, the function $F(t/T)$ on top of Fig.6.6-1 can be decomposed into the sinusoidal functions of the Fourier series—the first three of which are shown below $F(t/T)$—and a line spectrum showing the proper amplitudes $a(k)$ according to Eq.(1) may be caused by an emitted wave having the time variation of $F(t/T)$. The usual spectrometers show only the amplitude or power of the sinusoidal components, but not their phase. Depending on the phases one can compose many functions from these spectral components. The assumption is generally made that the phases of the sinusoidal components displayed by a spectrometer are random. Hence, the emitted wave can be viewed of consisting of samples of (colored) noise or of the sinusoidal components of this noise.

The laser permits the generation of coherent light. Several lasers producing coherent waves with frequencies $f_k = k/T$, $k = 1, 2, \ldots$, and synchronized phases can thus in principle produce a wave with the time variation of $F(t/T)$ in Fig.6.6-1. The absolute value of the constants $a(k)$ in Eq.(1) can be implemented by grey filters, while a negative sign of a constant $a(k)$ can be implemented by reflection of the respective sinusoidal component at a metallic mirror.

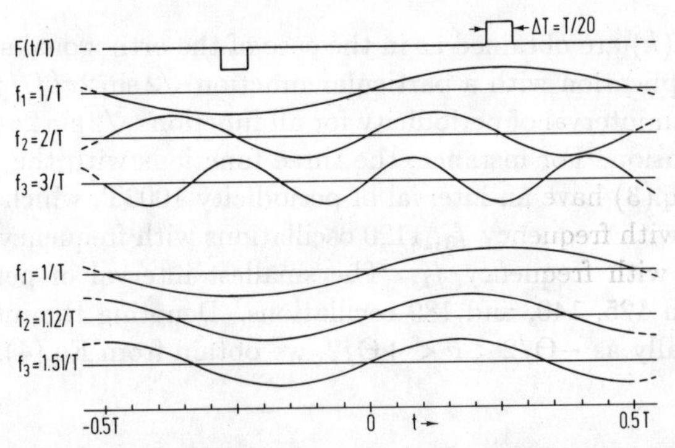

FIG.6.6-1. The sampled Walsh function $F(t) = \mathrm{Wal}(1, t/T) = \mathrm{Sal}(1, t/T)$ with a finite width $\Delta T = T/20$ of the samples, the first three functions for its expansion into an orthogonal Fourier series (frequencies $1/T$, $2/T$, $3/T$), and the first three functions for its expansion into a linearly independent Fourier series based on the Balmer series of the hydrogen atom (frequencies $1/T$, $1.12/T$, $1.51/T$).

Atomic spectra do not typically have frequencies f_k that are integer multiples of a lowest frequency i/T. The frequencies of the hydrogen spectrum are defined by

$$f = R_{\mathrm{H}} c \left(\frac{1}{n^2} - \frac{1}{l^2} \right), \quad l = n+1,\ n+2, \ldots;\ n = 1,\ 2,\ 3,\ 4,\ 5 \tag{2}$$

where R_{H} is the Rydberg constant of hydrogen and c the velocity of light. For $n = 2$ one obtains the Balmer series of spectral lines; its first three sinusoidal functions ($l = 3, 4, 5$) are plotted in Fig.6.6-1 with the frequencies

$$\begin{aligned} f_1 &= 3.021 \times 10^{14}\ \mathrm{Hz} = 1/T \\ f_2 &= 3.384 \times 10^{14}\ \mathrm{Hz} = 1.120/T \\ f_3 &= 4.568 \times 10^{14}\ \mathrm{Hz} = 1.512/T \\ T &= 2.189 \times 10^{-15}\ \mathrm{s} \end{aligned} \tag{3}$$

This system of functions is clearly not orthogonal in the time interval $-T/2 < t < T/2$, but it is linearly independent. For an expansion into a system of linearly independent sinusoidal functions we can rewrite Eq.(1) as follows:

$$F(\theta) = \sqrt{2} \sum_{k=1}^{\infty} a(k) \sin 2\pi (f_k/f_1)\theta, \quad f_k = k/T \tag{4}$$

The coefficients $a(k)$ are obtained as in the case of the orthogonal series expansion by multiplication with a particular function $\sqrt{2}\sin 2\pi(f_i/f_1)\theta$ and integration over an interval of periodicity for all functions $\sqrt{2}\sin 2\pi(f_k/f_1)\theta$ used in the expansion. For instance, the three functions with the frequencies defined by Eq.(3) have an interval of periodicity 1000T, which permits 1512 oscillations with frequency f_3, 1120 oscillations with frequency f_2, and 1000 oscillations with frequency f_1. The smallest interval of periodicity equals $125T$, with 125, 140, and 189 oscillations. Denoting the interval of periodicity generally as $-\Theta/2 < \theta < +\Theta/2$, we obtain from Eq.(4):

$$\sqrt{2}\int_{-\Theta/2}^{\Theta/2} F(\theta)\sin 2\pi(f_i/f_1)\theta\, d\theta = \sqrt{2}\sum_{k=1}^{\infty} a(k)\int_{-\Theta/2}^{\Theta/2} \sin 2\pi(f_k/f_1)\theta \sin 2\pi(f_i/f_1)\theta\, d\theta$$

$$a(i) = \frac{1}{2}\sqrt{2}\frac{1}{f_i/f_1}\int_{-\Theta/2}^{\Theta/2} F(\theta)\sin 2\pi(f_i/f_1)\theta\, d\theta \qquad (5)$$

The extension of the interval of integration from $-1/2 < \theta < +1/2$ to $-\Theta/2 < \theta < +\Theta/2$ transforms the linearly independent system of functions into an orthogonal system. However, this is not a complete orthogonal system, and the mean-square error of the approximation will not vanish as in the case of the Fourier series. If only the frequencies of the Balmer series are used, the ratio of the highest frequency f_∞ to the lowest frequency f_1 is only $f_\infty/f_1 = 1.8$, and the approximation of the function $F(t/T)$—periodically continued—would be extremely poor. However, the lowest frequency permitted by Eq.(2) is

$$f_{\rm L} = R_{\rm H}c\left(\frac{1}{5^2} - \frac{1}{6^2}\right) = 4.02 \times 10^{13}\ {\rm Hz}$$

(first line of Pfund series), while the highest permitted frequency is

$$f_{\rm H} = R_{\rm H}c(1-0) = 3.29 \times 10^{15}\ {\rm Hz}$$

(band limit of Lymann series), for a ratio of $f_{\rm H}/f_{\rm L} = 81.8$. This range implies that an approximation is possible which is equivalent to a Fourier series with about 81 terms. Hence, if one wants to compose nonsinusoidal functions from atomic spectra, one must look for spectra with a large ratio $f_{\rm H}/f_{\rm L}$ between the highest and the lowest frequency.

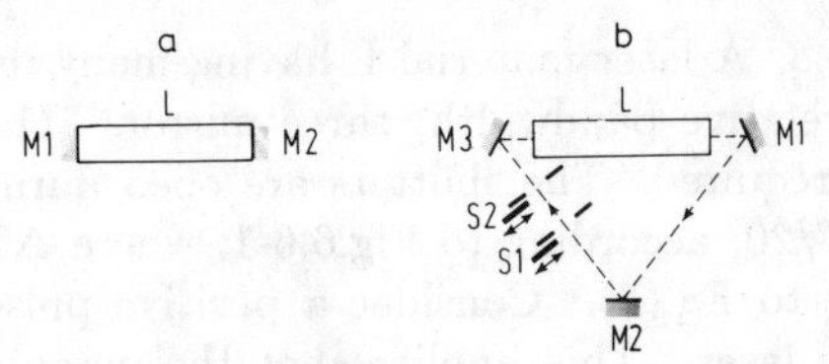

FIG.6.6-2. Principle of a laser producing waves with sinusoidal time variation (a) and of a laser producing waves with the time variation of $F(t/T)$ in Fig.6.6-1

Instead of the ratio f_H/f_L one often uses the relative bandwidth η,

$$\eta = \frac{f_H - f_L}{f_H + f_L} = \frac{f_H/f_L - 1}{f_H/f_L + 1} \tag{6}$$

The difference $f_H - f_L$ is the absolute frequency bandwidth of a function or signal with the highest nominal frequency f_H and the lowest nominal frequency f_L. A sinusoidal function occupies the bandwidth $f_H - f_L = 0$, and has the relative bandwidth $\eta = 0$. As the relative bandwidth increases, a function looks less and less like a sinusoidal function. The largest possible value of η equals 1. For $f_H/f_L = 81.8$ we obtain $\eta = 80.8/82.8 = 0.976$, which is extremely large and signifies that the spectral components of the hydrogen atom can compose a signal that has essentially no resemblance with a sinusoidal function. The ratio $f_H/f_L = 1.8$ of the Balmer series, on the other hand, yields $\eta = 0.286$, which implies that any signal composed of the components of the Balmer series does have some resemblance to a sinusoidal function.

Let us assume that we have a material with an emission spectrum whose relative bandwidth is about as large as that of the hydrogen atom, but whose energy levels are metastable and which in addition has at least one broad energy level. Such a material is inherently able of producing laser radiation at many frequencies that cover a large relative bandwidth.

Consider next how such a material could be used to produce a wave with the time variation of $F(t/T)$ in Fig.6.6-1, where the period T is in the order of 10^{-15} s according to Eq.(3). Refer to Fig.6.6-2a, which shows the basic arrangement of a laser. A lasing material L is first excited to a broad energy level and decays to a somewhat lower metastable energy level. The return to the ground state produces a sinusoidal wave with such a frequency that a standing wave is produced between the mirrors M1 and M2, which stimulates emission having the frequency and phase of the standing wave[1].

The modification of this laser principle for the production of a wave with the time variation of $F(t/T)$ in Fig.6.6-1 rather than a sinusoidal wave

[1] The discussion is in terms of the three level laser.

is shown in Fig.6.6-2b. A laser material L having many metastable energy levels with a large relative bandwidth, three mirrors M1 to M3, and two shutters S1, S2 are required. The shutters are open during time intervals of duration $\Delta T \approx T/20$, according to Fig.6.6-1, where ΔT is in the order of 10^{-16} s according to Eq.(3). Consider a positive pulse coming out at the right end of the laser. This implies that the many sinusoidal waves produced happen to sum up so that the electric field strength has the form of the positive pulse of $F(t/T)$ in Fig.6.6-1; we will, of course, have to explain why the sinusoidal waves should have the proper amplitudes and phases to sum in this way. The metallic mirror M1 reverses the amplitude, M2 reverses a second time, and M3 a third time, so that the amplitude of the pulse entering the lasing material has the opposite sign of the amplitude of the pulse that came out on the right side. If the propagation time of the light along the path via the three mirrors and through the lasing material equals $T/2$—or rather a large odd multiple of $T/2$—we have met a first condition for the generation of a wave with the time variation of $F(t/T)$. A second condition is that the shutter S1 is opened for the time ΔT at the times $t = 0, T/2, T, 3T/2, \ldots$. The third and last condition is that the light must be made to travel in the direction of the arrows in Fig.6.6-2b. This is accomplished by the shutter S2, which opens for the time ΔT a little later than shutter S1. As a result, a light pulse traveling in the direction of the arrows can pass both shutters, while a pulse traveling in the opposite direction cannot.

We hope this example shows that one can build a laser for nonsinusoidal waves in principle. The primary practical obstacle for an actual construction are again the fast shutters, just as in the case of the diffraction grating. The sophisticated lasing material is also a problem, but not one as great as the shutters[2].

Let us sum up the results of Sections 6.1 to 6.6. We have seen that the wave equation—and by implication other partial differential equations with time and space variables—favor sinusoidal functions, but this proves no more than that these equations assume a space-time continuum with the usual topology of real numbers. The experimental examples showed only that time-invariant diffraction gratings and lasers are easier to build than fast time-varying ones. The universal use of sinusoidal functions in these two cases thus reflects only our state of technology but tells us nothing about space-time.

[2]The laser discussed here is not the same as the lasers for ultrashort light pulses that produce pulses as short as 0.03 ps, containing only about 15 sinusoidal cycles (Letokhov 1969; Bradley 1977).

7 Discrete Topologies and Difference Equations

7.1 Finite Differences and Differentials

At the end of Section 1.4 we had pointed out two reasons why the concept of the continuum has proved to be so resistant to change since Aristotle. First, no convincing reason could be given why the continuous space-time should be abandoned after centuries of good service, until the development of information theory showed that a continuum cannot be verified by any physical observation. Furthermore, we have seen that such characteristic results of the continuum theory as differential calculus, Pythagorean distance, and standing sinusoidal waves are only particular examples which can be generalized by studying unusual coordinate systems. The emphasis during this investigation was always on studying something as different as possible from the continuum and the concepts derived from it, in order to free ones mind from the education in terms of the continuum.

We now turn to the second problem emphasized at the end of Section 1.4 and try to derive an observable result that contradicts the assumption of the continuum. To do so without first developing the equivalent of differential calculus for general, discrete rings we must restrict ourselves to the area where the development of mathematical foundations is most advanced. We have noted in Chapter 6 that the calculus of finite differences for the limit ring $2^N \to \infty$ is much further developed than the dyadic calculus for the ring $2^N = 2$, while essentially no mathematical methods have been developed for the intermediate rings. Hence, we will concentrate on the ring $2^N \to \infty$.

Since we have previously opposed something infinitely long as unobservable and thus outside any experimental science, we are making here an undesirable compromise. However, the limit ring $2^N \to \infty$ is closer to reality than the continuum and the real numbers axis. To emphasize this point once more, consider Fig.7.1-1a. The distance between the points P1 and P2 is to be measured with a ruler that has marks at the multiples of the normalized unit $\Delta\xi = \Delta x/X$. The point P1 is located closest to the mark $\xi = n\Delta\xi$. We say the distance between the points P1 and P2 equals

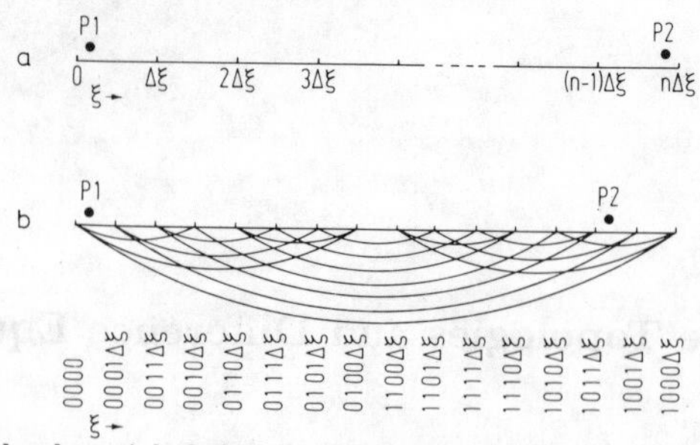

FIG.7.1-1. Usual ruler with finite distance $\Delta\xi$ between the marks (a) and a similar ruler with connections between the marks according to dyadic topology (b).

$n\Delta\xi$. For small values of $\Delta\xi$, n becomes very large and approaches *denumerable infinite* for $\Delta\xi \to 0$. The topology introduced by this ruler is the usual topology of the integers. On the other hand, if we assume marks with differential distance $d\xi$ between each other, the number of marks between P1 and P2 will be *nondenumerable infinite*; the usual topology of the real numbers is introduced. Hence, the limit ring $2^N \to \infty$ is an improvement over the real numbers axis if we want to use mathematical methods closer to reality.

At this point one may raise the question whether the results of differential calculus and difference calculus do not become equal for sufficiently small values of $\Delta\xi$. From the standpoint of abstract mathematics the answer is clearly no, since denumerable and nondenumerable infinite are not the same. This, however, does not prove that there is a difference in those cases that are of interest in physics. The way to find an answer for those cases is to use finite differences Δx^i instead of differentials dx^i, and difference equations instead of differential equations. One must find such solutions of the difference equations that become in general[1] independent of Δx^i for small values of Δx^i. If these solutions become for $\Delta x^i \to 0$ equal to the solutions of the corresponding differential equation, then one knows that the use of differentials instead of differences was permissible.

It would be wrong to assume that the transition from Δx^i to dx^i before or after the calculation yields the same result. If we solve a difference equation $f(x, \Delta x) = 0$ and let the difference Δx approach zero afterwards, we can have denumerably many intervals of length Δx cover a certain distance; if we let Δx become very small only, we can actually cover any finite distance with a finite number of intervals of length Δx. On the other hand, if one makes the transition from Δx to dx before the calculation, one has always nondenumerably many differentials cover any finite distance. The

[1]See Section 10.5 for the restriction "in general".

distinction between small differences Δx and differentials dx shows up mathematically in theorems like that by Hölder (Nörlund, 1924), which states that the gamma function satisfies the algebraic difference equation

$$\Gamma(X+1) = X\Gamma(X), \qquad X = x/\Delta x \tag{1}$$

but no algebraic differential equation.

The use of differentials in physics has been questioned for a long time[2]. Pauli (1933) as well as Landau and Peierls (1931) pointed out that the assumption of arbitrarily accurate position and time measurements was probably unjustified in relativistic quantum mechanics, since the Compton effect limits the accuracy of the position measurement of a particle. However, Pauli concluded that there was no such limitation in nonrelativistic quantum mechanics. Today, as a result of the development of information theory, we must reject any assumption of measurements with unlimited accuracy, since they imply gathering and processing of infinite information. In particular, nondenumerably infinite information is implied by "differentially accurate" measurements. March (1948, 1951) postulated that the distance of two particles could only be measured with an error that was at least equal to an *elementary unit of length*. Such elementary units are difficult to reconcile with the special theory of relativity. No such difficulties arise if one assumes the distance $\Delta x = \Delta\xi X$ between adjacent marks of the ruler in Fig.7.1-1a to be finite but otherwise as small as one wants. There is no implication that Δx is of the order of 10^{-13} cm; it is still finite if it is 10^{-100} cm long. Many papers on quantized or cellular space-time have been published over the last 30 years. We list in chronological sequence Snyder (1947a, b), Flint (1948), Schild (1949), Hellund and Tanaka (1954), Hill (1955), Das (1960; 1966a, b, c), Yukawa (1966), Atkinson and Halpern (1967), Cole (1970, 1971, 1972a, b; 1973a, b), Hasebe (1972), Welch (1976), and Kadishevsky (1978).

Consider the difference quotient of the time variable. A point shall be next to the mark $\xi(\theta_0)$ in Fig.7.1-1a at the time θ_0, and at the mark $\xi(\theta)$ at the time θ. The average velocity is defined by

$$\beta_{\text{av}} = \frac{\xi(\theta_0) - \xi(\theta)}{\theta_0 - \theta} \tag{2}$$

The limit of this average velocity β_{av} is the (instantaneous) velocity β:

$$\beta = \frac{\xi(\theta + \Delta\theta) - \xi(\theta)}{\Delta\theta} \tag{3}$$

[2]We have pointed out in Section 1.1 that Zeno of Elea already objected the notion of infinite divisibility of space and time.

If $\xi(\theta+\Delta\theta)-\xi(\theta)$ is smaller than $\Delta\xi$, one obtains $\xi(\theta+\Delta\theta)-\xi(\theta)=0$ and $\beta=0$, even though a velocity $\beta>0$ had been obtained for a larger time difference $\Delta\theta$. This could not happen in differential mathematics. One might think that the problem could always be avoided by making $\Delta\xi$ sufficiently small and $\Delta\theta$ sufficiently large. However, it will turn out that $\Delta\xi$ and $\Delta\theta$ cannot always be chosen independently. As a result, we will encounter cases where the instantaneous velocity is completely unspecified—except that it does not exceed the velocity of light—while there is a perfectly defined average velocity. This is a typical result of the calculus of finite differences, which is quite foreign to differential calculus.

7.2 Dyadic Difference Quotient

Let us investigate what the transition from the limit ring $2^N\to\infty$ to the dyadic ring $2^N=2$ implies. Consider the ruler in Fig.7.1-1b with 16 marks connected according to dyadic topology as discussed in connection with Fig.5.6-8. We define a scalar function $f(\xi)$ that is measured at the marks ξ. We had defined with Eq.(6.3-20) the dyadic difference quotient as the average difference at the point $m=\xi/\Delta\xi$. Alternately, one may define a difference quotient along a path from a mark ξ_0 to ξ. There are a number of paths between any two marks of the ruler of Fig.7.1-1b, while there is only one path in Fig.7.1-1a. Denote a particular path $L_j(\xi_0,\xi)$ and its distance $D_j(\xi_0,\xi)$. The difference quotient

$$Q_j=\frac{f(\xi_0)-f(\xi)}{D_j(\xi_0,\xi)} \tag{1}$$

is obtained. Ordinary subtraction is used for $f(\xi_0)-f(\xi)$ since the scalar $f(\xi)$ may represent the temperature at the point ξ, and temperature differences are obtained by ordinary subtraction.

As an example consider the following two paths from P1 to P2 in Fig.7.1-1b:

$$L_1(0000,1011\Delta\xi)=0000,\ 0001\Delta\xi,\ 0011\Delta\xi,\ 1011\Delta\xi$$
$$L_2(0000,1011\Delta\xi)=0000,\ 0100\Delta\xi,\ 1100\Delta\xi,\ 1110\Delta\xi,\ 1010\Delta\xi,\ 1011\Delta\xi$$

The paths have the lengths $D_1=3\Delta\xi$ and $D_2=5\Delta\xi$.

The limit of Q_j is obtained for the shortest possible distance $D_j=\Delta\xi$ which occurs between the point $\xi_0=m\Delta\xi$ and $\xi=(m+2^i)\Delta\xi$, where $i=0$, 1, 2, ... We denote this limit by $\Delta_2 f(m)/\Delta m$:

$$\frac{\Delta_2 f(m)}{\Delta m}=\frac{f(m\Delta\xi)-f\left[(m\oplus 2^i)\Delta\xi\right]}{\Delta\xi} \tag{2}$$

This is clearly one particular component of the average dyadic difference $Df(m)/Dm$ defined in Eqs.(6.3-1) and (6.3-20). However, one cannot only calculate $Df(m)/Dm$ from $\Delta_2 f(m)/\Delta m$ for every point m, but one can also derive $\Delta_2 f(m)/\Delta m$ from $Df(m)/Dm$ if $Df(m)/Dm$ is given for all points of the dyadic coordinate system. Hence, there is no essential difference between Eqs.(2) and (6.3-20).

Consider the difference quotient for the time variable. The time difference $\theta_0 - \theta$ of Eq.(7.1-2) can either remain as it is, if a conventional clock is used, or be replaced by a subtraction modulo 2 if a clock is used as discussed in Section 5.7. The coordinate difference $\xi(\theta_0) - \xi(\theta)$ must be replaced by $D_j[\xi(\theta_0), \xi(\theta)]$ measured along one of the paths $L_j[\xi(\theta_0), \xi(\theta)]$ from $\xi(\theta_0)$ to $\xi(\theta)$. Using a conventional clock one obtains as generalization of Eq.(7.1-2) for dyadic topology:

$$\beta_{\mathrm{av}} = \frac{D_j[\xi(\theta_0), \xi(\theta)]}{\theta_0 - \theta} \tag{3}$$

while the generalization of Eq.(7.1-3) yields:

$$\beta = \frac{\xi(\theta + \Delta\theta)/\Delta\xi \oplus \xi(\theta)/\Delta\xi}{\Delta\xi} \tag{4}$$

If modulo 2 addition is used for the time θ, one must replace $\theta_0 - \theta$ in Eq.(3) by $\theta_0/\Delta\theta \oplus \theta/\Delta\theta$, while $\theta + \Delta\theta$ in Eq.(4) is replaced by $\theta/\Delta\theta \oplus 1$.

If the path $L_j(\xi_0, \xi)$ between the points P1 and P2 in Fig.7.1-1b is taken along the straight line via $0001\Delta\xi$, $0011\Delta\xi$, $0010\Delta\xi$, ... rather than via the circular connections, one obtains $D(\xi_0, \xi) = \xi_0 - \xi$. In this case there is no difference between any measurements with the ruler of either Fig.7.1-1a or b. Equations (3) and (4) become equal to Eqs.(7.1-2) and (7.1-3). Similarly, Eqs.(1) and (2) assume the form the respective equations would have on the limit ring $2^N \to \infty$:

$$Q = \frac{f(\xi_0) - f(\xi)}{\xi_0 - \xi}$$

$$\frac{\Delta_2 f(m)}{\Delta m} = \frac{f(m\Delta\xi) - f[(m \pm 1)\Delta\xi]}{\Delta\xi} = \frac{f(\xi) - f(\xi \pm \Delta\xi)}{\Delta\xi}$$

We have already discussed in Section 5.6-7 that shifting in eigencoordinates yields the same result for modulo 2 and usual addition and subtraction. We see now that the usual ruler and the mathematical operations based on it may be regarded as a special case of a modulo 2 ruler and mathematical operations based on it. In the following chapters we will develop mathematical methods and derive physical results for rulers and clocks based on the limit ring $2^N \to \infty$, as represented by Fig.7.1-1a, but we recognize that these methods and results will also be special cases of dyadic calculus as exemplified by Fig.7.1-1b.

7.3 Observable Effects of Discrete Topologies

Let us consider the question of which physical problems one could investigate with reasonable hope of finding observable deviations from the results of the theory based on the usual topology of real numbers. Topology in physics is primarily associated with the general theory of relativity. The results obtained in Chapter 3 for unbounded but finite coordinate systems point in this direction. However, the finite number of marks of coordinate systems has led to finite resolutions $\Delta\xi$ and $\Delta\theta$, and the distinction between finite differences and differentials will most likely show up at very short distances and time intervals. Hence, we will turn to quantum physics. Our investigation will lead to results that deviate from those obtained for the usual real-number-topology at distances of about 10^{-10} to 10^{-12} cm, while the question of topology at short distances in the general theory of relativity becomes important at distances some 20 orders of magnitude smaller, which are far beyond our experimental range.

Let us give an overview that will make it easier to read the following more mathematical chapters. In Chapter 8 we will investigate under which conditions solution of partial difference equations exist that become independent of the time difference $\Delta\theta$ or Δt for small values of Δt. The difference equations of Schrödinger and Klein-Gordon are rewritten into difference equations, and the eigensolutions for force-free particles are obtained.

A method for obtaining eigensolutions of difference equations with variable coefficients is developed in Chapter 9. Its application to the nonrelativistic particle in a Coulomb field shows that eigenvalues and eigenfunctions of the difference equation are essentially equal to those of the differential equation.

The Klein-Gordon difference equation with Coulomb field is investigated in Chapter 10. Eigenvalues and eigenfunctions known from the differential equation are obtained, but the eigenfunctions turn out to converge asymptotically only if the fine structure constant $\alpha = e^2/\hbar c$ is not zero. Convergent eigenfunctions are obtained by means of conformal mapping. The convergent and the asymptotically convergent eigenfunctions are the same at distances from the center of the field that are large compared with the Compton wavelength. Differences occur at smaller distances. In particular, near the center of the Coulomb field one can no longer treat the difference Δr of the distance measurement as negligibly small without causing divergences.

The determination of the eigensolutions of the Dirac difference equations with Coulomb field in Chapter 11 yields at large distances the same eigenfunctions as known from the differential equations. Again they converge asymptotically only, and they are made convergent by additional terms.

The discussion of a number of problems required for mathematical rigor is postponed to Chapter 12 in order to make the preceding chapters more readable.

The physical interpretation of the mathematical results is generally avoided. This was done because interpretations have a tendency to cause unnecessary controversies. The claim that there is neither a three-dimensional space nor a space-time continuum should provide all the controversy one needs, and any further points of contention are thus avoided whenever possible. The results of the investigation in Chapters 8 to 12 rest on the axiom of using finite differences on the limit ring $2^N \to \infty$ instead of differentials, and the correctness of the mathematical manipulations. Mathematical mistakes are easy to agree upon. An axiom cannot be proved right, it can only be made plausible. All we have said so far about finite differences Δx and Δt is only needed to make the replacement of the differentials dx and dt plausible.

The main objection against the following investigation is that it is based on quantum mechanics before the development of quantum field theory. The primary reason for avoiding quantum field theory is the role of the concept of a field. Within the usual model of space-time a field fills the space between interacting matter, such as Sun and Earth, electrically charged particles, etc. This field is differentiable with respect to time and space coordinates, and it permits us to do away with the action-at-a-distance. A cause at point x can produce an effect at point $x + dx$ without action-at-a-distance, but not at a point $x + \Delta x$. One is not forced to accept the classical concept of action-at-a-distance. Since the point $x + \Delta x$ is as close to point x as one can measure, one only has to accept that one cannot observe how the cause bridges the gap Δx to produce an effect at $x + \Delta x$. It is the result of information theory, that information must always be finite, which prevents us from deciding whether there is an action-at-a-distance or not, when Δx is sufficiently small but still finite. The prospect appears to be that the concept of a field will need as much rethinking as the concept of a space-time continuum.

8 Schrödinger and Klein-Gordon Difference Equations

8.1 Time Dependence of the Solutions of Partial Difference Equations

Consider a linear partial difference equation of second order with the independent variables t and $\mathbf{r}$. The vector $\mathbf{r}$ represents any number of variables of the used coordinate system or "space variables" for short[1]; t is the time variable. Let $\Omega_i(\mathbf{r})$ denote an operator for the space variables:

$$\Omega_1(\mathbf{r})F(\mathbf{r},t+\Delta t) = \Omega_0(\mathbf{r})F(\mathbf{r},t) + \Omega_{-1}(\mathbf{r})F(\mathbf{r},t-\Delta t) \tag{1}$$

The operators $\Omega_i(\mathbf{r})$ shall be nonsingular. A representation of the operators by matrices and explicit difference operators will be given later. The meaning of singular will become apparent then.

Equation (1) may be solved if the operators $\Omega_i(\mathbf{r})$ have a common system of eigenfunctions $\varphi(\mathbf{r})$. Space and time variables may be separated in this case[2]:

$$\Omega_i(\mathbf{r})\varphi_n(\mathbf{r}) = \omega_{in}\varphi_n(\mathbf{r}), \quad i = 1,\ 0,\ -1 \tag{2}$$

$$\varphi_n(\mathbf{r},t) = \varphi_n(\mathbf{r})\varphi_n(t) \tag{3}$$

The function $F(\mathbf{r},t)$ is expanded in a series of eigenfunctions:

$$F(\mathbf{r},t) = \sum_{n=0}^{\infty} c_n\varphi_n(\mathbf{r},t) \tag{4}$$

Substitution of Eqs.(2) and (4) into Eq.(1) yields:

---

[1] We will generally use the term *space variable* instead of the more precise but cumbersome term *variables of the used coordinate system*.

[2] We are using the same letter φ for the different functions $\varphi_n(\mathbf{r},t)$, $\varphi_n(\mathbf{r})$, $\varphi_n(t)$ here, and later on for $\varphi_n(x)$, $\varphi_n(r)$, $\varphi_n(r,\vartheta,\varphi)$, which is not a good practice but avoids writing additional indices.

$$\sum_{n=0}^{\infty} c_n \omega_{1n} \varphi_n(\mathbf{r}) \varphi_n(t+\Delta t) = \sum_{n=0}^{\infty} c_n \omega_{0n} \varphi_n(\mathbf{r}) \varphi_n(t) + \sum_{n=0}^{\infty} c_n \omega_{-1n} \varphi_n(\mathbf{r}) \varphi_n(t-\Delta t) \quad (5)$$

Let us assume that the eigenfunctions $\varphi_n(\mathbf{r})$ form an orthogonal system[3]. One may multiply with the conjugated complex function $\varphi_k^*(\mathbf{r})$ and integrate over the interval of orthogonality. An ordinary difference equation with constant coefficients is obtained for $k = n$:

$$\omega_{1n} \varphi_n(t+\Delta t) = \omega_{0n} \varphi_n(t) + \omega_{-1n} \varphi_n(t+\Delta t) \quad (6)$$

The general solution of a difference equation with constant coefficients is obtained by the following substitution (Milne-Thompson, 1951):

$$\varphi_n(t) = \varphi_n(\tau) = b_n Q_n^\tau, \quad \tau = t/\Delta t \quad (7)$$

The characteristic equation is obtained from Eqs.(6) and (7):

$$\omega_{1n} Q_n^2 - \omega_{0n} Q_n - \omega_{-1n} = 0 \quad (8)$$

If Eq.(8) has two different roots one obtains the following two independent solutions of Eq.(6):

$$\varphi_{1n}(t) = b_{1n} Q_{1n}^\tau, \qquad \varphi_{2n}(t) = b_{2n} Q_{2n}^\tau \quad (9)$$

Two equal roots yield the solutions:

$$\varphi_{1n}(t) = b_{1n} Q_n^\tau, \qquad \varphi_{2n}(t) = b_{2n} \tau Q_n^\tau \quad (10)$$

The normalized variable $\tau = t/\Delta t$ may be made arbitrarily large or small for a fixed value of t by proper choice of Δt. The solutions which do not depend on the choice of Δt are of particular interest.

Let $O(1)$, *order 1*, denote a term that becomes independent of Δt for sufficiently small values of Δt. Furthermore, $O(\Delta t)^q$, *order* $(\Delta t)^q$, denotes a term that vanishes proportionate to $(\Delta t)^q$. The solutions of Eq.(9) and (10) become independent of Δt for sufficiently small values of Δt, if the following relation is satisfied:

---

[3]See Section 12.9 about the orthogonality of eigenfunctions.

$$1 - |O(\Delta t)| \leq |Q_{sn}| \leq 1 + |O(\Delta t)|, \quad sn = n,\ 1n,\ 2n \tag{11}$$

$|Q_{sn}| = 1+O(\Delta t)$ yields solutions that increase exponentially with t if Q_{sn} is real, while $|Q_{sn}| = 1 - O(\Delta t)$ yields exponentially decreasing solutions. For complex values of Q_{sn} one obtains oscillations with exponentially increasing or decreasing amplitudes. Solutions with constant amplitude are obtained for $1 - |O(\Delta t)| < Q_{sn} < 1 + |O(\Delta t)|$. These *stationary* solutions approach the following form for sufficiently small values of Δt:

$$\varphi_n(t) = b_n e^{iE_n t/\hbar} \tag{12}$$

The factor τ in the second solution of Eq.(10) may always be made independent of Δt by combining Δt with the constant b_{2n}.

Equations (8) and (12) yield for stationary solutions and vanishing values of Δt the relation:

$$\omega_{1n} e^{iE_n \Delta t/\hbar} - \omega_{0n} - \omega_{-1n} e^{-iE_n \Delta t/\hbar} = 0 \tag{13}$$

To permit stationary solutions, ω_{0n} must be real if ω_{1n} and ω_{-1n} are conjugate complex, and imaginary if ω_n and ω_{-1n} are real and equal.

Equation (11) is the necessary and sufficient condition for the existence of solutions of Eq.(1) that become independent of Δt for small values of Δt. It is called the *independence relation.* A similar relation is known in numerical analysis under the name stability condition (Courant *et al.*, 1928; Douglas, 1956; Richtmyer, 1957). The conditions under which the solutions of difference equations of the type of Eq.(1) converge to the solutions of the corresponding differential equations were investigated by Douglas (1956). Stability conditions for the Schrödinger and the Klein-Gordon equation were derived by Harmuth (1957) and used by Kelly (1965) as well as by Dawes and Marburger (1969).

8.2 Schrödinger Equation

The Schrödinger differential equation contains differential quotients of first and second order:

$$-i\hbar \frac{\partial \Psi}{\partial t} = \frac{\hbar^2}{2m} \nabla^2 \Psi - V(\mathbf{r}, t)\Psi \tag{1}$$

We replace them by the difference operators of Eqs.(6.2-6) and (6.2-12):

$$\frac{d^2\Psi}{dx^2} \to \frac{\Delta^2\Psi}{(\Delta x)^2} = \frac{1}{(\Delta x)^2}[\Psi(x+\Delta x) - 2\Psi(x) + \Psi(x-\Delta x)] \tag{2}$$

$$\frac{d\Psi}{dt} \to \frac{\Delta\Psi}{\Delta t} = \frac{1}{2\Delta t}[\Psi(t+\Delta t) - \Psi(t-\Delta t)] \tag{3}$$

In addition to the symmetric difference operator of Eq.(3) there exists also a right and a left difference operator of first order:

$$\frac{\Delta_r\Psi}{\Delta t} = \frac{1}{\Delta t}[\Psi(t+\Delta t) - \Psi(t)]$$

$$\frac{\Delta_l\Psi}{\Delta t} = \frac{1}{\Delta t}[\Psi(t) - \Psi(t-\Delta t)]$$

It will be shown in Section 12.4 that the right and the left difference operator lead to exponentially increasing or decreasing solutions for a force-free particle. The symmetric difference operator, on the other hand, yields stationary solutions. This confirms one's intuitive feeling that neighbors should be treated equally, which the right and the left difference operator do not do. We thus replace the differential equation Eq.(1) by the following difference equation:

$$\Psi(\mathbf{r}, t+\Delta t) = \Omega_0(\mathbf{r}, t)\Psi(\mathbf{r}, t) + \Psi(\mathbf{r}, t-\Delta t)$$

$$\Omega_0(\mathbf{r}, t) = i\frac{\hbar\Delta t}{m}\nabla\!\!\!\!\nabla^2 - 2i\frac{\Delta t}{\hbar}V(\mathbf{r}, t) \tag{4}$$

The symbol $\nabla\!\!\!\!\nabla$ denotes a difference operator that may be written explicitly for Cartesian coordinates with the help of Eq.(2).

Equation (4) is of the type of Eq.(8.1-1) if $V(\mathbf{r}, t)$ is time invariant. Its general solution may be obtained in this case by means of the eigenfunctions of the operator $\Omega_0(\mathbf{r}, t) = \Omega_0(\mathbf{r})$. This operator is the same as one obtains by elimination of the time variable from Eq.(1), except that differentials are replaced by finite differences.

Equation (4) yields $\omega_{1n} = \omega_{-1n}$, independent of $V(\mathbf{r})$ and the system of coordinates used. Hence, the eigenvalues ω_{0n} must be purely imaginary for stationary solutions according to Eq.(8.1-13).

Equation (4) may be solved for a force-free particle in one-dimensional Cartesian coordinates by means of the eigenfunctions $\varphi_n(x) = \varphi_\kappa(x) = \exp(i\kappa x)$. One obtains:

$$\Omega_0(x)\varphi_n(x) = i\frac{\hbar\Delta t}{m(\Delta x)^2}\left(e^{i\kappa\Delta x} - 2 + e^{-i\kappa\Delta x}\right) = \omega_{0n}\varphi_n(x)$$

$$\omega_{0n} = \omega_{0\kappa} = -2iB = -4i\frac{\hbar\Delta t}{m(\Delta x)^2}\sin^2\frac{\kappa\Delta x}{2} \tag{5}$$

Equation (8.1-8) yields:

$$Q_{1n} = -iB + (1-B^2)^{1/2}, \qquad Q_{2n} = -iB - (1-B^2)^{1/2}$$
$$Q_{1n}Q_{2n} = -1 \tag{6}$$

One obtains for $B^2 \leq 1$:

$$|Q_{1n}| = |Q_{2n}| = 1$$

The condition $1 < B^2 < 1 + |O(\Delta t)^2|$ yields:

$$|Q_{2n}| < 1 + |O(\Delta t)|, \qquad |Q_{1n}| > 1 - |O(\Delta t)|$$

The equality $B^2 = 1 + |O(\Delta t)^2|$ yields

$$|Q_{1n}| = 1 - |O(\Delta t)|, \qquad |Q_{2n}| = 1 + |O(\Delta t)|$$

and the inequality $B^2 > 1 + |(O(\Delta t)^2|$ yields:

$$|Q_{2n}| > 1 + |O(\Delta t)|, \qquad |Q_{1n}| < 1 - |O(\Delta t)|$$

Hence, stationary solutions exist for $B^2 < 1 + |O(\Delta t)^2|$; non-stationary solutions independent of Δt exist for $B^2 = 1 + |O(\Delta t)^2|$. The condition $B^2 > 1 + |O(\Delta t)^2|$ yields only solutions which can be made arbitrarily large or small by proper choice of Δt.

Let B be substituted from Eq.(5) and note that $\sin^2(\kappa x/2)$ cannot become larger than one for large values of κ. The necessary and sufficient condition for the existence of solutions that become independent of Δt for small values of Δt assumes the form:

$$\frac{m(\Delta x)^2}{\Delta t} \geq 2\hbar\left[1 + |O(\Delta t)^2|\right]$$
$$\Delta t \leq \frac{m(\Delta x)^2}{2\hbar}\left[1 - |O(\Delta t)^2|\right] \tag{7}$$

We have here a coupling between the time resolution Δt and the coordinate or space resolution Δx. For any choice of either Δt or Δx, the resolution of the other variable must be chosen so that Eq.(7) is satisfied.

For bounded values of κ and sufficiently small values of Δx one obtains instead of Eq.(7) the relation:

$$\Delta t \leq \frac{2m}{\hbar \kappa^2} \tag{8}$$

The time variation of the eigenfunctions of the force-free particle follows from Eq.(8.1-12) and (6):

$$\varphi_{1n}(t) = \varphi_{1\kappa}(t) = e^{-iE_n t/\hbar + O(\Delta t)}, \qquad \varphi_{2n}(t) = \varphi_{2\kappa}(t) = e^{+iE_n t/\hbar + O(\Delta t)}$$

$$E_n = E_\kappa = \frac{1}{2}\frac{\hbar^2 \kappa^2}{m}\left[1 - \frac{1}{2\cdot 3!}(\kappa \Delta x)^2 + \dots\right] \tag{9}$$

These results are the same for $\kappa \Delta x \ll 1$ and $E\Delta t/\hbar \ll 1$ as obtained from the differential equation, except for the conditions of Eqs.(7) and (8).

The operator $\Omega_0(\mathbf{r})$ in Eq.(4) may be written as line matrix with a finite or denumerably infinite number of terms:

$$\Omega_0(x) = i\frac{\hbar \Delta t}{m(\Delta x)^2}(\dots, 0, 1, -2, 1, 0, \dots) \tag{10}$$

Equation (5) may then be written as matrix equation:

$$i\frac{\hbar \Delta t}{m(\Delta x)^2}(\dots, 0, 1, -2, 1, 0, \dots)\begin{pmatrix} \cdot \\ \cdot \\ \varphi_n(x - 2\Delta x) \\ \varphi_n(x - \Delta x) \\ \varphi_n(x) \\ \varphi_n(x + \Delta x) \\ \varphi_n(x + 2\Delta x) \\ \cdot \\ \cdot \end{pmatrix} = \omega_{0n}\varphi_n(x)$$

The substitution $\varphi_n(x) = \exp(i\kappa x)$ yields again $\omega_{0n} = -2iB$.

Let us return to Eq.(7) and discuss its physical meaning, since this type of condition will be encountered repeatedly and it has no equivalent in a theory based on the concept of the continuum.

Refer to Fig.8.2-1a. A particle shall be closest to the mark x_0 at the time t and closest to the mark x_1 at the time $t + \Delta t$. The instant velocity

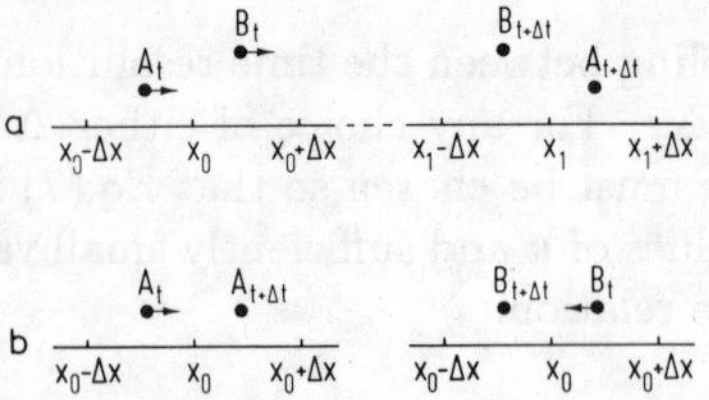

FIG.8.2-1. Uncertainty of velocity measurement using a ruler with resolution Δx and a clock with resolution Δt.

of this particle equals $v = (x_1 - x_0)/\Delta t$. The uncertainty of this velocity is obtained as follows. Particle A_t in Fig.8.2-1a is as far left as possible and still be closest to the mark x_0. At the time $t + \Delta t$ the particle, now denoted $A_{t+\Delta t}$, is as far right as possible and still be closest to mark x_1. On the other hand, particle B_t is as far right as possible at time t and as far left as possible at time $t + \Delta t$. Both particles would be observed to have the velocity $v = (x_1 - x_0)/\Delta t$, but the largest possible difference of their velocity—or the uncertainty of their velocity—equals $2\Delta x/\Delta t$.

Consider now the particle A in Fig.8.2-1b. It is observed at both times t and $t + \Delta t$ closest to the mark x_0. Its velocity is thus $v = 0$. However, the particle could actually have moved from the extreme left position (A_t) to the extreme right position ($A_{t+\Delta t}$) and still have been observed closest to x_0. The uncertainty of the velocity equals $\Delta x/\Delta t$. However, the movement could also have been in the opposite direction, as shown by the particle B in Fig.8.2-1b, and the uncertainty of the velocity $v = 0$ is thus again $2\Delta x/\Delta t$.

The uncertainty of velocity means that the momentum $p = mv$ of a particle of known mass m has the uncertainty $\Delta p = m\Delta v = 2m\Delta x/\Delta t$. The uncertainty of the location ΔX of a particle always equals the distance Δx between the marks of the rulers in Fig.8.2-1. Using Eq.(7) we may write the product $\Delta p \Delta X$ as follows:

$$\Delta p \Delta X = \frac{2m(\Delta x)^2}{\Delta t} \geq 4\hbar \tag{11}$$

Hence, the condition for the existence of solutions independent of Δx and Δt for small values of Δx and Δt turns out to be Heisenberg's uncertainty relation. The interpretation of Δp and ΔX is, of course, different from that used in the differential theory, since no finite differences Δx and Δt occur in a continuum theory; hence, Δp and ΔX are typically interpreted as mean-square-deviations of an observation. Statistics thus permits one to introduce the finite accuracy of measurements into a theory based on the concept of a space-time continuum with differential resolution dx and dt.

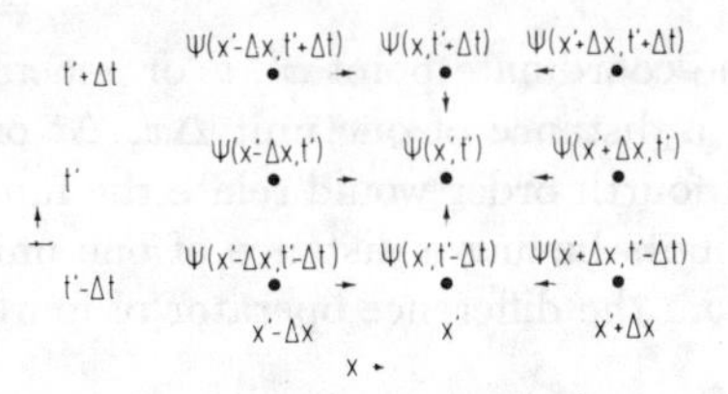

FIG.8.3-1. The point x', t' and its eight neighbors with distance Δx and/or Δt in a coordinate system with finite resolution.

8.3 Physical Meaning of a Difference Equation

Let us give a physical interpretation of the difference equation (8.2-4) which we derived from the differential equation (8.2-1). A difference equation is said to be of first order in a particular variable if two different values of that variable occur. For instance, all three difference operators of first order contain the variable Ψ at two different times: The symmetric difference operator of Eq.(8.2-3) contains the times $t+\Delta t$ and $t-\Delta t$, the following right difference operator contains the times $t+\Delta t$ and t, and the left difference operator contains t and $t-\Delta t$. A second order difference equation contains three different values of the particular variable, e.g. the values $x+\Delta x$, x, and $x-\Delta x$, as the difference operator of second order in Eq.(8.2-2).

Since Eq.(8.2-4) contains the times $t+\Delta t$, t, and $t-\Delta t$, it is a difference equation of second order in time. Note that Eq.(8.2-1) would be called a differential equation of first order in time, since both $\Psi(t)$ and $\partial\Psi/\partial t$ refer to the time t, while the times $t+dt$ and $t-dt$ do not occur.

The operator ∇^2 in Eq.(8.2-4) is of second order, since it contains $x+\Delta x$, x, and $x-\Delta x$ according to Eq.(8.2-2) for one variable, and corresponding terms for more space variables.

Refer to Fig.8.3-1 for the interpretation of Eq.(8.2-4) for the case $\Psi(\mathbf{r},t)$ $=\Psi(x,t)$. In a space-time coordinate system we have nine marks with spatial coordinates $x'-\Delta x$, x', $x'+\Delta x$, and time coordinates $t'-\Delta t$, t', $t'+\Delta t$. An operator $\Omega_0(\mathbf{r},t)=\Omega_0(x)$ of second order means a relationship between the functional variables $\Psi(x'-\Delta x,t)$, $\Psi(x',t)$, $\Psi(x'+\Delta x,t)$ at three adjacent points in x for any time t. Putting it differently, the neighbors $\Psi(x'-\Delta x,t)$ and $\Psi(x'+\Delta x,t)$ influence $\Psi(x',t)$. This is indicated by the six horizontal arrows in Fig.8.3-1.

For the time variable, Eq.(8.2-4) shows again a relationship of functional values $\Psi(\mathbf{r},t+\Delta t)=\Psi(x,t'+\Delta t)$, $\Psi(x,t')$, and $\Psi(x,t'-\Delta t)$ at three points adjacent in t for any coordinate x. This means that the neighbors in time $\Psi(x',t'-\Delta t)$ and $\Psi(x',t'+\Delta t)$ influence $\Psi(x',t')$. This influence is indicated by the two vertical arrows in Fig.8.3-1.

The meaning of a difference equation of second order is thus how a functional value at the coordinate point x', t' or generally x'^i is related to the neighbors having a distance of one unit Δx, Δt or generally Δx^i. A difference equation of fourth order would relate the functional value at the point x'^i to the neighbors having a distance of one unit Δx^i or two units $2\Delta x^i$, as is evident from the difference operator of fourth order[1]:

$$\begin{aligned}\frac{\Delta^4\Psi}{(\Delta x)^4} = \frac{1}{\Delta x}\Bigg\{\frac{1}{\Delta x}\Bigg[\frac{1}{\Delta x}\Bigg(&\frac{\Psi(x+2\Delta x)-\Psi(x+\Delta x)}{\Delta x} - \frac{\Psi(x+\Delta x)-\Psi(x)}{\Delta x}\Bigg)\\ &-\frac{1}{\Delta x}\Bigg(\frac{\Psi(x+\Delta x)-\Psi(x)}{\Delta x} - \frac{\Psi(x)-\Psi(x-\Delta x)}{\Delta x}\Bigg)\Bigg]\\ &-\frac{1}{\Delta x}\Bigg[\frac{1}{\Delta x}\Bigg(\frac{\Psi(x+\Delta x)-\Psi(x)}{\Delta x} - \frac{\Psi(x)-\Psi(x-\Delta x)}{\Delta x}\Bigg)\\ &-\frac{1}{\Delta x}\Bigg(\frac{\Psi(x)-\Psi(x-\Delta x)}{\Delta x} - \frac{\Psi(x-\Delta x)-\Psi(x-2\Delta x)}{\Delta x}\Bigg)\Bigg]\Bigg\}\\ = \frac{1}{\Delta x)^4}[\Psi(x+2\Delta x) - 4\Psi(x+\Delta x) + 6\Psi(x)&\\ &-4\Psi(x-\Delta x) + \Psi(x-2\Delta x)] \qquad (1)\end{aligned}$$

Note that the difference operator of second order defined by Eq.(6.3-21) contains neighbors with distance of either one unit $\Delta\eta$ or two units $2\Delta\eta$. Hence, the dyadic difference operator of second order resembles the difference operator of fourth order of Eq.(1).

8.4 Klein-Gordon Equation

The Klein-Gordon equation for a boson of charge e,

$$\begin{aligned}\left(\hbar^2\frac{\partial^2}{\partial t^2} - 2ie\hbar\Phi\frac{\partial}{\partial t} - ie\hbar\frac{\partial\Phi}{\partial t} + e^2\Phi^2\right)\Psi = &\\ (-\hbar^2c^2\nabla^2 + ie\hbar c\mathbf{A}\,\mathrm{grad} + ie\hbar c\,\mathrm{div}\,\mathbf{A} + e^2\mathbf{A}^2 + m^2c^2)\,\Psi& \qquad (1)\end{aligned}$$

is transformed by the substitutions

$$\frac{\partial y}{\partial t} \to \frac{1}{2\Delta t}[y(t+\Delta t) - y(t-\Delta t)]$$

$$\frac{\partial^2 y}{\partial t^2} \to \frac{1}{(\Delta t)^2}[y(t+\Delta t) - 2y(t) + y(t-\Delta t)]$$

---

[1] The difference operator of third order offers a variety of choices as did the difference operator of first order in Eq.(8.2-3). A logical, but not the only choice for a symmetric difference operator is

$\Delta^3\Psi/(\Delta x)^3 = (\Delta x)^{-3}[\Psi(x+2\Delta x) - 3\Psi(x+\Delta x) + 3\Psi(x-\Delta x) - \Psi(x-2\Delta x)]$.

into the following difference equation:

$$
\begin{aligned}
&\Omega_1(\mathbf{r},t)\Psi(\mathbf{r},t+\Delta t) = \Omega_0(\mathbf{r},t)\Psi(\mathbf{r},t) + \Omega_{-1}(\mathbf{r},t)\Psi(\mathbf{r},t-\Delta t) \\
&\Omega_1(\mathbf{r},t) = 1 + i\frac{e\Delta t}{\hbar}\Phi(\mathbf{r},t) \\
&\Omega_0(\mathbf{r},t) = 2 + \left(\frac{e\Delta t}{\hbar}\right)^2 \Phi^2(\mathbf{r},t) + (c\Delta t)^2 \nabla^2 \\
&\qquad - i\frac{ec}{\hbar}(\Delta t)^2\left[2\mathbf{A}(\mathbf{r},t)\,\mathrm{grad} + \mathrm{div}\,\mathbf{A}(\mathbf{r},t)\right] - \left(\frac{e\Delta t}{\hbar}\right)^2 \mathbf{A}^2(\mathbf{r},t) \\
&\qquad - \left(\frac{mc^2\Delta t}{\hbar}\right)^2 - i\frac{e\Delta t}{2\hbar}\left[\Phi(\mathbf{r},t+\Delta t) - \Phi(\mathbf{r},t-\Delta t)\right] \\
&\Omega_{-1}(\mathbf{r},t) = -1 + i\frac{e\Delta t}{\hbar}\Phi(\mathbf{r},t)
\end{aligned}
\tag{2}
$$

The form of the operators $\Omega_1(\mathbf{r},t)$ and $\Omega_{-1}(\mathbf{r},t)$ indicates that the general solution of Eq.(2) can be represented by a sum of eigenfunctions only if $\Phi(\mathbf{r},t)$ is a constant and if $\mathbf{A}(\mathbf{r},t)$ is time invariant. For time invariant potentials $\Phi(\mathbf{r},t)$ and $\mathbf{A}(\mathbf{r},t)$, one may represent the general solution approximately by eigenfunctions if the potential $\Phi(\mathbf{r})$ is either large, $e\Delta t\Phi(\mathbf{r})/\hbar \gg 1$, or small, $e\Delta t\Phi(\mathbf{r})/\hbar \ll 1$. One obtains for a small potential $\omega_{1n} = -\omega_{-1n} = 1$, and for a large one $\omega_{1n} = +\omega_{-1n} = 1$. The difference operator $\Omega_0(\mathbf{r},t) = \Omega_0(\mathbf{r})$ is in both cases significantly different from the differential operator obtained by elimination of the time variable from Eq.(1). The reason is the term $2ie\hbar\Phi\partial/\partial t$ in Eq.(1). A time differential of first order creates terms in the operators $\Omega_1(\mathbf{r},t)$ and $\Omega_{-1}(\mathbf{r},t)$, rather than in the operator $\Omega_0(\mathbf{r},t)$. In the non-relativistic Schrödinger equation (8.2-1) the potential V is multiplied by Ψ rather than by $\partial\Psi/\partial t$; as a result, V becomes a term of the operator $\Omega_0(\mathbf{r},t)$.

Equation (2) assumes the following simplified form for the force-free particle in one dimension:

$$
\begin{aligned}
&\Psi(x,t+\Delta t) = \Omega_0(x)\Psi(x,t) - \Psi(x,t-\Delta t) \\
&\Omega_0(x) = 2 + (c\Delta t)^2\frac{\nabla^2}{(\Delta x)^2} - \left(\frac{mc^2\Delta t}{\hbar}\right)^2
\end{aligned}
\tag{3}
$$

The eigenvalues ω_{1n} and ω_{-1n} have the values $+1$ and -1. The eigenvalue ω_{0n} is determined by the equation

$$\Omega_0(x)\varphi_n(x) = \left(\frac{c\Delta t}{\Delta x}\right)^2 [\varphi_n(x+\Delta x) - 2\varphi_n(x) + \varphi_n(x-\Delta x)]$$
$$+ \left[2 - \left(\frac{mc^2\Delta t}{\hbar}\right)^2\right] \varphi_n(x)$$
$$= \omega_{0n}\varphi_n(x)$$

which is satisfied by the eigenfunctions $\varphi_n(x) = \varphi_\kappa(x) = \exp(i\kappa x)$ and the following eigenvalues:

$$\omega_{0n} = \omega_{0\kappa} = 2B = -4\left(\frac{c\Delta t}{\Delta x}\right)^2 \sin^2\frac{\kappa\Delta x}{2} + 2 - \left(\frac{mc^2\Delta t}{\hbar}\right)^2 \tag{4}$$

One obtains from Eqs.(8.1-8) and (4):

$$Q_{1n} = B + (B^2-1)^{1/2}, \qquad Q_{2n} = B - (B^2-1)^{1/2}$$
$$Q_{1n}Q_{2n} = 1 \tag{5}$$

The relation

$$-1 - |O(\Delta t)^2| < B < 1 + |O(\Delta t)^2| \tag{6}$$

yields

$$-1 - |O(\Delta t)| < |Q_{1n}|, \qquad |Q_{2n}| < 1 + |O(\Delta t)|$$

while the equalities $B = \pm[1 + |O(\Delta t)^2|]$ yield:

$$|Q_{1n}| = 1 + |O(\Delta t)| \qquad |Q_{2n}| = 1 + |O(\Delta t)|$$

There are no solutions independent of Δt for $B > 1 + |O(\Delta t)^2|$ and $B < -1 - |O(\Delta t)^2|$.

Equations (4) and (5) yield two conditions for the existence of solutions with arbitrary values of κ, that are independent of Δt:

$$1 - \frac{1}{2}\left(\frac{mc^2\Delta t}{\hbar}\right)^2 \leq 1 + |O(\Delta t)^2| \tag{7}$$

$$\left(\frac{\Delta x}{c\Delta t}\right)^2 \geq \left[1 + \left(\frac{mc\Delta x}{2\hbar}\right)^2\right] [1 + |O(\Delta t)^2|] \tag{8}$$

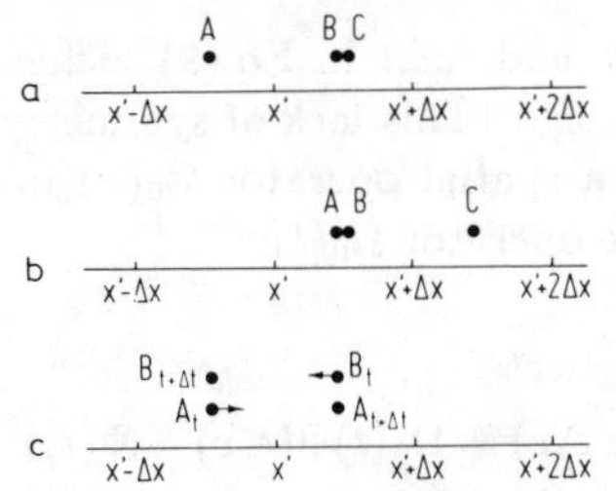

FIG.8.4-1. The instant velocity $(x_1 - x_0)/\Delta t$ of a boson remains undetermined within the range $0 \leq |v| \leq c$ due to the relation $(\Delta t)^2 \leq (\Delta x/c)^2$.

The first condition is always satisfied, the second one approaches the form

$$(\Delta t)^2 \leq \left(\frac{\Delta x}{c}\right)^2 \tag{9}$$

for small values of Δx and Δt.

We have pointed out in connection with Eq.(7.1-3) that the instantaneous velocity may not be defined if the unit time interval cannot be chosen sufficiently large[1]. Consider a force-free boson at the mark x' of a ruler at time t. At time $t + \Delta t$ it can have propagated at most the distance $\Delta x = c\Delta t$ according to Eq.(9), and be at the mark $x' + \Delta x$ for a positive velocity or at mark $x' - \Delta x$ for a negative velocity. Consider Fig.8.4-1a. A boson "closest to the mark x'" can be as far left as A or as far right as B. After the time Δt has elapsed, it shall be closest to the mark $x' + \Delta x$, which means it must have propagated to the right at least as far as C. The lower limit for the velocity of the boson is zero—if it moved from B to C—while the upper limit is c—if it moved from A to C. Nothing more can be said about the instantaneous velocity than that it must have been in the range $0 \leq |v| \leq c$.

Figure 8.4-1b shows a corollary. A boson A is closest to the mark x' at the time t. At the time $t+\Delta t$ it cannot have propagated to the right further than C due to Eq.(9). If it is observed closest to the mark $x' + \Delta x$, it may be as far left as B or as far right as C. Its velocity remains undetermined within the range $0 \leq |v| \leq c$.

If a boson is closest to the mark x' both at time t and $t + \Delta t$ it may have moved as much to the right as particle A or as much to the left as particle B in Fig.8.4-1c. In this case the square of the velocity remains undetermined within the range $0 \leq v^2 \leq c^2$.

---

[1] The normalized variables used there are connected with the nonnormalized variables here through the relations $\theta = t/T$, $\xi = x/X$, $\Delta\theta = \Delta t/T$, $\Delta\xi = \Delta x/X$.

An exchange of Δx and $ic\Delta t$ in Eq.(9) yields the same relation but with reversed inequality sign. This lack of symmetry between Δx and $ic\Delta t$ is caused by the use of a spatial operator $\Omega_0(x)$ in Eq.(3). One may also write Eq.(3) with a time operator $\Omega_0(t)$:

$$\Psi(t, \mathbf{r} + \Delta\mathbf{r}) = \Omega_0(t)\Psi(t, \mathbf{r}) - \Psi(t, \mathbf{r} - \Delta\mathbf{r})$$

$$\begin{aligned}
\Omega_0(t)\varphi_n(t) &= \left(\frac{\Delta x}{c\Delta t}\right)^2 [\varphi_n(t + \Delta t) - 2\varphi_n(t) + \varphi_n(t - \Delta t)] \\
&\quad + \left[2 + \left(\frac{mc\Delta x}{\hbar}\right)^2\right]\varphi_n(t) \\
&= \omega_{0n}\varphi_n(t) \\
\varphi_n(t) &= e^{iE_nt/\hbar} \\
\omega_{0n} &= -4\left(\frac{\Delta x}{c\Delta t}\right)^2 \sin^2\frac{E_n\Delta t}{2\hbar} + 2 + \left(\frac{mc\Delta x}{\hbar}\right)^2
\end{aligned} \tag{10}$$

The following inequalities replace Eqs.(7) and (8) for arbitrary values of E:

$$1 + \frac{1}{2}\left(\frac{mc\Delta x}{\hbar}\right)^2 \leq 1 + |O(\Delta x)^2| \tag{11}$$

$$\left(\frac{c\Delta t}{\Delta x}\right)^2 \geq \left[1 - \left(\frac{mc^2\Delta t}{2\hbar}\right)^2\right][1 + |O(\Delta x)^2|] \tag{12}$$

Equations (11) and (12) are symmetric in Δx and $ic\Delta t$ to Eqs.(7) and (8). One can satisfy Eq.(11) for the equality sign only. Hence, there are no stationary solutions but only solutions that increase or decrease exponentially with x, if they are to become independent of Δx for sufficiently small values of Δx.

The independence relation for a non-relativistic particle in three dimensions is derived in Section 12.5. For a relativistic particle one obtains from Eq.(2) again Eq.(7), but Eq.(8) is replaced by the following relation:

$$\begin{aligned}
\left(\frac{\Delta r}{c\Delta t}\right)^2 &\geq \left[\left(\frac{\Delta r}{\Delta x}\right)^2 + \left(\frac{\Delta r}{\Delta y}\right)^2 + \left(\frac{\Delta r}{\Delta z}\right)^2 + \left(\frac{mc\Delta r}{2\hbar}\right)^2\right][1 + |O(\Delta t)^2|] \\
(\Delta r)^2 &= (\Delta x)^2 + (\Delta y)^2 + (\Delta z)^2
\end{aligned} \tag{13}$$

Equation (13) is reduced to Eq.(8) if Δx is small compared with Δy and Δz.

The time variation of the eigenfunctions of the force-free boson for small values of Δt follows from Eq.(5):

$$Q_{1n}^{\tau} = e^{iE_n t/\hbar}, \quad Q_{2n}^{\tau} = e^{-iE_n t/\hbar}$$

$$E_n = E_\kappa = \frac{\hbar}{\Delta t}\sin^{-1}(1-B^2)^{1/2}$$
$$= \left(m^2c^4 + \hbar^2c^2\kappa^2\right)^{1/2}\left[1 + |O(\Delta t)^2| + |O(\Delta x)^2|\right] \quad (14)$$

Equations (7) and (8) are replaced by the relation

$$\Delta t \leq \frac{2\hbar}{mc^2}\left(1 + \frac{\kappa^2\hbar^2}{m^2c^2}\right)^{-1/2} \quad (15)$$

for small values of Δx and bounded values of κ. Similarly, one obtains the relation

$$E^2 \geq m^2c^4$$

for small values of Δt and bounded values of E instead of Eqs.(11) and (12).

9 Schrödinger Difference Equation with Coulomb Field

9.1 Separation of Variables for a Centrally Symmetric Field

Consider the three-dimensional Cartesian coordinate system of Fig.5.4-5. A difference equation defines the values of a function at the grid points. Let now the Cartesian coordinate system be replaced by a spherical coordinate system as shown in Fig.9.1-1. A difference equation written for this coordinate system will define the values of a function at its grid points. Generally, the grid points of a Cartesian coordinate system will not coincide with those of a spherical coordinate system. This lack of an one-to-one correspondence of points creates a problem when one wants to transfer from one coordinate system to the other. No such problem occurs in differential mathematics, since there are enough points in a continuum to have for any point of a Cartesian coordinate system an exactly corresponding point of a spherical coordinate system.

There are two ways around this problem. First, one may write the differential equation for the desired coordinate system, and then replace the differential quotients by difference quotients. This is a simple and quick approach, but it is obviously unsatisfactory to have to fall back on the differential equations.

One may avoid going through the differential equations by means of the rules derived in Section 12.7. After lengthy calculations one obtains the operator $\Omega_0(\mathbf{r},t) = \Omega_0(\mathbf{r})$ of Eq.(8.2-4) in spherical coordinates:

$$\begin{aligned}
\Omega_0(r,\vartheta,\varphi) &= \Omega_{01}(r,\vartheta,\varphi) + \Omega_{02}(r,\vartheta,\varphi) \\
\Omega_{01}(r,\vartheta,\varphi) &= i\frac{\hbar\Delta t}{m}\left[\frac{2}{r}\frac{\Delta_\mathrm{s}}{\Delta r} + \frac{\Delta^2}{(\Delta r)^2} + \frac{1}{r^2}\left(\operatorname{ctn}\vartheta\frac{\Delta_\mathrm{s}}{\Delta\vartheta} + \frac{\Delta^2}{(\Delta\vartheta)^2}\right)\right. \\
&\qquad\left. + \frac{1}{r^2\sin^2\vartheta}\frac{\Delta^2}{(\Delta\varphi)^2} - \frac{2m}{\hbar^2}V(r,\vartheta,\varphi)\right] \\
\Omega_{02}(r,\vartheta,\varphi) &= i\frac{\hbar\Delta t}{m}O(\Delta r,\Delta\vartheta,\Delta\varphi)
\end{aligned}$$

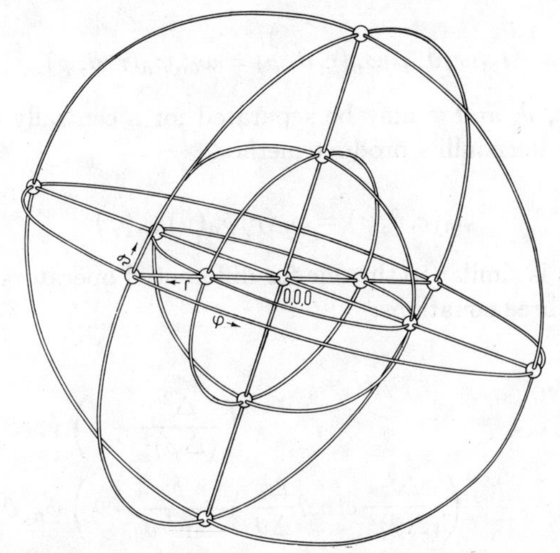

FIG.9.1-1. Spherical coordinate system with finite resolution Δr, $\Delta\vartheta$, $\Delta\varphi$.

The operator Ω_{01} is equal to the operator one obtains from the differential operator ∇^2 in spherical coordinates by the substitution of differences for the differentials. The operator Ω_{02} contains difference operators up to the sixth order, which are multiplied by Δr, $\Delta\vartheta$, $\Delta\varphi$, and their powers. The differences Δr, $\Delta\vartheta$, and $\Delta\varphi$ are assumed to have arbitrary values. Hence, it is reasonable to try the following substitution for the solution of Eq.(8.2-4):

$$\Psi(\mathbf{r},t) = \Psi_1(\mathbf{r},t) + \sum_{i=1}^{\infty}\sum_{j=1}^{\infty}\sum_{k=1}^{\infty}(\Delta r)^i(\Delta\vartheta)^j(\Delta\varphi)^k\Psi_{ijk}(\mathbf{r},t) \qquad (1)$$

The terms with the same factor $(\Delta r)^i(\Delta\vartheta)^j(\Delta\varphi)^k$ must satisfy the difference equation by themselves, since the value of Δr, $\Delta\vartheta$, and $\Delta\varphi$ is arbitrary. For $\Psi_1(\mathbf{r},t)$ one obtains according to Eq.(8.2-4):

$$\Psi_1(\mathbf{r},t+\Delta t) = \Omega_{01}(r,\vartheta,\varphi)\Psi_1(\mathbf{r},t) + \Psi_1(\mathbf{r},t-\Delta t) \qquad (2)$$

Let the expansion in Eq.(1) converge. One may then neglect the terms $\Psi_{ijk}(\mathbf{r},t)$ for sufficiently small values of Δr, $\Delta\vartheta$, and $\Delta\varphi$. Hence, one obtains the same result in this approximation whether one makes the transition from differential to difference operators in Cartesian or spherical coordinates.

Equation (2) is reduced to ordinary difference equations if one can find eigenfunctions $\varphi_n(r,\vartheta,\varphi)$ and eigenvalues ω_{0n} of the operator $\Omega_{01}(r,\vartheta,\varphi)$:

$$\Omega_{01}(r,\vartheta,\varphi)\varphi_n(r,\vartheta,\varphi) = \omega_{0n}\varphi_n(r,\vartheta,\varphi) \tag{3}$$

The variables r, ϑ, and φ may be separated for a centrally symmetric potential $V(r)$ by Bernoulli's product method:

$$\varphi_n(r,\vartheta,\varphi) = \varphi_n(r)\psi_n(\vartheta)\chi_n(\varphi) \tag{4}$$

The calculation is similar to the one for differential operators. Equation (3) is replaced by three equations:

$$\left(\frac{\Delta^2}{(\Delta\varphi)^2} - \delta\right)\chi_n(\varphi) = 0 \tag{5}$$

$$\left(\frac{\Delta^2}{(\Delta\vartheta)^2}\operatorname{ctn}\vartheta\frac{\Delta_s}{\Delta\vartheta} + \frac{\delta}{\sin^2\vartheta} - \lambda\right)\psi_n(\vartheta) = 0 \tag{6}$$

$$\left[i\frac{\hbar\Delta t}{m}\left(\frac{\Delta^2}{(\Delta r)^2} + \frac{2}{r}\frac{\Delta_s}{\Delta r}\frac{2m}{\hbar^2}V(r) + \frac{\lambda}{r^2}\right) - \omega_{0n}\right]\varphi_n(r) = 0 \tag{7}$$

The substitution of $\chi_n(\varphi) = \exp(\pm im\varphi)$ into Eq.(5) yields the eigenvalues:

$$\delta = \pm\frac{4}{(\Delta\phi)^2}\sin^2\frac{m\Delta\varphi}{2}$$

For $m\Delta\varphi \ll 1$ one obtains:

$$\delta = \pm m^2 \tag{8}$$

Let us turn to Eq.(6), which we call the *difference equation of the finite spherical functions.* Its solution is very tedious; it will be found in Sections 12.6 and 12.7. Here it is only of interest to know that λ can assume the values $-l(l+1)$ with $l = 1, 2, \ldots$ for arbitrary values of $\Delta\vartheta$.

9.2 Discrete Eigenvalues in a Coulomb Field

Equation (9.1-7) assumes the following form for a Coulomb potential $V(r) = -Ze^2/r$:

$$\left(\frac{\Delta^2}{(\Delta r)^2} + \frac{2}{r}\frac{\Delta_s}{\Delta r} + \kappa^2 + \frac{2b}{r} - \frac{l(l+1)}{r^2}\right)\varphi_n(r) = 0$$

$$\kappa^2 = i\frac{\omega_{0n}m}{\hbar\Delta t} = \frac{2mW}{\hbar}, \quad b = \frac{Ze^2m}{\hbar^2} \tag{1}$$

Using the difference operators defined by Eqs.(8.2-2) and (8.2-3), but with the variable r substituted for x and t, one obtains a difference equation of second order with variable coefficients:

$$R(R+1)\varphi_n(R+1) + \left[gR^2 + 2b\Delta rR - l(l+1)\right]\varphi_n(R) + R(R-1)\varphi_n(R-1) = 0$$

$$g = \kappa^2(\Delta r)^2 - 2, \quad R = \frac{r}{\Delta r}$$

$$\varphi_n(r) = \varphi_n(R), \quad \varphi_n(r \pm \Delta r) = \varphi_n(R \pm 1) \tag{2}$$

Equation (2) is reduced to the hypergeometric difference equation for $l = 0$ (Wallenberg and Guldberg, 1911).

To obtain eigenfunctions of Eq.(2) we use a method based on the Laplace transform (Nörlund, 1910, 1915, 1924; Milne-Thomson, 1951; Wallenberg and Guldberg, 1911). It is not necessary to use this detour via differential calculus. One may obtain the same solution by purely algebraic means. The Laplace transform is used only because it shortens the calculation drastically.

Let the contour integral

$$\varphi_n(R) = \frac{1}{2i\pi}\int_{\ell} s^{R+1} w(s)\, ds \tag{3}$$

be taken in the complex plane from point ℓ_1 along a contour ℓ to the point ℓ_2. Partial integration yields:

$$\varphi_n(R) = \frac{1}{2i\pi}\left(\frac{1}{R}s^R w\Big|_{\ell_1}^{\ell_2} - \frac{1}{R}\int_{\ell} s^R w'\, ds\right)$$

Let ℓ_1 and ℓ_2 be chosen so that the term

$$s^R w(s)\Big|_{\ell_1}^{\ell_2} \tag{4}$$

vanishes. One obtains:

$$-R\varphi_n(R) = \frac{1}{2i\pi}\int_{\ell} s^R w'\, ds \tag{5}$$

Further partial integration yields

$$R(R+1)\varphi_n(R) = \frac{1}{2i\pi}\int_\ell s^{R+1}w''\,ds \tag{6}$$

provided

$$s^{R+1}w'(s)\Big|_{\ell_1}^{\ell_2} \tag{7}$$

vanishes.

Making the corresponding assumptions about beginning and end of the contour of integration one obtains the following relations for $\varphi_n(R-1)$ and $\varphi_n(R+1)$:

$$(-1)^k(R+1)\cdots(R+k)\varphi_n(R+1) = \frac{1}{2i\pi}\int_\ell s^{R+k}w^{(k)}\,ds$$

$$(-1)^k(R-1)\cdots(R+k-2)\varphi_n(R-1) = \frac{1}{2i\pi}\int_\ell s^{R+k-2}w^{(k)}\,ds \tag{8}$$

We rewrite Eq.(2) into such a form that the coefficients become multiples of the terms on the left side of Eqs.(5), (6), and (8):

$$\begin{aligned}[(R+2)(R+1)-2(R+1)\varphi_n(R+1)]& \\ +\left[gR(R+1)+(2b\Delta r-g)R-l(l+1)\right]\varphi_n(R)& \\ +R(R-1)\varphi_n(R-1)=0&\end{aligned} \tag{9}$$

Substitution of the integral transforms for $\varphi_n(R)$, $\varphi_n(R+1)$, and $\varphi_n(R-1)$ yields the following differential equation:

$$(s^2+gs+1)sw''+(2s-2b\Delta r+g)sw'-l(l+1)w=0 \tag{10}$$

This equation has one singular point at $s=\infty$. The other singular points are located at the roots of the equation[1]:

$$(s^2+gs+1)s=0 \tag{11}$$

One obtains:

[1] We are following a standard method for the solution of differential equations with variable coefficients. A detailed explanation may be found in textbooks under the heading Fuchs-type differential equations.

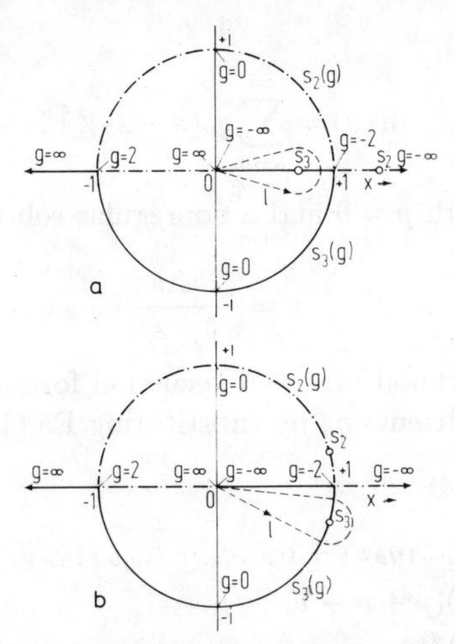

FIG.9.2-1. Location of the singular points s_2 (dashed-dotted line) and s_3 (solid line) in the complex plane as function of g for the nonrelativistic Schrödinger equation, the Klein-Gordon equation, and the iterated Dirac equation. The line of integration ℓ is shown in (a) for the bound particle with $W < 0$, and in (b) for the free particle with $W > 0$.

$$s_1 = 0, \quad s_2 = -\frac{1}{2}g + \left(\frac{1}{4}g^2 - 1\right)^{1/2}, \quad s_3 = \frac{1}{s_2} = -\frac{1}{2}g - \left(\frac{1}{4}g^2 - 1\right)^{1/2}$$

$$s_2 + s_3 = -g, \quad s_2 - s_3 = 2\left(\frac{1}{4}g^2 - 1\right)^{1/2} = 2i\kappa\Delta r\left(1 - \frac{1}{4}(\kappa\Delta r)^2\right)^{1/2}$$

$$g = \kappa^2(\Delta r)^2 - 2 = \frac{2mW(\Delta r)^2}{\hbar} - 2 \tag{12}$$

The parameter g is always real, since the eigenvalues ω_{0n} of stationary solutions of the Schrödinger equation are purely imaginary as shown by Eq.(8.1-13). The loci of s_2 and s_3 for real values of g are shown in Fig.9.2-1a. The value of g is approximately -2 for small values of Δr. Let W in Eq.(12) be negative. The value of g is in this case smaller than -2, and the root s_3 is located on the real axis between 0 and +1. A positive value of W makes g larger than -2, and s_3 is now located on the unit circle ($-2 < g < +2$) or on the negative real axis to the left of -1 ($g > +2$).

A formal solution of Eq.(10) is obtained by a series expansion of $w(s)$ at the point s_3. This power series converges for small values of Δr inside a circle that passes through the point s_2:

$$w(s) = \sum_{\nu=0}^{\infty} q_\nu (s - s_3)^{p+\nu} \tag{13}$$

A regular solution with $p = 0$ and a nonregular solution with

$$p = \frac{2b\Delta r}{s_3 - s_2} \tag{14}$$

exists in the neighborhood of s_3. A recursion formula with three terms is obtained for the coefficients q_ν by substituting Eq.(13) into Eq.(10):

$$\begin{aligned}
\alpha_{\nu,\nu-1}q_{\nu-1} + \alpha_{\nu,\nu}q_\nu + \alpha_{\nu,\nu+1}q_{\nu+1} &= 0 \\
\alpha_{\nu,\nu-1} &= (p+\nu)(p+\nu-1) \\
\alpha_{\nu,\nu} &= (p+\nu)^2(2s_3 - s_2) + (p+\nu)[s_3 - p(s_3 - s_2)] - l(l+1) \\
\alpha_{\nu,\nu+1} &= s_3(s_3 - s_2)(p+\nu+1)(\nu+1)
\end{aligned} \tag{15}$$

Let the line of integration of the contour integral of Eq.(3) begin at the origin, run around s_3 and return to the origin, as shown by the line ℓ in Fig.9.1-2a. The point s_2 remains outside this loop. More details of this integration may be found in the works of Wallenberg and Guldberg (1911), and Nörlund (1910, 1915, 1924). One obtains a factorial series:

$$\varphi_n(R) = s_3^R \frac{\Gamma(R)}{\Gamma(R+p+1)} \sum_{\nu=0}^{\infty} (-1)^\nu q_\nu \frac{(p+1)\cdots(p+\nu)}{(R+p+1)\cdots(R+p+\nu)} \tag{16}$$

The fraction $(p+1)\cdots(p+\nu)/(R+p+1)\cdots(R+p+\nu)$ shall always be replaced by 1 for $\nu = 0$. This convention avoids the need for writing a special term for $\nu = 0$.

The power series of Eq.(13) does not converge everywhere along the line ℓ of integration. As a result, the factorial series of Eq.(16) diverges for all values of R. A convergent factorial series is obtained by conformal mapping. The circle of convergence of the power series of Eq.(13) is mapped onto a loop that runs through the origin; except for $s = s_3$ and $s = 0$ there must be no singularity either inside or on the loop. The resulting factorial series does still not converge in the whole interval $0 \leq R < \infty$. Eigenfunctions $\varphi_n(R)$ that are bounded and convergent in the whole interval $0 \leq R < \infty$ are obtained, if the factorial series of Eq.(16) terminates and thus contains a finite number of terms only.

The series of Eq.(16) can terminate only if p is a negative integer:

$$p = -n, \quad n = 1, 2, \ldots \tag{17}$$

The term q_ν with $\nu = n - 1$ in Eq.(16) is the last one that does not vanish.

One obtains from Eqs.(12), (14), and (17):

$$s_3 - s_2 = -\frac{2b\Delta r}{n} = -\frac{2Zme^2\Delta r}{n\hbar^2} \tag{18}$$

The right side of Eq.(18) is real and negative, and the same must hold for the difference $s_3 - s_2$. One may infer from Fig.9.2-1a that this is only possible if s_3 and s_2 are located on the positive real axis. This is a necessary condition for the existence of solutions in the form of terminating factorial series. One may also see from Fig.9.2-1a that this condition implies $g < -2$. Equations (1) and (2) then yield $W < 0$.

Two values are obtained for κ from Eqs.(12), (14), and (17):

$$\begin{aligned}
\kappa_1^2 &= \frac{4}{(\Delta r)^2}\left[1 + \left(\frac{b\Delta r}{2n}\right)^2 + \cdots\right] \\
\kappa_2^2 &= -\left(\frac{b}{n}\right)^2\left[1 - \left(\frac{b\Delta r}{2n}\right)^2 + \cdots\right] \\
&= -\left(\frac{Zme^2}{n\hbar^2}\right)^2\left[1 - \left(\frac{Zme^2\Delta r}{2n\hbar^2}\right)^2 + \cdots\right]
\end{aligned} \tag{19}$$

The value of κ_1 depends essentially on Δr, but the value of κ_2 becomes independent of Δr for small values of Δr. The energy eigenvalues associated with κ_2 follow from Eq.(1):

$$W = \frac{\hbar^2}{2m}\kappa_2^2 = -\frac{Z^2e^4m}{2n^2\hbar^2}\left[1 - \left(\frac{Zme^2}{2n\hbar^2}\Delta r\right)^2 + \cdots\right] \tag{20}$$

The factor s_3^R in Eq.(16) may be rewritten:

$$s_3^R = e^{-|\kappa_2|r} \tag{21}$$

The solution in the point $s = s_2$ contains the factor $\exp(+|\kappa_2|r)$ instead of $\exp(-|\kappa|r)$. This solution grows beyond all bounds for $r \to \infty$.

Equation (17) is only a necessary condition for the termination of the factorial series of Eq.(16). A negative integer $p = -n$ produces in general an infinite coefficient $q_\nu = q_n$. Let us rewrite the recursion formula of Eq.(15) as a system of linear equations for greater clarity:

$$
\begin{aligned}
0 &= \alpha_{00}q_0 + \alpha_{01}q_1 \\
0 &= \alpha_{10}q_0 + \alpha_{11}q_1 + \alpha_{12}q_2 \\
&\vdots \\
0 &= \qquad \alpha_{\nu,\nu-1}q_{\nu-1} + \alpha_{\nu,\nu}q_\nu + \alpha_{\nu,\nu+1}q_{\nu+1} \\
&\vdots \\
0 &= \qquad \alpha_{n-1,n-2}q_{n-2} + \alpha_{n-1,n-1}q_{n-1} + 0q_n
\end{aligned} \tag{22}
$$

These are n equations with $n-1$ unknowns q_1 to q_{n-1}. Such a system of equations can exist only if the determinant of the coefficients vanishes:

$$
D = \begin{vmatrix}
\alpha_{00} & \alpha_{01} & & & & & \\
\alpha_{10} & \alpha_{11} & \alpha_{12} & & & & \\
\vdots & & & \ddots & & & \\
& & \alpha_{\nu,\nu-1} & \alpha_{\nu,\nu} & \alpha_{\nu,\nu+1} & & \\
\vdots & & & & & \ddots & \\
& & & & \alpha_{n-2,n-3} & \alpha_{n-2,n-2} & \alpha_{n-2,n-1} \\
& & & & & \alpha_{n-1,n-2} & \alpha_{n-1,n-1}
\end{vmatrix} = 0 \tag{23}
$$

The values of this determinant can be computed readily for small values of Δr. One obtains from Eq.(15) for $p = -n$ and $\Delta r \to 0$:

$$
\begin{aligned}
\alpha_{\nu,\nu-1} &= (n-\nu)(n+1-\nu) \\
\alpha_{\nu,\nu} &= (n-\nu)(n-\nu-1) - l(l+1) \\
\alpha_{\nu,\nu+1} &= 0
\end{aligned}
$$

$$
D = \prod_{\nu=0}^{n-1} \alpha_{\nu,\nu} = 0 \tag{24}
$$

The roots of D are $l = n-\nu-1$ and $l = -(n-\nu)$. Only the first root is of interest since l must be a non-negative integer. Let ν run from 0 to $n-1$. The possible values of l run then from $n-1$ to 0. Hence, l may assume the following values only:

$$
l = 0,\ 1,\ \ldots,\ n-1 \tag{25}
$$

Equations (17) and (25) are the necessary and sufficient conditions for the termination of the factorial series of Eq.(16).

One may also conclude from Eq.(23) that there are no logarithmic terms in the solution of the differential equation (10) in the point $s = s_3$.

Let us turn to the terms of order $O(\Delta r)^q$, where $q = 1, 2, \ldots$. The roots of the determinant D for $\Delta r \neq 0$ are the roots of an equation of order n. For $n = 1$ one obtains $l = 0$; for $n = 2$ one obtains

$$l_1 = \frac{Ze^2 m\Delta r}{\hbar^2}, \quad l_2 = 1 - \frac{Ze^2 m\Delta r}{\hbar^2}$$

The eigenvalues l must be integers. Hence, only the terms of order $O(1)$ in Eq.(16) terminate, the terms of order $O(\Delta r)^q$ form an asymptotically convergent series. In Eq.(17) one must replace $p = -n$ by $p = -n + O(\Delta r)$. We will be satisfied here with the asymptotically convergent series, since a nonrelativistic equation is primarily of historic and tutorial interest, but we will derive a convergent solution in the case of the Klein-Gordon equation.

Substitution of Eqs.(17) and (21) into Eq.(16) yields the following polynomials:

$$\varphi_n(R) = e^{-|\kappa_2|r} \sum_{\nu=0}^{n-1} (-1)^\nu q_\nu \binom{R-1}{n-\nu+1}$$

$$\varphi_1(r) = \varphi_1(R) = q_0 e^{-|\kappa_{21}|r}$$

$$\varphi_2(r) = \frac{q_0}{\Delta r} e^{-|\kappa_{22}|r} \left(r - \Delta r - \frac{\Delta r q_1}{q_0} \right)$$

$$\varphi_3(r) = \frac{q_0}{(\Delta r)^2} e^{-|\kappa_{22}|r} \left(\frac{(r-\Delta r)(r-2\Delta r)}{2} + \frac{q_1 \Delta r (r - \Delta r)}{q_0} + \frac{q_2 (\Delta r)^2}{q_0} \right) \tag{26}$$

The coefficients $\alpha_{\nu,\nu+1}$ of Eq.(15) may be written for small values of Δr as follows:

$$\alpha_{\nu,\nu+1} = \frac{2(n-\nu-1)(\nu+1)b\Delta r}{n} \tag{27}$$

The recursion formula of Eq.(15) is reduced to two terms in this approximation:

$$q_{\nu+1} = -\frac{n}{2b\Delta r} \frac{(n-\nu)(n-\nu-1) - l(l+1)}{(n-\nu-1)(\nu+1)} q_\nu \tag{28}$$

The polynomials in Eq.(26) are identical in this approximation with those obtained from the differential Schrödinger equation:

$$\varphi_1(r) = q_0 e^{-Zme^2/\hbar}$$
$$\varphi_2(r) = \frac{q_0}{\Delta r} e^{-Zme^2/2\hbar}\left(r - \frac{\hbar^2}{Zme^2}[2 - l(l+1)]\right), \quad l = 0,\ 1 \tag{29}$$

Let us summarize our results. The use of finite differences and difference equations has led to mathematical methods completely different from those used for the solution of the differential Schrödinger equation with Coulomb field. Nevertheless, the terms of order $O(1)$ of the eigenvalues and eigenfunctions are the same as those obtained from the differential equation. Two results are noteworthy: Asymptotically convergent series are encountered that do not represent the function exactly but only with some uncertainty that depends on the resolution Δr of the coordinate system. Second, it is known that the solution of a differential equation can be quantized by termination of a power series, only if the recursion formula for the coefficients of the power series has two terms (Sommerfeld, 1939); this limitation does not exist for factorial series, as is clear from the fact that the recursion formula of Eq.(15) has three terms.

9.3 Time Variable Solutions

The eigenfunctions $\varphi_n(t, r, \vartheta, \varphi) = \varphi_n(t)\varphi_n(r, \vartheta, \varphi)$ have a time variation that is obtained from Eqs.(8.1-7) and (8.1-8) by substituting $\omega_{1n} = \omega_{-1n} = 1$ and one of the two values of ω_{0n}:

$$\omega_{0n,1} = -i\frac{\hbar\Delta t\kappa_1^2}{m} = -4i\frac{\hbar\Delta t}{m(\Delta r)^2}\left[1 + \left(\frac{b\Delta r}{2n}\right)^2 + \cdots\right]$$
$$\omega_{0n,2} = -i\frac{\hbar\Delta t\kappa_2^2}{m} = i\frac{\hbar\Delta t}{m}\left(\frac{b}{n}\right)^2\left[1 - \left(\frac{b\Delta r}{2n}\right)^2 + \cdots\right] \tag{1}$$

$$Q_{1n,j} = \frac{1}{2}\omega_{0n,j} + \left(\frac{1}{4}\omega_{0n,j}^2 + 1\right)^{1/2}$$
$$Q_{2n,j} = \frac{1}{2}\omega_{0n,j} - \left(\frac{1}{4}\omega_{0n,j} + 1\right)^{1/2}, \qquad j = 1,\ 2 \tag{2}$$

The independence relation of Eq.(8.1-11) assumes the form:

$$\left|\left(\frac{1}{2}\omega_{0n,j}\right)^2\right| \leq 1 + |O(\Delta t)^2|$$

For $\omega_{0n,j} = \omega_{0n,1}$ one obtains in the limit of small differences Δr and Δt

$$\frac{m(\Delta r)^2}{\Delta t} \geq 2\hbar \tag{3}$$

while $\omega_{0n,j} = \omega_{0n,2}$ yields:

$$\Delta t \leq \frac{\hbar^3 n^2}{Z^2 m e^4} \tag{4}$$

Only the condition of Eq.(4) needs to be satisfied if one is strictly interested in the time variation of the eigenfunctions belonging to the energy eigenvalues of the bound particle of Eq.(9.1-2). In order to make the general solution $\Psi(\mathbf{r}, t) = \sum c_n \varphi_n(t) \varphi_n(r, \theta, \varphi)$ independent of Δt, one must in addition satisfy the condition of Eq.(3).

Conditions very similar to Eqs.(3) and (4) have been obtained previously as Eqs.(8.2-7) and (8.2-8) for the nonrelativistic free particle. The ratio $m(\Delta r)^2/2\hbar\Delta t$ is in both cases only important if one wants to make all the eigenfunctions independent of Δt. This is required, e.g. for the solution of initial value problems. For particular eigenfunctions one only has to make Δt and Δr sufficiently small.

The time variation of the eigenfunctions $\varphi_n(t)$ belonging to the energy eigenvalues of Eq.(9.2-20) are readily obtained from Eqs.(8.1-9) and (2) for $j = 2$:

$$\varphi_{1n,2}(t) = b_{1n,2} Q_{1n,2}^{\tau} = b_{1n,2} e^{iEt/\hbar}$$

$$\varphi_{2n,2}(t) = b_{2n,2} Q_{2n,2}^{\tau} = (-1)^{\tau} e^{-iEt/\hbar}$$

$$E = \frac{\hbar}{\Delta t \sin(W\Delta t/\hbar)} = W\left(1 + \frac{1}{3!}\frac{W\Delta t}{\hbar} + \cdots\right), \quad \tau = \frac{t}{\Delta t} \tag{5}$$

The solutions for $W > 0$ will not be considered here since the calculation is similar to the one for the Klein-Gordon difference equation, which will be carried out in Section 10.6.

10 Klein-Gordon Difference Equation with Coulomb Field

10.1 Separation of Variables and Initial Value Problem

Consider the operators of Eq.(8.4-2). Let the scalar potential $\Phi(\mathbf{r}, t)$ be time-invariant and let the vector potential $\mathbf{A}(\mathbf{r}, t)$ be zero. The resulting operators are transformed into spherical coordinates in analogy to the transformation in Section 9.1. The terms of order $O(\Delta r)$, $O(\Delta\vartheta)$, and $O(\Delta\varphi)$ are left out from the beginning. One obtains:

$$
\begin{aligned}
\Omega_1(r,\vartheta,\varphi) &= 1 + i\frac{e\Delta t}{\hbar}\Phi(r,\vartheta,\varphi) \\
\Omega_0(r,\vartheta,\varphi) &= 2 + \left(\frac{e\Delta t}{\hbar}\right)^2 \Phi^2(r,\vartheta,\varphi) - \left(\frac{mc^2\Delta t}{\hbar}\right)^2 + (c\Delta t)^2\Bigg[\frac{2}{r}\frac{\Delta_{\mathrm{s}}}{\Delta r} \\
&\quad + \frac{\Delta^2}{(\Delta r)^2} + \frac{1}{r^2}\left(\operatorname{ctn}\vartheta\frac{\Delta_{\mathrm{s}}}{\Delta\vartheta} + \frac{\Delta^2}{(\Delta\vartheta)^2}\right) + \frac{1}{r^2\sin^2\vartheta}\frac{\Delta^2}{(\Delta\varphi)^2}\Bigg] \\
\Omega_{-1}(r,\vartheta,\varphi) &= -1 + i\frac{e\Delta t}{\hbar}\Phi(r,\vartheta,\varphi)
\end{aligned}
\tag{1}
$$

The operators Ω_1 and Ω_{-1} have the same constant value $+1$ for the Schrödinger equation as shown by Eq.(8.2-4). The variable terms $ie\Delta t\Phi/\hbar$ of Ω_1 and Ω_{-1} in Eq.(1) make it impossible to write the general solution by a superposition of eigenfunctions unless the potential Φ is constant. This feature of the Klein-Gordon equation makes the analytical solution of initial value problems very difficult. Numerical solutions can be obtained in principle from the difference equation by advancing in steps Δt, but a peculiar difficulty is encountered. If one chooses the differences Δr, $\Delta\vartheta$, $\Delta\varphi$, and Δt arbitrarily, one may obtain solutions with an error that increases exponentially with t, regardless of how small the differences Δr, $\Delta\vartheta$, $\Delta\varphi$, and Δt were chosen. Certain conditions between these differences must be satisfied in order to obtain *stable solutions* (Courant et al., 1928; Richtmyer, 1957). These conditions can be calculated if the operators of Eq.(1) have a common system of eigenfunctions (Douglas, 1956). This brings us back to where we started. A stepwise, numerical solution without knowledge of its stability is of little value. Hence, we cannot solve the initial value problem of the operators of Eq.(1) in any satisfying way. Particular solutions of the Klein-Gordon difference equation, on the other hand, can be found by the substitution:

$$\Psi(\mathbf{r},t) = \sum_{n=1}^{\infty} c_n \varphi_n(r,\vartheta,\varphi)\varphi_n(t) \tag{2}$$

Assume the functions $\varphi_n(t)$ to be orthogonal in a certain interval. One obtains from Eqs.(8.4-2), (1), and (2):

$$\varphi_n(t+\Delta t)\varphi_n(r,\vartheta,\varphi) = \varphi_n(t)\Omega_0'(r,\vartheta,\varphi)\varphi_n(r,\vartheta,\varphi) - \varphi_n(t-\Delta t)\varphi_n(r,\vartheta,\varphi)$$
$$\Omega_0'(r,\vartheta,\varphi) = \Omega_0(r,\vartheta,\varphi) - i\frac{e\Delta t}{\Delta r}\Phi(r,\vartheta,\varphi)\frac{\varphi_n(t+\Delta t) - \varphi_n(t-\Delta t)}{\varphi_n(t)} \tag{3}$$

Assume that $[\varphi_n(t+\Delta t) - \varphi_n(t-\Delta t)]/\varphi_n(t)$ is constant:

$$\varphi_n(t+\Delta t) = K\varphi_n(t) + \varphi_n(t-\Delta t) \tag{4}$$

In this case one may determine eigenvalues ω_{0n}:

$$\Omega_0'(r,\vartheta,\varphi)\varphi_n(r,\vartheta,\varphi) = \omega_{0n}\varphi_n(r,\vartheta,\varphi) \tag{5}$$

Equation (3) is then reduced to an ordinary difference equation with constant coefficients:

$$\varphi_n(t+\Delta t) = \omega_{0n}\varphi_n(t) - \varphi(t-\Delta t) \tag{6}$$

Equation (6) is equal to Eq.(8.1-6), but Eq.(6) only applies if the assumption of Eq.(4) is satisfied. There may be other solutions of Eq.(3) for which the assumption of Eq.(4) is not satisfied and which are eliminated by this assumption.

From Eqs.(4) and (6) follows the relation $K = \omega_{0n}^2 - 4$. Equation (6) has two solutions:

$$\varphi_{1n}(t) = b_{1n}Q_{1n}^{\tau} = b_{1n}e^{iEt/\hbar}$$
$$\varphi_{2n}(t) = b_{2n}Q_{2n}^{\tau} = b_{2n}e^{-iEt/\hbar}$$
$$Q_{1n} = e^{iE\Delta t/\hbar} = \frac{1}{2}\omega_{0n} + \left(\frac{1}{4}\omega_{0n} - 1\right)^{1/2}$$
$$Q_{2n} = e^{-iE\Delta t/\hbar} = \frac{1}{2}\omega_{0n} - \left(\frac{1}{4}\omega_{0n} - 1\right)^{1/2}, \qquad \tau = \frac{t}{\Delta t} \tag{7}$$

It follows from Eq.(8.1-13) that ω_{0n} must be real for stationary solutions. The independence condition of Eq.(8.1-11) yields:

$$-2 - \left|O(\Delta t)^2\right| \le \omega_{0n} \le +2 + \left|O(\Delta t)^2\right| \tag{8}$$

Equation (5) may be transformed from a partial difference equation with three variables into three ordinary difference equations by the substitution:

$$\varphi_n(r,\vartheta,\varphi) = \varphi_n(r)\psi_n(\vartheta)\chi_n(\varphi)$$

The equations for $\chi_n(\varphi)$ and $\psi_n(\vartheta)$ are the same as Eqs.(9.1-5) and (9.1-6). The equation for $\varphi_n(r)$ has the following form:

$$(c\Delta t)^2\left[\frac{\Delta^2}{(\Delta r)^2} + \frac{2}{r}\frac{\Delta_s}{\Delta r} + \frac{2}{(c\Delta t)^2} + \left(\frac{e}{\hbar c}\right)^2\Phi^2(r) - \left(\frac{mc}{\hbar}\right)^2 \right.$$
$$\left. + \frac{\lambda}{r^2} - \frac{\omega_{0n}}{(c\Delta t)^2} + \frac{2e}{c^2\Delta t\hbar}\Phi(r)\sin\frac{E\Delta t}{\hbar}\right]\varphi_n(r) = 0 \tag{9}$$

10.2 Discrete Eigenvalues of Bosons in a Coulomb Field

Equation (10.1-9) assumes for a boson with charge e in a field $\Phi(r) = -Ze/r^2$ the following form, if the symmetric difference quotient $\Delta_s/\Delta r$ is used as before:

$$R(R+1)\varphi_n(R+1) + [gR^2 + 2b\Delta rR - l(l+1) + Z^2\alpha^2]\varphi_n(R)$$
$$+R(R-1)\varphi_n(R-1) = 0$$
$$g = -2 - \kappa^2(\Delta r)^2 = -2 - \left(\frac{m^2c^2}{\hbar^2} - \frac{E^2}{\hbar^2c^2}\right)(\Delta r)^2, \quad \alpha = \frac{e^2}{\hbar c}$$
$$\kappa^2 = \left(\frac{mc}{\hbar}\right)^2 + \frac{\omega_{0n} - 2}{(c\Delta t)^2}, \quad b = -\frac{Z\alpha}{c\Delta t}\sin\frac{E\Delta t}{\hbar} \approx -\frac{Z\alpha E}{\hbar c}, \quad R = \frac{r}{\Delta r}$$
$$E = mc^2\left(1 - \frac{\kappa^2\hbar^2}{m^2c^2}\right)^{1/2} + O(\Delta r, \Delta t), \quad \varphi_n(R) = \varphi_n(r) \tag{1}$$

This equation is the same as Eq.(9.2-2) for a nonrelativistic particle except for the term $Z^2\alpha^2$.

If a solution of Eq.(1) is known that converges for $R > R_0$, one can continue it to the left into the region $R \leq R_0$ by rewriting Eq.(1):

$$\begin{aligned}\varphi_n(R) = -R^{-1}(R+1)^{-1}\{[g(R+1)^2 + 2b\Delta r(R+1) - l(l+1) + Z^2\alpha^2] \\ \times \varphi_n(R+1) + (R+1)(R+2)\varphi_n(R+2)\}\end{aligned} \quad (2)$$

This equation has the singular points $R = 0$ and $R = -1$. Similarly, a solution that converges for $R < R_0$ may be continued to the right into the region $R \geq R_0$:

$$\begin{aligned}\varphi_n(R) = -R^{-1}(R-1)^{-1}\{[g(R-1)^2 + 2b\Delta r(R-1) - l(l+1) + Z^2\alpha^2] \\ \times \varphi(R-1) + (R-1)(R-2)\varphi_n(R-2)\}\end{aligned} \quad (3)$$

The singular points of Eq.(3) are $R = 0$ and $R = +1$. The position of the singular points depends on the coefficients of $\varphi_n(R+1)$ and $\varphi_n(R-1)$ only; it is independent of the potential $\Phi(r)$ in Eq.(10.1-9).

Solutions $\varphi_n(R)$ that converge for $R > R_0$ or $R < R_0$ may be found again with the help of the Laplace transform as for the nonrelativistic Schrödinger equation in Section 9.2. The continuation of the solution beyond the abscissa of convergence of a factorial series was of no concern there, since the factorial series terminated and became a polynomial.

The following differential equation is obtained from Eq.(1) by means of the Laplace transform of Eq.(9.2-3):

$$(s^2 + gs + 1)sw'' + (2s + g - 2b\Delta r)sw' - [l(l+1) + Z^2\alpha^2]w = 0 \quad (4)$$

The singular points of this equation in the finite plane are determined by the roots of the term $(s^2 + gs + 1)s$:

$$\begin{aligned}s_1 &= 0 \\ s_2 &= -\frac{1}{2}g + \left(\frac{1}{4}g^2 - 1\right)^{1/2} \\ s_3 &= -\frac{1}{2}g - \left(\frac{1}{4}g^2 - 1\right)^{1/2} = s_2^{-1}\end{aligned} \quad (5)$$

Since ω_{0n} is real, it follows from Eq.(1) that κ^2 and g are real. Let g vary from $-\infty$ to $+\infty$. The points s_2 and s_3 move then along the solid and dashed-dotted lines in Fig.9.2-1a.

The series expansion

$$w(s) = \sum_{\nu=0}^{\infty} q_\nu (s - s_3)^{p+\nu} \tag{6}$$

yields, in addition to $p = 0$, the initial power

$$p = \frac{2b\Delta r}{s_3 - s_2} \tag{7}$$

and the recursion formula:

$$\begin{aligned}
\alpha_{\nu,\nu-1} q_{\nu-1} + \alpha_{\nu,\nu} q_\nu &+ \alpha_{\nu,\nu+1} q_{\nu+1} = 0 \\
\alpha_{\nu,\nu-1} &= (p+\nu-1)(p+\nu) \\
\alpha_{\nu,\nu+1} &= s_3(s_3 - s_2)(\nu+1)(p+\nu+1) \\
\alpha_{\nu,\nu} &= (2s_3 - s_2)(p+\nu)(p+\nu-1) + (3s_3 - s_2 - 2b\Delta r)(p+\nu) \\
&\qquad - l(l+1) + Z^2\alpha^2 \\
&\qquad s_2 \neq s_3
\end{aligned} \tag{8}$$

Consider the values of s_3 on the positive real axis, $0 < s_3 < 1$. One obtains for $\varphi_n(R)$ an asymptotically convergent factorial series like Eq.(9.2-16):

$$\varphi_n(R) = s_3^R \frac{\Gamma(R)}{\Gamma(R+p+1)} \sum_{\nu=0}^{\infty} (-1)^\nu q_\nu \frac{(p+1)\cdots(p+\nu)}{(R+p+1)\cdots(R+p+\nu)} \tag{9}$$

Small values of Δr and Δt yield:

$$\begin{aligned}
s_2 &\approx 1 + |\kappa|\Delta r \approx e^{|\kappa|\Delta r} \\
s_3 &\approx 1 - |\kappa|\Delta r \approx e^{-|\kappa|\Delta r} \\
p \approx p_0 &= \frac{Z\alpha mc}{|\kappa|\hbar}\left(1 - \frac{\kappa^2\hbar^2}{m^2c^2}\right)^{1/2}
\end{aligned} \tag{10}$$

An attempt to terminate the series of Eq.(9) by choosing negative integers for p fails, since the determinant corresponding to Eq.(9.2-23) does not vanish this time. One may verify this statement by substituting s_3 for s_2 in Eq.(12.6-19) in Section 12.6. The following values have to be chosen for the coefficients of the resulting equation:

$$d_{10} = d_{11} = -d_{-11} = 1, \quad d_{02} = -l(l+1) + Z^2\alpha^2, \quad d_{12} = d_{-12} = 0$$
$$\text{not needed: } d_{-10} = 1, \quad d_{01} = 2b\Delta r$$

For small values of Δr one obtains the equation:

$$(p+\nu)(p+\nu+1) - l(l+1) + Z^2\alpha^2 = 0$$

One cannot satisfy this equation with integer values of p. If Δr is not assumed to be vanishing small, one can terminate the factorial series of Eq.(9), but the resulting equation—corresponding to Eq.(9.2-23)—then yields certain values for Δr. Such solutions are against the philosophy of this investigation, since we are trying to find solutions that become independent of Δr for small values of Δr. Not surprisingly, these solutions for certain values of Δr are not in any way similar to those known from the differential equation. For instance, the condition $p = -n$ yields the eigenvalues

$$E = mc^2 \left(1 - \frac{Z^2\alpha^2}{n^2 + Z^2\alpha^2}\right)^{1/2}$$

for the energy, which are different from those of the differential Klein-Gordon equation. Hence, we will not consider the polynomial solutions any further, but will try to find solutions in the form of factorial series whose terms of order $O(1)$ converge for sufficiently small values of Δr in the half plane $R \geq 0$.

As a first step, let us look for the conditions that make the factorial series of Eq.(9) approach asymptotically a polynomial. To this end we expand $\alpha_{\nu,\nu-1}$, $\alpha_{\nu,\nu}$, and $\alpha_{\nu,\nu+1}$ of Eq.(8) in powers of Δr and Δt:

$$\begin{aligned}
\alpha_{\nu,\nu-1} &= (p_0+\nu-1)(p_0+\nu) + O(\Delta r, \Delta t)\\
\alpha_{\nu,\nu} &= (p_0+\nu)(p_0+\nu+1) - l(l+1) + Z^2\alpha^2 + O(\Delta r, \Delta t)\\
\alpha_{\nu,\nu+1} &= -2\kappa\Delta r(\nu+1)(p_0+\nu+1)[1 + O(\Delta r, \Delta t)]
\end{aligned} \tag{11}$$

The symbol $O(\Delta r, \Delta t)$ stands for terms that vanish proportionately either to Δr or Δt. We call these terms small if the conditions

$$|\kappa|\Delta r \ll 1, \quad |\kappa|c\Delta t \ll 1 \tag{12}$$

are satisfied.

The recursion formula of Eq.(8) may be rewritten as a system of linear equations just like Eq.(9.2-22). Cramer's rule yields the coefficient $q_{\mu+1}$:

$$\frac{(-1)^{\mu+1}}{\prod_{\nu=0}^{\mu}\alpha_{\nu,\nu+1}}\begin{vmatrix} \alpha_{00} & \alpha_{01} & & & & & \\ \alpha_{10} & \alpha_{11} & \alpha_{12} & & & & \\ \vdots & & & \ddots & & & \\ & \alpha_{\nu,\nu-1} & & \alpha_{\nu,\nu} & & \alpha_{\nu,\nu+1} & \\ \vdots & & & & & \ddots & \\ & & & \alpha_{\mu-1,\mu-2} & \alpha_{\mu-1,\mu-1} & & \alpha_{\mu-1,\mu} \\ & & & & \alpha_{\mu,\mu-1} & & \alpha_{\mu,\mu} \end{vmatrix} = \frac{q_{\mu+1}}{q_0} \tag{13}$$

The terms $\alpha_{\nu,\nu+1}$ are of the order $O(\Delta r)$. Hence, one may readily compute the determinant in first approximation:

$$\frac{q_{\mu+1}}{q_0} = \frac{(-1)^{\mu+1}\prod_{\nu=0}^{\mu}[(p_0+\nu)(p_0+\nu+1)-l(l+1)+Z^2\alpha^2]+O(\Delta r,\Delta t)}{\prod_{\nu=0}^{\mu}\alpha_{\nu,\nu+1}} \tag{14}$$

Let the last factor of the product of Eq.(14) be zero:

$$(p_0+\mu)(p_0+\mu+1)-l(l+1)+Z^2\alpha^2=0 \tag{15}$$

The $\mu+1$ coefficients $q_0, \Delta r q_1, \ldots, (\Delta r)^\mu q_\mu$ are then of order $O(1)$, while the coefficients $(\Delta r)^\nu q_\nu$ for $\nu > \mu$ are of order $O(\Delta r, \Delta t)$. In this case, the factorial series of Eq.(9) looks like a polynomial for small values of Δr and Δt, provided one considers a finite number of terms only. The series still diverges, no matter how small Δr and Δt are.

The change of notation

$$\mu = n - l - 1 \tag{16}$$

yields with the help of Eq.(15):

$$p_0 = -n + l + \frac{1}{2} \pm \left[\left(l + \frac{1}{2}\right)^2 - Z^2\alpha^2\right]^{1/2}$$
$$p_{01} = -n + l + \frac{1}{2} - \left[\left(l + \frac{1}{2}\right)^2 - Z^2\alpha^2\right]^{1/2} \approx -n + \frac{Z^2\alpha^2}{2l+1}$$
$$p_{02} = -n + l + \frac{1}{2} + \left[\left(l + \frac{1}{2}\right)^2 - Z^2\alpha^2\right]^{1/2} \approx -n' - \frac{Z^2\alpha^2}{2l+1}$$
$$n' = n - 2l - 1 \tag{17}$$

Let μ run from 0 to $n-1$. The index l then assumes the values

$$l = 0,\, 1, \ldots,\, n-1 \tag{18}$$

The coefficients q_0, $\Delta r q_1, \ldots, (\Delta r)^{n-l-1} q_{n-l-1}$ are of the order $O(1)$. One obtains from Eqs.(1), (10), and (17):

$$\kappa = \left|\frac{mc}{\hbar}\left(1 + \frac{p_0^2}{Z^2\alpha^2}\right)^{-1/2}\right|$$
$$\kappa_1 \approx \frac{Z\alpha mc}{n\hbar}\left[1 + \frac{Z^2\alpha^2}{n^2}\left(\frac{n}{2l+1} - \frac{1}{2}\right)\right]$$
$$\kappa_2 \approx \left|\frac{Z\alpha mc}{n'\hbar}\left[1 - \frac{Z^2\alpha^2}{n'^2}\left(\frac{n'}{2l+1} + \frac{1}{2}\right)\right]\right|$$
$$E = \left|mc^2\left(1 + \frac{Z^2\alpha^2}{p_0^2 + Z^2\alpha^2}\right)^{-1/2}\right|$$
$$E_1 \approx mc^2\left[1 - \frac{Z^2\alpha^2}{2n^2} - \frac{Z^4\alpha^4}{n^4}\left(\frac{n}{2l+1} - \frac{3}{8}\right)\right]$$
$$E_2 \approx mc^2\left[1 - \frac{Z^2\alpha^2}{2n'^2} + \frac{Z^4\alpha^4}{n'^4}\left(\frac{n'}{2l+1} + \frac{3}{8}\right)\right]$$
$$n' = n - 2l - 1 \tag{19}$$

The values shown in Eq.(19) for κ and E are the same as obtained from the differential Klein-Gordon equation. The negative sign of the square roots and thus the value κ_2 and E_2 are usually not considered, in order to avoid poles in the eigenfunctions of the differential equation.

10.3 Asymptotic Solution

The recursion formula of Eq.(10.2-8) yields the coefficients $q_1, q_2, \ldots$:

$$\frac{q_1}{q_0} = \frac{1}{\kappa \Delta r}\left(\frac{p_0(p_0+1) - l(l+1) + Z^2\alpha^2}{2(p_0+1)} + O(\Delta r, \Delta t)\right)$$
$$\frac{q_2}{q_0} = \frac{1}{(\kappa \Delta r)^2}\left(\frac{(p_0+1)(p_0+2) - l(l+1) + Z^2\alpha^2}{4(p_0+2)} \right.$$
$$\left. \times \frac{p_0(p_0+1) - l(l+1) + Z^2\alpha^2}{2(p_0+1)} + O(\Delta r, \Delta t)\right) \tag{1}$$

With the help of the following relations holding for the gamma function

$$\Gamma(R) = (R-1)\cdots(R-n+l+1)\Gamma(R-n+l+1)$$
$$\lim_{R\to\infty} \frac{\Gamma(R-n+l+1)}{\Gamma(R+p_0+1)} = R^{-(p_0+n-l)}$$
$$s_3^{-R} \approx e^{-\kappa r} \tag{2}$$

one may rewrite the factorial series of Eq.(10.2-9) as follows:

$$\varphi_n(r) = q_0' e^{-\kappa r} r^{u+n-l-1}\left(1 - \frac{p_0(p_0+1) - l(l+1) + Z^2\alpha^2}{2\kappa r} + \ldots\right)$$
$$u = -\frac{1}{2} \mp \left[\left(l+\frac{1}{2}\right)^2 - Z^2\alpha^2\right]^{1/2} \tag{3}$$

Equation (3) contains a polynomial with $n-l$ terms of order $O(1)$ and an infinite series of terms of order $O(\Delta r, \Delta t)$, which have been left out. The polynomial is the same as derived from the differential Klein-Gordon equation. It may be used as approximation for the factorial series of Eq.(10.2-9) for $r \neq 0$, since $R = r/\Delta r$ may be made arbitrarily large for sufficiently small values of Δr in this case. Nothing can be said so far about the accuracy of this approximation beyond the fact that the series is asymptotically convergent. The polynomial cannot be used for $R = r = 0$. In the differential theory one permits the solution

$$u = -\tfrac{1}{2} + \left[\left(l+\tfrac{1}{2}\right)^2 - Z^2\alpha^2\right]^{1/2}$$

since it causes only a "weak pole" at $r = 0$, while

$$u = -\tfrac{1}{2} - \left[\left(l + \tfrac{1}{2}\right)^2 - Z^2\alpha^2\right]^{1/2}$$

is excluded for causing a "strong pole". We cannot do so here, since a slowly divergent series is still divergent. Both the Schrödinger difference equation (9.2-2) and the Klein-Gordon difference equation (1) have a singularity at $R = r = 0$ regardless of the potentials $V(r)$ or $\Phi(r)$. As a result, the abscissa of convergence of their solutions in the form of a factorial series equals $R = 0$. This assures asymptotic convergence for $R > 0$, but for $R = 0$ one may either have convergence or divergence. In the case of the Schrödinger equation we obtained convergence automatically for the terms of order $O(1)$, but the term $Z^2\alpha^2$ prevents this automatic convergence on the convergence abscissa in the case of the Klein-Gordon equation. The problem is very similar to that encountered with power series. They converge inside their circle of convergence, and they diverge outside their circle of convergence, but they may or may not converge on all or part of the circle of convergence. Hence, we must try to find a solution of the Klein-Gordon difference equation that converges—at least asymptotically—on the abscissa of convergence $R_0 = 0$. We will be only partly successful in this endeavor, since we will be able to show how to go about it, but the resulting series converges so slowly that we will not be able to derive final conclusions. However, the problem is reduced to a clear-cut problem of mathematics, and we may thus hope that a mathematician will be able to solve it.

Let us reflect once more on the difference in the solution of the Schrödinger and the Klein-Gordon equation. A factorial series

$$s(R) = \sum_{\nu=0}^{\infty} q_\nu \frac{(p+1)\cdots(p+\nu)}{(R+p+1)\cdots(R+p+\nu)} \tag{4}$$

can be terminated either by a negative integer value $p = -n$ or by a vanishing coefficient $q_\nu = q_n = 0$; in the second case, the recursion formula for the coefficients q_ν must have two terms only. The factorial series of the Schrödinger equation was terminated for terms of order $O(1)$ by the choice $p = -n$ in Eq.(9.2-17), while the factorial series of the Klein-Gordon equation was terminated by the choice $q_{\mu+1} = 0$ implied by Eq.(10.2-15). The difficulties we are encountering with the solution of the Klein-Gordon equation flows from this different termination of the factorial series.

10.4 Convergent Solution

The power series of Eq.(10.2-6) converges inside a circle around the point s_3 which runs through s_2 for small values of Δr. One may readily recognize this from Fig.9.2-1a. The power series converges absolutely inside a somewhat smaller circle. This smaller circle may be transformed by conformal mapping into a loop around s_3, which runs through $s = 0$. The point s_2 will be outside this loop. The transformed series of Eq.(10.2-6) converges absolutely and uniformly along a line of integration ℓ that is inside this loop except at the point $s = 0$, and that runs from $s = 0$ around s_3 and back to $s = 0$. One may thus integrate term by term and one obtains a factorial series that converges for $R > R_0$, where R_0 is the still to be determined abscissa of convergence.

The transformation $\zeta = (s/s_3)^P$, where P is real, maps the point s_3 of the $s-$plane into the point 1, 0 of the $\zeta-$plane, and the point s_2 into the point s_2^{2P}. The point s_2^{2P} is outside the circle $|\zeta - 1| = 1$ around the point 1, 0 if the condition

$$P > \frac{\ln 2}{2 \ln s_2} \tag{1}$$

is satisfied. We had previously used the relation $R = r/\Delta r$ to connect the normalized variable R with a non-normalized distance r. In analogy we write now:

$$P = \frac{\rho}{\Delta r} \tag{2}$$

One may then rewrite Eq.(1) with the help of Eq.(10.2-10) in the form

$$\kappa\rho > \frac{1}{2} \ln 2 \approx 0.347 \tag{3}$$

We see that the conformal mapping introduces a length ρ that must be larger than zero for a finite value of the wavenumber κ.

The inverse transformation

$$s = s_3 \zeta^{1/P} \tag{4}$$

maps the circle $|\zeta - 1| = 1$ into a loop in the $s-$plane. In order to get some idea how this loop looks, let us substitute the following:

$$\zeta = r_1 e^{i\chi_1}, \quad s = r_2 e^{i\chi_2}, \quad s_3 = r_3 e^{i\chi_3}$$

One obtains:

$$r_2 = r_3 r_1^{1/P}, \quad \chi_2 - \chi_3 = \frac{\chi_1}{P}$$

Let χ_1 run from $-\pi/2$ to $+\pi/2$, and r_1 from 0 to 2 and back to 0. The difference $\chi_2 - \chi_3$ then runs from $-\pi/2P$ to $+\pi/2P$, and r_2 runs from 0 to

$2^{1/P}r_3$ and back to 0. The term χ_3 equals zero if s_3 is located on the real axis.

The power series in the point s_2 may be transformed in the same manner as the one in point s_3. A complex number P must be used if s_2 is located on the positive real axis. This solution varies for large values of r like $e^{+\kappa r}$, as one may recognize from Eqs.(10.2-9) and (10.2-10), and it will not be considered here.

The transformation of Eq.(4) brings the power series of Eq.(10.2-6) into the following form:

$$w(s) = w(\zeta^{1/P}) = (\zeta^{1/P}-1)^p \sum_{\nu=0}^{\infty} s_3^{p+\nu} q_\nu (\zeta^{1/P}-1)^\nu = (\zeta-1)^p \sum_{\nu=0}^{\infty} A_\nu (1-\zeta)^\nu \tag{5}$$

The coefficients A_ν can be obtained from the coefficients q_ν by purely algebraic operations. One expands $1-\zeta^{1/P}$ into a binomial series:

$$1-\zeta^{1/P} = 1-[1-(1-\zeta)]^{1/P} = \frac{1-\zeta}{P} + \frac{P-1}{2!}\left(\frac{1-\zeta}{P}\right)^2 + \cdots = XP^{-1}(1-\zeta) \tag{6}$$

This series converges absolutely for $-1 < 1-\zeta < +1$. For $1-\zeta = +1$ the series converges only if Δr is not zero, $\Delta r \neq 0$. The expression

$$\left(\frac{\zeta^{1/P}-1}{\zeta-1}\right)^p = P^{-p}X^p = P^{-p}\left[1-\left(1-\frac{1}{X}\right)\right]^{-p} \tag{7}$$

may also be expanded into a binomial series, which converges absolutely for $|1-1/X| < 1$, and thus certainly for $0 < 1-\zeta \leq +1$ too. Using the expansion

$$X^{-1} = c_0 + c_1(1-\zeta) + c_2(1-\zeta)^2 + \cdots \tag{8}$$

and the relation $XX^{-1} = 1$ one finally obtains $[(\zeta^{1/P}-1)/(\zeta-1)]^p$ as a series expansion in $1-\zeta$. The series of Eq.(8) converges absolutely if Eqs.(6) and (7) converge absolutely. This is certainly the case for $0 < 1-\zeta \leq +1$. The same statements about convergence apply to the series:

$$\left(\frac{\zeta^{1/P}-1}{\zeta-1}\right)^p = P^{-p}\left[1 + d_1(1-\zeta) + d_2(1-\zeta)^2 + \cdots\right] \tag{9}$$

The terms $q_\nu(\zeta^{1/P}-1)^\nu$ also require series expansions:

$$\left(\zeta^{1/P}-1\right)^{\nu}=(-1)^{\nu}P^{-\nu}(1-\zeta)\left[1+e_{\nu 1}(1-\zeta)+e_{\nu 2}(1-\zeta)^2+\cdots\right] \quad (10)$$

These series expansions again converge absolutely for $-1<1-\zeta<+1$. Multiplication of the series of Eqs.(9) and (10), and comparison of the coefficients according to Eq.(5) yields after a few days of computation the coefficients A_ν of the convergent power series:

$$\begin{aligned}
A_0 &= P^{-p}s_3^p q_0 \\
A_1 &= A_0\left(\frac{1}{2}p\frac{P-1}{P}-\frac{1}{P}s_3\frac{q_1}{q_0}\right) \\
A_2 &= A_0\left[\frac{1}{8}p(p-1)\left(\frac{P-1}{P}\right)^2+\frac{1}{6}p\frac{(2P-1)(P-1)}{P^2}\right. \\
&\qquad \left. -\frac{1}{2}(p+1)\frac{P-1}{P^2}s_3\frac{q_1}{q_0}+\frac{1}{P^2}s_3^2\frac{q_2}{q_0}\right]
\end{aligned} \quad (11)$$

The coefficients A_ν for $\nu>2$ are rather lengthy to write. Hence, A_3 and A_4 are only presented for the limit $P\gg 1$, $\kappa\Delta r\ll 1$:

$$\begin{aligned}
A_1 &= A_0\left(\frac{1}{2}p_0-\frac{q_1}{q_0}\right) \\
A_2 &= A_0\left[\frac{1}{8}p_0\left(p_0+\frac{5}{8}\right)-\frac{1}{2}(p_0+1)\frac{q_1}{Pq_0}+\frac{q_2}{P^2q_0}\right] \\
A_3 &= A_0\left[\frac{1}{48}p_0(p_0^2+5p_0+6)-\frac{1}{24}(3p_0^2+11p_0+8)\frac{q_1}{Pq_0}+\frac{1}{2}(p_0+2)\frac{q_2}{P^2q_0}\right. \\
&\qquad \left. -\frac{q_3}{P^3q_0}\right] \\
A_4 &= A_0\left[p_0\left(\frac{p_0^3}{384}+\frac{5p_0^2}{192}+\frac{97p_0}{1152}+\frac{231}{2880}\right)-\frac{1}{48}(p_0^3+8p_0^2+19p_0+12)\frac{q_1}{Pq_0}\right. \\
&\qquad \left. +\left(\frac{1}{8}p_0^2+\frac{17}{24}p_0+\frac{11}{12}\right)\frac{q_2}{P^2q_0}-\frac{1}{2}(p_0+3)\frac{q_3}{P^3q_0}+\frac{q_4}{P^4q_0}\right]
\end{aligned} \quad (12)$$

Let us not forget that we did all this for the contour integral of Eq.(9.2-3). The conformal mapping of Eq.(4) brings it into the following form:

$$\varphi_n(R) = \frac{s_3^R}{2i\pi P} \int_\ell \zeta^{R/P-1} w(\zeta^{1/P})\, d\zeta \tag{13}$$

Having spent so much time on the convergence of the series $w(\zeta^{1/P})$ we may now reap the benefits and integrate term by term. One obtains a factorial series (Nörlund, 1914, 1924; Wallenberg and Guldberg, 1911):

$$\varphi_n\left(\frac{r}{\rho}\right) = e^{-\kappa r} \frac{\Gamma(r/\rho)}{\Gamma(r/\rho + p + 1)} \sum_{\nu=0}^{\infty} A_\nu \frac{(p+1)\cdots(p+\nu)}{(r/\rho + p + 1)\cdots(r/\rho + p + \nu)}$$

$$\frac{r}{\rho} = \frac{R}{P}, \quad s_3^R \approx e^{-\kappa r}, \quad \varphi_n\left(\frac{r}{\rho}\right) = \varphi\left(\frac{R}{P}\right) = \varphi_n(R)$$

$$\frac{A_1}{A_0} = \frac{1}{2}p_0 - \frac{1}{\kappa\rho} \frac{p_0(p_0+1) - l(l+1) + Z^2\alpha^2}{2(p_0+1)} + O(\Delta r, \Delta t)$$

$$\begin{aligned} \frac{A_2}{A_0} &= \frac{1}{8}p_0\left(p_0 + \frac{5}{3}\right) - \frac{1}{\kappa\rho} \frac{(p_0+1)[p_0(p_0+1) - l(l+1) + Z^2\alpha^2]}{4(p_0+1)} \\ &+ \frac{1}{\kappa^2\rho^2} \frac{p_0(p_0+1) - l(l+1) + Z^2\alpha^2}{8(p_0+1)(p_0+2)} \\ &\times [(p_0+1)(p_0+2) - l(l+1) + Z^2\alpha^2] + O(\Delta r, \Delta t) \end{aligned} \tag{14}$$

It is shown in Section 12.8 that this factorial series converges absolutely for $r > 0$, if ρ satisfies the condition of Eq.(3). One may thus separate it into two series, the first containing the terms of order $O(1)$ and the second one the terms of order $O(\Delta r, \Delta t)$. Both series converge. Hence, one may neglect the second series for sufficiently small values of Δr and Δt, provided the first series is not equal to zero. The point $r = 0$ must be excepted since we know nothing about the convergence there.

Using Eq.(10.3-2) one may rewrite the factorial series of Eq.(14) for large values of r,

$$r \gg \rho|p_0 + 1| \approx \rho \left[1 + \frac{Z\alpha mc}{\kappa\hbar}\left(1 - \frac{\kappa^2\hbar^2}{m^2c^2}\right)^{1/2}\right] \tag{15}$$

into the simpler form:

$$\varphi_n\left(\frac{r}{\rho}\right) \approx A_0' e^{-\kappa r} r^{u+n-l-1}\Big(1 - \frac{1}{2\kappa r}\Big[(1+\kappa\rho)p_0(p_0+1)$$
$$-l(l+1) + Z^2\alpha^2\Big] + \dots\Big)$$
$$u = -\frac{1}{2} \mp \left[\left(l+\frac{1}{2}\right)^2 - Z^2\alpha^2\right]^{1/2} \tag{16}$$

The asymptotic series of Eq.(10.3-3) and thus the eigenfunctions of the differential Klein-Gordon equation do not contain the term $\kappa\rho p_0(p_0+1)$ in the brackets but are otherwise identical. The condition $\kappa\rho > 0.347$ of Eq.(3) shows that $\kappa\rho$ cannot be ignored in the sum $1+\kappa\rho$. However, the term in brackets multiplied by $1/2\kappa r$ in Eq.(16) will be small compared with the first summand 1 for

$$r \gg \frac{1}{2\kappa} \tag{17}$$

provided n and thus $p_0(p_0+1) \approx n(n+1)$ as well as $\kappa\rho$ are not too large. Equation (17) yields with the help of Eq.(10.2-19):

$$r \gg \frac{n}{2Z\alpha}\frac{\hbar}{mc} \tag{18}$$

Hence, deviations of the eigenfunctions of the difference equation from those of the differential equation may be expected to show at distances of less than about 100 Compton wavelengths from the center of the Coulomb field. The Compton wavelength equals 8.9×10^{-13} cm for the pions π^+ or π^-, and 2.6×10^{-13} cm for the kaon K^+. Hence, the deviations between the differential theory and the finite differences theory shows up at typical nuclear distances, but not at atomic distances of about[1] 10^{-8} cm. One should be careful not to draw any more conclusions from Eq.(16) since it contains the approximation $r/\rho \gg |p+1|$ of Eq.(15), but $p_0(p_0+1)/2 \approx p(p+1)/2$ must not be small compared to r/ρ. The factorial series of Eq.(14) is free from this approximation, but it is cumbersome to work with.

The factorial series of Eq.(14) converges to the right of the abscissa of convergence $R_0 = 0$, which means at all distances $r > 0$. It *may* also

[1]In Section 2.3, particularly in the two sentences following Eq.(2.3-9), we had concluded that information flow is not a restriction at atomic distances but becomes a problem at nuclear distances. The detailed calculation carried through here yields the same conclusion.

converge *on* the abscissa of convergence, that is for $r = 0$, but a special investigation is required. This investigation has so far not been successful. What has been done is summarized in Section 10.5, and the difficulties encountered will become apparent there. Presumably, the requirement of convergence for $r = 0$ will put some constraint on the mysterious parameter ρ with the dimension of a length. For $\rho = 0$, the convergent solution, Eq.(16), of the difference equation becomes equal to the solution of the differential equation, but this value of ρ is excluded by Eq.(3).

10.5 Convergence at the Origin

One obtains from Eq.(10.4-14) for $r = 0$:

$$\varphi_n(0) = \lim_{r\to 0} \frac{\rho}{r} \frac{1}{\Gamma(p+1)} \sum_{\nu=0}^{\infty} A_\nu \left[1 - \frac{1}{1!}\frac{r}{\rho} \sum_{j=1}^{\nu} \frac{1}{p+j} + \frac{1}{2!}\frac{r^2}{\rho^2}\left(\sum_{j=1}^{\nu}\frac{1}{p+j}\right)^2 + \cdots\right] \tag{1}$$

The following two equations must hold in order to obtain a defined value for $\varphi_n(0)$:

$$\sum_{\nu=0}^{\infty} A_\nu = 0 \tag{2}$$

$$\varphi_n(0) = -\frac{1}{\Gamma(p+1)} \sum_{\nu=0}^{\infty} \left(A_\nu \sum_{j=1}^{\nu} \frac{1}{p+j} \right) \tag{3}$$

The sum $\sum 1/(p+j)$ in Eq.(3) varies for large values of ν like $\ln \nu$. Hence, the terms of Eq.(3) decrease for large values of ν like $A_\nu \ln \nu$, and Eq.(3) calls for a somewhat faster decrease of the coefficients A_ν than Eq.(2) in order to make the sums convergent.

The sum of Eq.(2) may be written in the following form with the help of Eqs.(10.4-5) to (10.4-10):

$$\sum_{\nu=0}^{\infty} A_\nu = \lim_{\zeta\to 0} \sum_{\nu=0}^{\infty} A_\nu (1-\zeta)^\nu$$

$$\sum_{\nu=0}^{\infty} A_\nu = \frac{A_0}{q_0} \lim_{\zeta\to 0}[1 + d_1(1-\zeta) + \cdots]$$
$$\times \sum_{\nu=0}^{\infty} s_3^{p+\nu}(-1)^\nu q_\nu \left(\frac{1-\zeta}{P}\right)^\nu [e_{\nu 0} + e_{\nu 1}(1-\zeta) + \cdots] \quad (4)$$

The following relations hold:

$$\lim_{\zeta\to 0} \sum_{\nu=0}^{\infty} d_\nu (1-\zeta)^\nu = \lim_{\zeta\to 0}\left(\frac{\zeta^{1/P}-1}{\zeta-1}\right)^p = 1$$

$$\lim_{\zeta\to 0} \sum_{\mu=0}^{\infty} (-1)^\nu P^{-\nu} e_{\nu\mu}(1-\zeta)^{\nu+\mu} = \lim_{\zeta\to 0}\left(\zeta^{1/P}-1\right)^\nu = (-1)^\nu$$

$$\lim_{\zeta\to 0} \sum_{\nu=0}^{\infty} s_3^{p+\nu} P^{-\nu}(-1)^\nu q_\nu (1-\zeta)^\nu [e_{\nu 0} + e_{\nu 1}(1-\zeta) + \cdots] =$$
$$= \sum_{\nu=0}^{\infty} (-1)^\nu P^{-\nu} s_3^{p+\nu} q_\nu \quad (5)$$

$$\sum_{\nu=0}^{\infty} A_\nu = \frac{A_0}{q_0} \sum_{\nu=0}^{\infty} (-1)^\nu s_3^{p+\nu} P^{-\nu} q_\nu \quad (6)$$

The term $(-1)^\nu s_3^{p+\nu} P^{-\nu} q_\nu$ assumes the following form for large values of ν according to Eq.(10.2-14):

$$\lim_{\nu\to\infty} (-1)^\nu s_3^{p+\nu} P^{-\nu} q_\nu = (-2\kappa\nu)^{-\nu} \quad (7)$$

Hence, the series of Eq.(5) converges absolutely if the condition[1]

$$2\kappa\rho > 1, \quad \Delta r \neq 0 \quad (8)$$

is satisfied. This condition restricts ρ more than Eq.(10.4-3). The terms of the sum in Eq.(3) have for large values of ν the form $(-2\kappa\rho)^{-\nu} \ln \nu$; hence, the sum is convergent if Eq.(7) is satisfied.

Consider the eigenfunctions for $\mu = 0$ or $l = n-1$ according to Eq.(10.2-16). Only the terms A_0 or q_0 of Eq.(6) have terms of order $O(1)$. As a result, one obtains from Eq.(6):

[1]Note that κ is always positive because of Eq.(10.2-19), and ρ is always positive because of Eq.(10.4-3).

$$q_0 + O(\Delta r) = 0 \tag{9}$$

If one ignores the terms of order $O(\Delta r)$ one obtains only the trivial solution $q_0 = 0$.

Let us next choose $\mu = 1$ in Eq.(10.2-16), which yields $l = n-2$, $n \geq 2$. Equation (6) now assumes the form:

$$q_0 - \frac{\Delta r}{\rho} q_1 + O(\Delta r) = 0 \tag{10}$$

If one ignores $O(\Delta r)$ one obtains with the help of Eq.(10.3-1) and (8):

$$\rho = \frac{1}{\kappa} \frac{p_0(p_0+1) - l(l+1) + Z^2\alpha^2}{2(p_0+1)} > \frac{1}{2\kappa} \tag{11}$$

According to Eq.(10.2-17), p_0 becomes complex for $Z^2\alpha^2 > (l+1/2)^2$. Since ρ must be real, no acceptable solution of Eq.(11) exists. For the range $Z^2\alpha^2 < (l+1/2)^2$ we use the series expansion of $p_0 = p_{01}$ in Eq.(10.2-17):

$$2\kappa\rho = \frac{p_{01}(p_{01}+1) - l(l+1) + Z^2\alpha^2}{p_{01}+1} \tag{12}$$

Substituting $l = n-2$, $p_{01} = -n + Z^2\alpha^2/(2n-3)$ and neglecting the term $Z^4\alpha^4$, one obtains

$$2\kappa\rho = -2 \tag{13}$$

while Eq.(8) calls for $2\kappa\rho > 1$. Hence, again no solution is obtained if the terms $O(\Delta r)$ are ignored. The same value $2\kappa\rho = -2$ is obtained if $p_0 = p_{02} = -n + 2l - 1 - Z^2\alpha^2/(2l+1)$, $l = n-2$, is substituted from Eq.(10.2-16) into Eq.(11).

Let us try one more step and choose $\mu = 2$, $l = n-3$, $n \geq 3$ in Eq.(10.2-17). Equation (6) assumes the form

$$q_0 - \frac{\Delta r}{\rho} q_1 + \left(\frac{\Delta r}{\rho}\right)^2 q_2 + O(\Delta r) = 0 \tag{14}$$

and one obtains by neglecting $O(\Delta r)$:

$$2\kappa\rho = u\left[1 \pm \left(1 - \frac{v}{u}\right)^{1/2}\right]$$

$$u = \frac{p_0(p_0+1) - l(l+1) + Z^2\alpha^2}{2(p_0+1)}$$

$$v = \frac{(p_0+1)(p_0+2) - l(l+1) + Z^2\alpha^2}{p_0+2} \tag{15}$$

For $p_0 = p_{01} = -n + Z^2\alpha^2/(2n-5)$ one obtains:

$$u = \frac{2n-3-2Z^2\alpha^2/(2n-5)}{n-1-Z^2\alpha^2/(2n-5)}, \quad v = -2$$

The right side of the first line of Eq.(15) becomes complex for any finite value of n, and it becomes -2 for $n \to \infty$.

The substitution $p_0 = p_{02} = n - 5 - Z^2\alpha^2/(2n-5)$ yields

$$u = -\frac{2n-7-2Z^2\alpha^2/(2n-5)}{n-4-Z^2\alpha^2/(2n-5)}, \quad v = -2$$

and the right side of Eq.(15) becomes negative for any value of n. A special case occurs for $n = 4$, but in the end one obtains -1 for the right side of the first line of Eq.(15).

Let us interpret these results. Previously, we have always been able to obtain results that became independent of Δr for sufficiently small values of Δr. Putting it differently, for $r \gg \Delta r$ we could ignore the terms of order $O(\Delta r)$, but for $r = 0$ one cannot ignore a finite value of Δr anymore. The parameter ρ must thus be a function of Δr as well as of Z, n, l, and α. Hence, close to the origin $r = 0$ the eigenfunction describes not only the observed system—represented by Z, n, l, and α—but also the coordinate system or observation equipment represented by its resolution Δr. We have here a very lucid example how the use of finite differences Δr leads to results different from those obtained with differentials dr. One may suspect from these preliminary results that the divergences of quantum field theory for $r \to 0$ have something to do with the use of differentials.

A mathematically correct investigation of the conditions under which the sum of Eq.(2) becomes zero requires the inclusion of at least terms of order $O(\Delta r)$ of the coefficients A_ν. Although there is in principle no difficulty in extending the series expansions in Section 10.4 from order $O(1)$ to order $O(\Delta r)$, anyone attempting this extension by brute force will find that this is no easy task. One can handle the problem by computer programming, but even this calls for a transformation to a faster convergent series. What is needed is a good mathematician who can approach the problem with a new idea.

10.6 Free Particles in a Coulomb Field

We turn to the solutions of the differential equation Eq.(10.2-4) for which s_2 and s_3 are located on the circle $|s| = 1$ as shown by Fig.9.2-1b. One obtains for small values of Δr the following relations instead of Eqs.(10.2-1) and (10.2-10):

$$s_2 \approx 1 + i\kappa'\Delta r, \quad s_3 \approx 1 - i\kappa'\Delta r$$

$$E = mc^2 \left(1 + \frac{\kappa'^2\hbar^2}{m^2c^2}\right)^{1/2}, \quad \kappa' = -i\kappa \approx \frac{mc}{\hbar}\left(\frac{E^2}{m^2c^4} - 1\right)^{1/2}$$

$$p = ip' \approx p_0 \approx ip_0' = -i\frac{Z\alpha mc}{\kappa'\hbar}\left(1 + \frac{\kappa'^2\hbar^2}{m^2c^2}\right)^{1/2} \tag{1}$$

No solutions in the form of polynomials can exist, since p is imaginary and the series of Eq.(10.2-9) cannot terminate. There are also no solutions that become polynomials asymptotically, since Eq.(10.2-15) becomes a complex equation whose real and imaginary parts do no vanish simultaneously. Using the relation

$$\lim_{R\to\infty} \frac{\Gamma(R)}{\Gamma(R + ip' + 1)} = \frac{1}{R^{1+ip'}} = \frac{K}{r}e^{-ip'\ln R} \tag{2}$$

one obtains from Eq.(10.2-9) the following asymptotic series in place of Eq.(10.3-3):

$$\varphi_n(r) = q_0' e^{ip_0'\ln R} e^{-i\kappa' r}\frac{1}{r}\left(1 + i\frac{ip_0'(1 + ip_0') - l(l+1) + Z^2\alpha^2}{2\kappa' r} + \cdots\right) \tag{3}$$

The term $\exp(-ip_0'\ln R)$ is a phase factor that depends essentially on the chosen value of Δr. The eigenfunctions $\varphi_n(r,t)$ are at large distances spherical waves according to Eqs.(10.1-7) and (3):

$$\varphi_n(r,t) \approx q_0'' e^{ip_0'\ln R}\frac{1}{r}e^{-i(\kappa' r \pm Et/\hbar)} \tag{4}$$

These waves have the phase velocity v_p and the group velocity v_g:

$$v_p = \frac{E}{\kappa'\hbar} = c\left(1 - \frac{m^2c^4}{E^2}\right)^{-1/2}$$

$$v_g = \frac{1}{\hbar}\frac{dE}{d\kappa'} = c\left(1 - \frac{m^2c^4}{E^2}\right)^{1/2}$$

The solution in the point s_3 yields in essence again Eq.(3) but $\exp(-i\kappa' r)$ is replaced by $\exp(+i\kappa' r)$.

A convergent series is again obtained by conformal mapping according to Eq.(10.4-4). The angle $s_2 0 s_3$ in Fig.9.2-1b must be increased from $2\kappa\Delta r$ to at least $\pi/3$ and no more than $5\pi/3$, in order to make the circle of convergence around s_3 pass through the origin and to leave s_2 outside this circle. One obtains thus the following condition in place of Eq.(10.4-3):

$$\frac{5\pi}{6} + k\pi > \kappa'\rho' > \frac{\pi}{6} + k\pi, \quad k = 0,\ 1, \ldots \tag{5}$$

The value of $\kappa'\rho'$ must now fall into certain intervals that are periodically repeated, while Eq.(10.4-3) required a certain lower bound for $\kappa\rho$. The circle of convergence in the ζ−plane around the point 1, 0 does not reach the origin if $\kappa'\rho'$ is not in one of the intervals $0 < \kappa'\rho' < \pi/6$, $5\pi/6 < \kappa'\rho' < \pi/6 + \pi$, etc. No convergent factorial series exists in this case.

The calculation from here on follows closely the one in Section 10.4 and is just as lengthy. One obtains in essence the factorial series of Eq.(10.4-14) but κ and p are replaced by $i\kappa'$ and ip':

$$\begin{aligned} \varphi_n\left(\frac{r}{\rho'}\right) &= q_0' e^{-i\kappa' r} \frac{\Gamma(r/\rho')}{\Gamma(r/\rho' + ip_0' + 1)} \left[1 + \frac{i}{r/\rho' + ip_0' + 1} \right. \\ &\times \left. \left(\frac{ip_0'(ip_0'+1) - l(l+1) + Z^2\alpha^2}{2\kappa'\rho'} - \frac{1}{2} p_0'(ip_0'+1) \right) + \cdots \right] \end{aligned} \tag{6}$$

Equation (6) deviates for large values of r/ρ' by the term $i\kappa'\rho' p_0'(ip_0'+1)$ from the asymptotic series in Eq.(3):

$$\begin{aligned} \varphi_n\left(\frac{r}{\rho'}\right) &= q_0'' e^{-ip_0' \ln(r/\rho')} e^{-i\kappa' r} \frac{1}{r} \\ &\times \left(1 + \frac{i}{2\kappa' r} \left[(1 + \kappa'\rho') ip_0'(ip_0'+1) - l(l+1) + Z^2\alpha^2 \right] + \cdots \right) \end{aligned} \tag{7}$$

The phase factor $\exp[-ip_0' \ln(r/\rho')]$ has now a defined value, while the factor $\exp(-ip_0' \ln R)$ in Eq.(3) was arbitrary.

The series of Eq.(7) converges absolutely for $r > 0$, since the real series of Eq.(10.4-14) converges absolutely for $r > 0$. For $r = 0$ one has the same kind of difficulties that are discussed in more detail in Section 10.5.

The parameter g becomes larger than $+2$ for $\kappa'^2 > 4/(\Delta r)^2$ as may be seen from Eq.(10.2-1). The point s_3 is then located on the negative real axis in Fig.9.2-1b and $\varphi_n(R)$ grows exponentially with R. The solution in the point s_2, on the other hand, assumes the form of the solution of the bound particle shown in Eqs.(10.3-3) and (10.4-14). The energy E of the boson must be within the limits

$$mc^2 < E < mc^2\left(1+\frac{4\hbar^2}{m^2c^2(\Delta r)^2}\right)^{1/2}$$

in order to obtain spherical waves as solutions. The larger the energy, the smaller must be Δr.

10.7 Independence Relation

Equations (10.1-8) and (10.2-1) yield the independence relation for the bound boson in a Coulomb field:

$$\Delta t \leq 2\frac{\hbar}{mc^2}\left(1-\frac{\kappa^2\hbar^2}{m^2c^2}\right)^{-1/2}, \quad \kappa < \frac{mc}{\hbar} \tag{1}$$

A similar relation holds for the free boson having a certain finite value of the wavenumber κ':

$$\Delta t \leq 2\frac{\hbar}{mc^2}\left(1+\frac{\kappa'^2\hbar^2}{m^2c^2}\right)^{-1/2} \tag{2}$$

This relation does not hold for arbitrary values of κ'. The parameter g must be smaller than $+2$ to obtain spherical waves as eigenfunctions according to Fig.9.2-1b. The substitution $g < 2$ in Eqs.(10.1-8) and (10.2-1) yields the sufficient condition

$$\left(\frac{\Delta r}{c\Delta t}\right)^2 \geq 1+\left(\frac{mc\Delta r}{2\hbar}\right)^2 \tag{3}$$

for the existence of spherical waves for arbitrary values of κ', which become independent of Δt for small values of Δt. This equation is equal to Eq.(8.4-8), except that Δx is replaced by Δr and the term $O(\Delta t)^2$ is left out. The transition from Cartesian to spherical coordinates had thus no effect, as one would demand from an equation with physical meaning.

The relations of Eqs.(1), (2), and (3) apply to the solutions of Eq.(10.1-1) which are composed of eigenfunctions that satisfy Eq.(10.1-6). The general solution of Eq.(10.1-1) with the potential $\Phi = -Ze/r^2$ may be represented by eigenfunctions in the limit $r \to \infty$. Equation (10.1-1) assumes the following form for $r \to \infty$:

$$\begin{aligned}
\Omega_1(r,\vartheta,\varphi) &= 1 \\
\Omega_0(r,\vartheta,\varphi) &= 2 - \left(\frac{mc^2\Delta t}{\hbar}\right)^2 + (c\Delta t)^2 \frac{\Delta^2}{(\Delta r)^2} \\
\Omega_{-1}(r,\vartheta,\varphi) &= -1 \\
\Psi(r,\vartheta,\varphi,t+\Delta t) &= \Omega_0(r,\vartheta,\varphi)\Psi(r,\vartheta,\varphi,t) - \Psi(r,\vartheta,\varphi,t-\Delta t)
\end{aligned} \tag{4}$$

The substitution

$$\Psi(t) = \sum_{n=0}^{\infty} c_n \varphi_n(r,\vartheta,\varphi)\varphi_n(t)$$

$$\Omega_0(r,\vartheta,\varphi)\varphi_n(r,\vartheta,\varphi) = \omega_{0n}\varphi_n(r,\vartheta,\varphi)$$

yields:

$$\left[(c\Delta t)^2 \frac{\Delta^2}{(\Delta r)^2} + 2 - \left(\frac{mc^2\Delta t}{\hbar}\right)^2\right] \varphi_n(r) = \omega_{0n}\varphi_n(r)$$

The eigenfunctions $\varphi_n(r) = \exp(i\kappa r)$ provide the following eigenvalues in analogy to Eq.(8.4-4):

$$\omega_{0n} = 2B = -4\left(\frac{c\Delta t}{\Delta r}\right)^2 \sin^2\frac{\kappa\Delta r}{2} + 2 - \left(\frac{mc^2\Delta t}{\hbar}\right)^2 \tag{5}$$

Following the line of argument that led from Eq.(8.4-4) to Eq.(8.4-8) one obtains the necessary condition for the existence of a general solution $\Psi(r,\vartheta,\varphi,t)$ of Eq.(4), which becomes independent of Δt for small values of Δt:

$$\left(\frac{\Delta r}{c\Delta t}\right)^2 \geq 1 + \left(\frac{mc\Delta r}{2\hbar}\right)^2 \tag{6}$$

We do not know whether this condition is also sufficient for finite values of r.

The nonrelativistic independence relation of the free particle in one-dimensional Cartesian coordinates shown by Eq.(8.2-7) is equal to the relation of Eq.(9.3-3), which holds for three-dimensional spherical coordinates. The differences $\Delta\vartheta$ and $\Delta\varphi$ without dimension are not contained in the independence relation, while the differences Δx, Δy, and Δz appear on an equal footing in the independence relation for three-dimensional Cartesian coordinates in Section 12.5. Similarly, Eq.(6) contains Δr only, while the relativistic independence relation in three-dimensional Cartesian coordinates contains Δx, Δy, and Δz as shown by Eq.(8.4-13).

11 Dirac Difference Equation with Coulomb Field

11.1 Iterated Dirac Equation

The iterated Dirac equation with scalar potential Φ and vector potential $\mathbf{A}$ has the following form:

$$\left[\frac{\partial^2}{\partial t^2} + i\frac{e}{\hbar}\frac{\partial\Phi}{\partial t} + 2i\frac{e}{\hbar}\Phi\frac{\partial}{\partial t} - \frac{e^2}{\hbar^2}\Phi^2 - \frac{1}{\hbar^2}(i\hbar c\,\mathrm{grad} + e\mathbf{A})^2 + \frac{m^2c^4}{\hbar^2}\right.$$
$$\left. - \frac{ec}{\hbar}\boldsymbol{\sigma}'\,\mathrm{curl}\,\mathbf{A} + i\frac{e}{\hbar}\boldsymbol{\alpha}\frac{\partial\mathbf{A}}{\partial t} + i\frac{e}{\hbar}\boldsymbol{\alpha}\mathbf{A}\frac{\partial}{\partial t} + i\frac{ec}{\hbar}\boldsymbol{\alpha}\,\mathrm{grad}\,\Phi\right]\Psi = 0 \quad (1)$$

Let us substitute $\Delta^2/(\Delta t)^2$ for $\partial^2/\partial t^2$ and $\Delta_s/\Delta t$ for $\partial/\partial t$. The differential operators grad and curl are formally replaced by the difference operators Grad and Curl:

$$\left[1 + i\frac{e\Delta t}{\hbar}(\Phi + \boldsymbol{\alpha}\mathbf{A})\right]\Psi(\mathbf{r}, t + \Delta t)$$
$$= \Omega_0\Psi(\mathbf{r}, t) - \left[1 - i\frac{e\Delta t}{\hbar}(\Phi + \boldsymbol{\alpha}\mathbf{A})\right]\Psi(\mathbf{r}, t - \Delta t)$$

$$\Omega_0 = 2 - \left(\frac{mc^2\Delta t}{\hbar}\right)^2 - i\frac{e(\Delta t)^2}{\hbar}\frac{\Delta_s\Phi}{\Delta t} + \left(\frac{e\Delta t}{\hbar}\right)^2\Phi^2 - (\Delta t)^2\left(ic\,\mathrm{Grad} + \frac{e}{\hbar}\mathbf{A}\right)^2$$
$$+ \frac{ec(\Delta t)^2}{\hbar}\boldsymbol{\sigma}'\,\mathrm{Curl}\,\mathbf{A} - i\frac{ec(\Delta t)^2}{\hbar}\boldsymbol{\alpha}\left(\frac{\Delta_s\mathbf{A}}{\Delta t} + c\,\mathrm{Grad}\,\Phi\right) \quad (2)$$

The vector $\mathbf{r}$ stands for any spatial variable. The vectors $\boldsymbol{\sigma}'$ and $\boldsymbol{\alpha}$ have the components σ_x, σ_y, σ_z and α_x, α_y, α_z. The following representation by matrices is used (Schiff, 1968)[1]:

[1] The matrix β will be required in Section 11.2. Note that boldface is used for the matrices $\boldsymbol{\alpha}$, $\boldsymbol{\sigma}'$, and $\boldsymbol{\sigma}$, which have components for x, y, and z, while lightface type is used for matrices like β and σ_x. Lightface α is the fine structure constant, lightface σ is defined in Eq.(15).

$$\sigma_x = \begin{pmatrix} 0 & 1 \\ 1 & 0 \end{pmatrix}, \quad \sigma_y = \begin{pmatrix} 0 & -i \\ i & 0 \end{pmatrix}, \quad \sigma_z = \begin{pmatrix} 1 & 0 \\ 0 & -1 \end{pmatrix}, \quad \mathbf{1} = \begin{pmatrix} 1 & 0 \\ 0 & 1 \end{pmatrix}$$

$$\boldsymbol{\alpha} = \begin{pmatrix} 0 & \boldsymbol{\sigma} \\ \boldsymbol{\sigma} & 0 \end{pmatrix}, \quad \boldsymbol{\sigma}' = \begin{pmatrix} \boldsymbol{\sigma} & 0 \\ 0 & \boldsymbol{\sigma} \end{pmatrix}, \quad \beta = \begin{pmatrix} \mathbf{1} & 0 \\ 0 & \mathbf{1} \end{pmatrix} \tag{3}$$

The terms $ie\Delta t\hbar^{-1}(\Phi+\boldsymbol{\alpha}\mathbf{A})$ in Eq.(2) in front of $\Psi(\mathbf{r}, t+\Delta t)$ and $\Psi(\mathbf{r}, t-\Delta t)$ cause the same difficulty as in the case of the Klein-Gordon equation. One must make the substitution

$$\Psi(\mathbf{r}, t) = \sum_{n=0}^{\infty} c_n \psi_n(\mathbf{r}) \psi_n(t)$$

and one obtains in analogy to Eqs.(10.1-3) to (10.1-7) the following equations:

$$\begin{aligned} \Omega_0' \psi_n(\mathbf{r}) &= \omega_{0n} \psi_n(\mathbf{r}) \\ \Omega_0' &= \Omega_0 - i(\omega_{0n}^2 - 4)\frac{e\Delta t}{\hbar}\Phi \end{aligned} \tag{4}$$

$$\begin{aligned} \psi_n(t+\Delta t) &= \omega_{0n}\psi_n(t) - \psi_n(t-\Delta t) \\ \psi_{1n}(t) &= b_{1n}e^{iEt/\hbar}, \quad \psi_{2n}(t) = b_{2n}e^{-iEt/\hbar} \\ \sin^2\frac{E\Delta t}{\hbar} &= 1 - \frac{1}{4}\omega_{0n}^2 \end{aligned} \tag{5}$$

Let Φ be centrally symmetric and let us change to spherical coordinates. Neglecting terms of order $O(\Delta r, \Delta\vartheta, \Delta\varphi)$ one obtains the following four equations:

$$\begin{aligned} &\Omega\psi_{jn} - \frac{\alpha}{e}\frac{\Delta_s\Phi(r)}{\Delta' r}\big[i\psi_{5-j,n}\sin\vartheta\cos\varphi - (-1)^j\psi_{5-j,n}\sin\vartheta\sin\varphi \\ &\qquad\qquad - (-1)^j\psi_{j+2,n}\cos\vartheta\big] = 0 \\ &\Omega = \frac{2-\omega_{0n}}{(c\Delta t)^2} - \left(\frac{mc}{\hbar}\right)^2 + \frac{\alpha^2}{e^2}\Phi^2(r) + \frac{2\alpha}{ec\Delta t}\Phi(r)\sin\frac{E\Delta t}{\hbar} + \frac{\Delta^2}{(\Delta r)^2} \\ &\qquad + \frac{2}{r}\frac{\Delta_s}{\Delta r} + \frac{1}{r^2}\left(\operatorname{ctn}\vartheta\frac{\Delta_s}{(\Delta\vartheta)^2}\right) + \frac{1}{r^2}\frac{1}{\sin^2\vartheta}\frac{\Delta^2}{(\Delta\varphi)^2} \\ &j = 1,\ 2,\ 3,\ 4, \quad \psi_{5,n} = \psi_{1,n}, \quad \psi_{6,n} = \psi_{2,n}, \quad \alpha = \frac{e^2}{\hbar c} \end{aligned} \tag{6}$$

Equation (6) distinguishes between the difference Δr, which belongs to the operator Grad in the term $(ic\ \mathrm{Grad}+\mathbf{A}e/\hbar)^2$ in Eq.(2), and the difference $\Delta' r$, which belongs to the operator Grad in the term $(\Delta_s \mathbf{A}/\Delta t + c\ \mathrm{Grad}\,\Phi)$. No such distinction is made in the differential theory, but the equality of Δr and $\Delta' r$ would have to be assumed in a finite difference theory.

Let us neglect terms of the order $O(\Delta\varphi)$. The substitution

$$\psi_{jn}(r,\vartheta,\varphi) = \psi_{jn}(r,\vartheta)e^{im_j\varphi} \tag{7}$$

brings then Eq.(6) into the following form:

$$\Omega'\psi_{jn}(r,\vartheta) - \frac{\alpha}{e}\frac{\Delta_s\Phi(r)}{\Delta' r}\left[i\psi_{5-j,n}(r,\vartheta)\sin\vartheta - (-1)^j\psi_{j+2,n}(r,\vartheta)\cos\theta\right] = 0$$

$$\Omega' = \Omega - \frac{1}{r^2}\left(\frac{1}{\sin^2\vartheta}\frac{\Delta^2}{(\Delta\varphi)^2} - \frac{m_j^2}{\sin^2\vartheta}\right) \tag{8}$$

The variable ϑ is separated out by the substitution

$$\psi_{jn}(r,\vartheta) = \psi_{jn}(r)\psi_{jn}(\vartheta) \tag{9}$$

and one obtains the relations:

$$\begin{aligned}
\psi_{4n}(\vartheta)\sin\vartheta + \psi_{3n}(\vartheta)\cos\vartheta &= \lambda_1\psi_{1n}(\vartheta)\\
\psi_{2n}(\vartheta)\sin\vartheta + \psi_{1n}(\vartheta)\cos\vartheta &= \lambda_3\psi_{3n}(\vartheta)\\
\psi_{3n}(\vartheta)\sin\vartheta - \psi_{4n}(\vartheta)\cos\vartheta &= \lambda_2\psi_{2n}(\vartheta)\\
\psi_{1n}(\vartheta)\sin\vartheta - \psi_{2n}(\vartheta)\cos\vartheta &= \lambda_4\psi_{4n}(\vartheta)
\end{aligned} \tag{10}$$

$$\left(\frac{\Delta^2}{(\Delta\theta)^2} + \mathrm{ctn}\,\vartheta\frac{\Delta_s}{\Delta\vartheta} + \frac{m_j^2}{\sin^2\vartheta}\right)\psi_{jn}(\theta) = l_j(l_j+1)\psi_{jn}(\vartheta) \tag{11}$$

Equation (11) is the difference equation of the finite spherical harmonics. The remaining equation with the variable r only has the following form:

$$\Omega''\psi_{jn}(r) - i\frac{\alpha}{e}\lambda_j\frac{\Delta_s\Phi(r)}{\Delta' r}\psi_{j+2,n}(r) = 0$$

$$\begin{aligned}
\Omega'' = \frac{2-\omega_{0n}}{(c\Delta t)^2} - \left(\frac{mc}{\hbar}\right)^2 + \frac{\alpha^2}{e^2}\Phi^2(r) + \frac{2\alpha}{ec\Delta t}\Phi(r)\sin\frac{E\Delta t}{\hbar}\\
+ \frac{\Delta^2}{(\Delta r)^2} + \frac{2}{r}\frac{\Delta_s}{\Delta r} - \frac{l_j(l_j+1)}{r^2}
\end{aligned} \tag{12}$$

The four relations of Eq.(10) are satisfied by spherical harmonics:

$$\psi_{jn}(\vartheta) = c_{jn} P_{l_j}^{m_j}(\cos\vartheta)$$
$$\psi_{1n}(r) = \psi_{2n}(r), \quad \psi_{3n}(r) = \psi_{4n}(r)$$

One obtains for $l_1 = l_2 = l+1$ and $l_3 = l_4 = l$:

$$c_{1n} = l - m + 1, \quad c_{3n} = i(l+m+1), \quad c_{2n} = 1, \quad c_{4n} = -i$$
$$\lambda_1 = \lambda_2 = i, \quad \lambda_3 = \lambda_4 = -i \tag{13}$$

On the other hand, $l_1 = l_2 = l-1$ and $l_3 = l_4 = l$ yield:

$$c_{1n} = l + m, \quad c_{2n} = -1, \quad c_{3n} = i(l-m), \quad c_{4n} = i$$
$$\lambda_1 = \lambda_2 = i, \quad \lambda_3 = \lambda_4 = -i \tag{14}$$

Equation (11) is satisfied by spherical harmonics, if terms of order $O(\Delta\vartheta)$ are neglected. This is shown in detail in Section 12.7.

Let us substitute a Coulomb potential for $\Phi(r)$:

$$\Phi(r) = -\frac{Ze}{r}, \quad \frac{\Delta_s\Phi(r)}{\Delta' r} = \frac{1}{(\Delta r)^2}\frac{Ze}{R^2-\sigma^2}$$
$$R = \frac{r}{\Delta r}, \quad \sigma = \frac{\Delta' r}{\Delta r} \tag{15}$$

Equation (12) yields for $j = 1$ and the coefficients defined by Eq.(13) the explicit difference equation:

$$R(R+1)(R^2-\sigma^2)\psi_{1n}(R+1) + \Big(gR^4 + 2b\Delta r R^3$$
$$-[\sigma^2 g + (l+1)(l+2) - Z^2\alpha^2]R^2 - 2b\Delta r\sigma^2 R + \sigma^2(l+1)(l+2) - Z^2\alpha^2\sigma^2\Big)\psi_{1n}(R)$$
$$+ R(R-1)(R^2-\sigma^2)\psi_{1n}(R-1) + Z\alpha R^2\psi_{3n}(R) = 0$$
$$\kappa^2 = \left(\frac{mc}{\hbar}\right)^2 + \frac{\omega_{0n}-2}{(c\Delta t)^2} = \left(\frac{mc}{\hbar}\right)^2\left(1 - \frac{E^2}{m^2c^4}\right)$$
$$g = -2 - \kappa^2(\Delta r)^2, \quad b = -\frac{Z\alpha}{c\Delta t}\sin\frac{E\Delta t}{\hbar} \tag{16}$$

The same equation holds for $j = 3$, if one interchanges ψ_{1n} with ψ_{3n} and replaces $(l+1)(l+2)$ by $l(l+1)$.

Let us rewrite the polynomial $R(R+1)(R^2-\sigma^2)$ in Eq.(16):

$$\begin{aligned}R(R+1)(R^2-\sigma^2) &= \\ &= (R+1)(R+2)(R+3)(R+4) - 9(R+1)(R+2)(R+3) \\ &\quad + (19-\sigma^2)(R+1)(R+2) - (8-2\sigma^2)(R+1)\end{aligned}$$

The other two polynomials in Eq.(16) may be rewritten correspondingly. One may then use the Laplace transforms

$$\begin{aligned}\psi_{1n}(R) &= \frac{1}{2i\pi}\int_{\ell} s^{R-1} w_1(s)\,ds \\ \psi_{3n}(R) &= \frac{1}{2i\pi}\int_{\ell} s^{R-1} w_3(s)\,ds \end{aligned} \tag{17}$$

to obtain from Eq.(16) the following differential equation:

$$\begin{aligned}&(s^2+gs+1)s^2 w_1'''' + \left[9s^2 - 2(b\Delta r - 3g)s + 3\right]s^2 w_1''' \\ &+ \Big((19-\sigma^2)s^2 + \left[(7-\sigma^2)g - 6b\Delta r - (l+1)(l+2) + Z^2\alpha^2\right]s + (1-\sigma^2)\Big)s w_1'' \\ &\quad + \left[(8-2\sigma^2)s - (\sigma^2-1)g - 2(1-\sigma^2)b\Delta r - (l+1)(l+2) + Z^2\alpha^2\right]s w_1' \\ &\qquad + \sigma^2\left[(l+1)(l+2) - Z^2\alpha^2\right]w_1 + Z\alpha(s^2 w_3'' + s w_3') = 0\end{aligned} \tag{18}$$

Interchanging w_1 and w_3, and replacing $(l+1)(l+2)$ by $l(l+1)$ yields the differential equation for $j = 3$.

Equation (18) has in the finite plane the following singular points:

$$\begin{gathered}s_1 = 0, \quad s_2 = -\frac{1}{2}g + \left(\frac{1}{4}g^2 - 1\right)^{1/2}, \quad s_3 = -\frac{1}{2}g - \left(\frac{1}{4}g^2 - 1\right)^{1/2} = s_2^{-1} \\ s_2 - s_3 = 3\kappa\Delta r\end{gathered} \tag{19}$$

The two series expansions

$$w_1 = \sum_{\nu=0}^{\infty} a_\nu (s-s_3)^{p+\nu}, \quad w_3 = \sum_{\nu=0}^{\infty} d_\nu (s-s_3)^{p+\nu} \tag{20}$$

are made to solve Eq.(18). Their substitution yields recursion formulas for a_ν and d_ν (The coefficients $\delta_{\nu,\nu-1}$, $\delta_{\nu,\nu}$, and $\delta_{\nu,\nu+1}$ are obtained from the coefficients $\alpha_{\nu,\nu-1}$, $\alpha_{\nu,\nu}$, and $\alpha_{\nu,\nu+1}$ by the substitution of $l(l+1)$ for $(l+1)(l+2)$.):

$$\begin{aligned}
&\alpha_{\nu,\nu-2}d_{\nu-2} + \delta_{\nu,\nu-1}d_{\nu-1} + \delta_{\nu,\nu}d_\nu + \delta_{\nu,\nu+1}d_{\nu+1} + \alpha_{\nu,\nu+2}d_{\nu+2} \\
&\qquad\qquad - \alpha'_{\nu,\nu-1}a_{\nu-1} - \alpha'_{\nu,\nu}a_\nu - \alpha'_{\nu,\nu+1}a_{\nu+1} = 0 \\
&\alpha_{\nu,\nu-2}a_{\nu-2} + \alpha_{\nu,\nu-1}a_{\nu-1} + \alpha_{\nu,\nu}a_\nu + \alpha_{\nu,\nu+1}a_{\nu+1} + \alpha_{\nu,\nu+2}a_{\nu+2} \\
&\qquad\qquad + \alpha'_{\nu,\nu-1}d_{\nu-1} + \alpha'_{\nu,\nu}d_\nu + \alpha'_{\nu,\nu+1}d_{\nu+1} = 0
\end{aligned}$$

$$\begin{aligned}
\alpha_{\nu,\nu-2} &= (p+\nu-2)\big[8 - 2\sigma^2 + (19-\sigma^2)(p+\nu-3) \\
&\quad + 9(p+\nu-3)(p+\nu-4) + (p+\nu-3)(p+\nu-4)(p+\nu-5)\big] \\
\alpha_{\nu,\nu-1} &= \sigma^2\left[(l+1)(l+2) - Z^2\alpha^2\right] + (p+\nu-1)\big\{(16 - 4\sigma^2 s_3^2 \\
&\quad - (\sigma^2-1)g - 2(1-\sigma^2)b\Delta r - (l+1)(l+2) - Z^2\alpha^2 \\
&\quad + \left[(57-3\sigma^2)s_3 + (7-\sigma^2)g - 6b\Delta r - (l+1)(l+2)\right](p+\nu-2) \\
&\quad + (36s_3 - 2b\Delta r + 6g)(p+\nu-2)(p+\nu-3) \\
&\quad + (2s_3 - s_2)(p+\nu-2)(p+\nu-3)(p+\nu-4)\big\} \\
\alpha_{\nu,\nu} &= (p+\nu)\Big((8-2\sigma^2)s_3^2 - \big[(\sigma^2-1)g + 2(1-\sigma^2)b\Delta r \\
&\quad + (l+1)(l+2) - Z^2\alpha^2\big]s_3 + \big\{(57-3\sigma^2)s_3^2 + 2s_3\big[(7-\sigma^2)g \\
&\quad - 6b\Delta r - (l+1)(l+2) + Z^2\alpha^2\big] + 1 - \sigma^2\big\}(p+\nu-1) \\
&\quad + \left[54s_3^2 - 3s_3(2b\Delta r - 6g) + 3\right](p+\nu-1)(p+\nu-2) \\
&\quad + 3s_3(2s_3 - s_2)(p+\nu-1)(p+\nu-2)(p+\nu-3)\Big) \\
\alpha_{\nu,\nu+1} &= (p+\nu+1)(p+\nu)\big\{(19-\sigma^2)s_3^2 + \big[(7-\sigma^2)g - 6b\Delta r \\
&\quad - (l+1)(l+2) + Z^2\alpha^2\big]s_3^2 + (1-\sigma^2)s_3 + \big[36s_3^3 - 3s_3^2(2b\Delta r - 6g) \\
&\quad + 6s_3\big](p+\nu-1) + s_3^2(4s_3 - 3s_2)(p+\nu-1)(p+\nu-2)\big\} \\
\alpha_{\nu,\nu+2} &= (p+\nu+2)(p+\nu+1)(p+\nu)\big[9s_3^4 - 2(b\Delta r - 3g)s_3^3 + 3s_3^2 \\
&\quad + s_3^3(s_3 - s_2)(p+\nu-1)\big]
\end{aligned}$$

$$\begin{aligned}
\alpha'_{\nu,\nu-1} &= Z\alpha(p+\nu-1)^2 \\
\alpha'_{\nu,\nu} &= 2Z\alpha s_3(p+\nu)\left(p+\nu-\frac{1}{2}\right) \\
\alpha'_{\nu,\nu+1} &= Z\alpha s_3(p+\nu)(p+\nu+1) \qquad (21)
\end{aligned}$$

The recursion formulas of Eq.(21) yield for $\nu = -2$ an equation for p:

$$p = \frac{2b\Delta r}{s_3 - s_2} \approx \frac{Z\alpha mc}{\kappa\hbar}\left(1 - \frac{\kappa^2\hbar^2}{m^2c^2}\right)^{1/2} \tag{22}$$

Eigenfunctions in the form of polynomials exist only if $Z^2\alpha^2$ equals zero. Hence, we try again to find asymptotically terminating series, that become equal to the eigenfunctions of the Dirac differential equation at large distances. To this end we write the recursion formulas of Eq.(21) as a system of linear equations:

$$\begin{aligned}
0 &= \alpha_{-10}a_0 + \alpha'_{-10}d_0 + \alpha_{-11}a_1 \\
0 &= -\alpha'_{-10}a_0 + \delta_{-10}d_0 \qquad\qquad + \alpha_{-11}d_1 \\
0 &= \alpha_{00}a_0 \;+ \alpha'_{00}d_0 \;+ \alpha_{01}a_1 \;+ \alpha'_{01}d_1 \;+ \alpha_{02}a_2 \\
&\vdots
\end{aligned} \tag{23}$$

Let a_μ and d_μ equal zero. One has then 2μ equations with 2μ variables, and the determinant of the coefficients must be zero. The terms $\alpha_{\nu,\nu+2}$ are of order $O(\Delta r)$. Hence, one may write in first approximation the determinant of the coefficients as a product of sub-determinants with two rows:

$$D \approx \begin{vmatrix} \alpha_{-10} & \alpha'_{-10} \\ -\alpha'_{-10} & \delta_{-10} \end{vmatrix} \begin{vmatrix} \alpha_{01} & \alpha'_{01} \\ -\alpha'_{01} & \delta_{01} \end{vmatrix} \cdots \begin{vmatrix} \alpha_{\mu-2,\mu-1} & \alpha'_{\mu-3,\mu-1} \\ -\alpha'_{\mu-3,\mu-1} & \delta_{\mu-2,\mu-1} \end{vmatrix} = 0 \tag{24}$$

Equation (24) yields two pairs of values for p, which are independent of the ratio $\sigma = \Delta' r/\Delta r$:

$$\begin{aligned}
p_1 &= -n + l + 1 - \left[(l+1)^2 - Z^2\alpha^2\right]^{1/2} \\
p_2 &= -n + l + 1 + \left[(l+1)^2 - Z^2\alpha^2\right]^{1/2} \\
p_3 &= -n + l - \left[(l+1)^2 - Z^2\alpha^2\right]^{1/2} \\
p_4 &= -n + l + \left[(l+1)^2 - Z^2\alpha^2\right]^{1/2} \\
n &= \mu + 1, \quad \mu = 1,\, 2, \ldots, \quad l = 0,\, 1, \ldots, n-1
\end{aligned} \tag{25}$$

Let us substitute $n' = n + 1/2$. One obtains the following energy eigenvalues from Eqs.(16), (22), and (25):

$$E_1 = mc^2 \left(1 + \frac{Z^2\alpha^2}{\left\{ n' - \frac{1}{2} - l + 1 + [(l+1)^2 - Z^2\alpha^2]^{1/2} \right\}^2} \right)^{-1/2}$$

$$E_2 = mc^2 \left(1 + \frac{Z^2\alpha^2}{\left\{ n' - \frac{1}{2} - l + 1 - [(l+1)^2 - Z^2\alpha^2]^{1/2} \right\}^2} \right)^{-1/2}$$

$$E_3 = mc^2 \left(1 + \frac{Z^2\alpha^2}{\left\{ n' + \frac{1}{2} - l + 1 + [(l+1)^2 - Z^2\alpha^2]^{1/2} \right\}^2} \right)^{-1/2}$$

$$E_4 = mc^2 \left(1 + \frac{Z^2\alpha^2}{\left\{ n' + \frac{1}{2} - l + 1 - [(l+1)^2 - Z^2\alpha^2]^{1/2} \right\}^2} \right)^{-1/2} \tag{26}$$

The energy E_1 is the eigenvalue used in the differential theory; E_2 is excluded in the differential theory due to the negative sign in front of the root in the denominator in order to avoid a "strong pole" for $r = 0$. The eigenvalues E_3 and E_4 do not occur explicitly in the differential theory. Each pair of energy eigenvalues E_1, E_2 and E_3, E_4 defines a system of orthogonal eigenfunctions, because E_3 and E_4 become E_1 and E_2 if n' is replaced by $n' - 1$.

The recursion formula of Eq.(21) is simplified for small values of Δr:

$$\begin{aligned}
\alpha_{\nu,\nu+2}a_{\nu+2} + \alpha_{\nu,\nu+1}a_{\nu+1} - Z\alpha d_{\nu+1} &= 0 \\
\alpha_{\nu,\nu+2}d_{\nu+2} + \delta_{\nu,\nu+1}d_{\nu+1} + Z\alpha a_{\nu+1} &= 0 \\
\alpha_{\nu,\nu+2} &\approx 2\kappa\Delta r(\nu+2)(p+\nu+2) \\
\alpha_{\nu,\nu+1} &\approx (p+\nu+1)(p+\nu+2) - (l+1)(l+2) + Z^2\alpha^2 \\
\delta_{\nu,\nu+1} &\approx (p+\nu+1)(p+\nu+2) - l(l+1) + Z^2\alpha^2
\end{aligned} \tag{27}$$

The condition

$$\begin{vmatrix} \alpha_{\mu-2,\mu-1} & \alpha'_{\mu-2,\mu-1} \\ -\alpha'_{\mu-2,\mu-1} & \delta_{\mu-2,\mu-1} \end{vmatrix} = 0$$

in Eq.(24) means that the two equations

$$\begin{aligned}\alpha_{\mu-2,\mu-1}a_{\mu-1} + \alpha'_{\mu-2,\mu-1}d_{\mu-1} &= 0\\ -\alpha'_{\mu-2,\mu-1}a_{\mu-1} + \delta_{\mu-2,\mu-1}d_{\mu-1} &= 0\end{aligned}$$

are linearly dependent in the terms of order $O(1)$. This yields the relation:

$$d_{\mu-1} = -\frac{\alpha_{\mu-2,\mu-1}a_{\mu-1}}{Z\alpha} \tag{28}$$

One deduces from this relation that there must also be a linear relation between a_0 and d_0. This linear relation is obtained from Eq.(28) by inverting the recursion formulas of Eq.(27):

$$\begin{aligned}(\delta_{\nu-1,\nu}\alpha_{\nu-1,\nu} + Z^2\alpha^2)a_\nu + \delta_{\nu-1,\nu}\alpha_{\nu-1,\nu+1}a_{\nu+1} - Z\alpha\alpha_{\nu-1,\nu+1}d_{\nu+1} &= 0\\ (\delta_{\nu-1,\nu}\alpha_{\nu-1,\nu} + Z^2\alpha^2)d_\nu + \alpha_{\nu-1,\nu}\alpha_{\nu-1,\nu+1}d_{\nu+1} + Z\alpha\alpha_{\nu-1,\nu+1}a_{\nu+1} &= 0\end{aligned} \tag{29}$$

The recursion formulas of Eq.(27) may be separated for a_ν and d_ν:

$$\begin{aligned}a_{\nu+1} &= \frac{1}{\kappa\Delta r}\frac{(p+\nu)^2 - (l+1)^2 + Z^2\alpha^2}{(\nu+1)(p+\nu+1)}\\ &\quad\times\left(a_\nu - \frac{1}{4\kappa\Delta r}\frac{(p+\nu-1)^2 - (l+1)^2 + Z^2\alpha^2}{\nu(p+\nu)}a_{\nu-1}\right)\\ d_{\nu+1} &= \frac{1}{\kappa\Delta r}\frac{(p+\nu)^2 - (l+1)^2 + Z^2\alpha^2}{(\nu+1)(p+\nu+1)}\\ &\quad\times\left(d_\nu - \frac{1}{4\kappa\Delta r}\frac{(p+\nu-1)^2 - (l+1)^2 + Z^2\alpha^2}{\nu(p+\nu)}d_{\nu-1}\right)\end{aligned} \tag{30}$$

The coefficients of the recursion formulas of Eq.(30) become constant for large values of ν. The coefficients a_ν and d_ν become proportionate to:

$$C_1(-2\kappa\Delta r)^{-\nu} + C_2\nu(-2\kappa\Delta r)^{-\nu} \tag{31}$$

The asymptotically convergent eigenfunctions of the iterated Dirac equation are defined by the factorial series of Eq.(10.2-9). The convergent solution is obtained by means of Eqs.(10.4-11) and (10.4-14). In order to find the abscissa of convergence of the factorial series, let us assume that the factorial series of Eq.(10.4-14) converges in the half plane $R > R_0$, and let it be continued by means of the difference equation Eq.(16) to the left. Let

Eq.(16) be solved for $\psi_{1n}(R-1)$ and let then $R-1$ be replaced by R. The right side of the equation contains the denominator $(R+1)R(R+1+\sigma)(R+1-\sigma)$. The singularity of $\psi_{1n}(R)$ located furthest to the right is either $R=0$ or $R=\sigma-1$. The abscissa of convergence is

$$R_0 = 0 \tag{32}$$

for $\sigma \leq 1$. The investigation of the convergence for $R=0$ runs along the lines discussed in Section 10.5. Due to Eq.(31) one obtains:

$$\lim_{\nu\to\infty}(-1)^\nu P^{-\nu} a_\nu = \frac{C_1+\nu C_2}{(2\kappa\rho)^\nu}$$
$$\lim_{\nu\to\infty}(-1)^\nu P^{-\nu} d_\nu = \frac{C_3+\nu C_4}{(2\kappa\rho)^\nu} \tag{33}$$

These relations imply $|2\kappa\rho|>1$ as necessary condition for the convergence at $r=0$. The investigation of the sufficient condition of convergence has been no more successful than for the Klein-Gordon equation. We are again finding a parameter ρ with the dimension of a length, but we can say no more than that it must be larger than $1/2\kappa$.

Let us use the values of Eq.(14) instead of those of Eq.(13). One must replace $(l+1)(l+2)$ by $(l-1)l$ in Eq.(16), while $l(l+1)$ remains unchanged in the equation for $j=3$. One obtains:

$$p_1 = -n+l-(l^2-Z^2\alpha^2)^{1/2}$$
$$p_2 = -n+l+(l^2-Z^2\alpha^2)^{1/2}$$
$$p_3 = -n+l-1-(l^2-Z^2\alpha^2)^{1/2}$$
$$p_4 = -n+l-1+(l^2-Z^2\alpha^2)^{1/2}$$
$$E_1 = mc^2\left(1+\frac{Z^2\alpha^2}{\left[n'-\frac{1}{2}-l+(l^2-Z^2\alpha^2)^{1/2}\right]^2}\right)^{-1/2}$$
$$E_2 = mc^2\left(1+\frac{Z^2\alpha^2}{\left[n'-\frac{1}{2}-l-(l^2-Z^2\alpha^2)^{1/2}\right]^2}\right)^{-1/2}$$
$$E_3 = mc^2\left(1+\frac{Z^2\alpha^2}{\left[n'+\frac{1}{2}-l+(l^2-Z^2\alpha^2)^{1/2}\right]^2}\right)^{-1/2}$$
$$E_4 = mc^2\left(1+\frac{Z^2\alpha^2}{\left[n'+\frac{1}{2}-l-(l^2-Z^2\alpha^2)^{1/2}\right]^2}\right)^{-1/2} \tag{34}$$

These are the same energy eigenvalues as in the differential theory for spin $-1/2$, but one has again two orthogonal systems of eigenfunctions, just like in Eq.(26).

11.2 Linearized Dirac Equation

There are two ways to derive the linearized Dirac difference equations. One may replace the differentials by finite differences in the linearized differential equations, or one may extract the root of the difference operators of second order. Let us try the following equation[1]:

$$(E + c\boldsymbol{\alpha}\mathbf{p} + \beta mc^2)\Psi(\mathbf{r},t) = 0$$

$$E^2 = -\hbar^2 \frac{\Delta^2}{(\Delta t)^2}, \quad p_{x_j}^2 = -\hbar^2 \frac{\Delta^2}{(\Delta x_j)^2}, \quad j = 1, 2, 3 \tag{1}$$

The substitution

$$E = i\hbar\left(\frac{\Delta_s}{\Delta t} + c_1 \Delta t \frac{\Delta^2}{(\Delta t)^2} + c_2(\Delta t)^2 \frac{\Delta^2}{(\Delta t)^2}\frac{\Delta_s}{\Delta t} + \cdots\right)$$

yields the following expression for the root of E^2:

$$E = \pm i\hbar\left(\frac{\Delta_s}{\Delta t} - \frac{1}{8}(\Delta t)^2 \frac{\Delta^2}{(\Delta t)^2}\frac{\Delta_s}{\Delta t} + \frac{3}{128}(\Delta t)^4 \frac{\Delta^4}{(\Delta t)^4}\frac{\Delta_s}{\Delta t} + \cdots\right) \tag{2}$$

This attempt leads to a difference equation of infinite order. If $\Psi(\mathbf{r},t)$ changes slowly with time one can neglect the terms $(\Delta t)^{2i}\Psi(\mathbf{r},t)/(\Delta t)^{2i}$ and only the term $i\hbar\Delta_s/\Delta t$ in Eq.(2) remains. We will return to this simplified, linearized operator of Eq.(2) a little later. The substitutions

$$\begin{aligned} E &= i\hbar\frac{\partial}{\partial t} \to i\hbar\frac{\Delta_s}{\Delta t} \\ p_{x_j} &= -i\hbar\frac{\partial}{\partial x_j} \to -i\hbar\frac{\Delta_s}{\Delta x_j} \end{aligned} \tag{3}$$

lead, on the other hand, to an equation which does not assume the form of the iterated equation Eq.(11.1-2) when squared. For instance, one obtains:

[1] The matrices $\boldsymbol{\alpha}$ and β are defined in Eq.(11.1-3).

$$\left(i\hbar\frac{\Delta_s}{\Delta t}\right)^2 = -i\hbar\left(\frac{\Delta^2}{(\Delta t)^2} + \frac{1}{4}(\Delta t)^2\frac{\Delta^4}{(\Delta t)^4}\right)$$

The eigenfunctions of the iterated equations and the linearized equations with the operators of Eq.(3) can be different even if Δt and Δx_j are arbitrarily small. Let us now return to Eq.(2). The operator E of Eq.(2) becomes equal to the operator E of Eq.(3) if the higher finite derivatives are neglected. The solution of an example for the operators of Eq.(3) will thus show at the same time the effect of neglecting the higher finite derivatives in Eq.(2). We consider a particle in a Coulomb field and substitute $E - e\Phi$ for E in Eq.(1). One obtains:

$$\begin{aligned}
\Psi(\mathbf{r}, t+\Delta t) &= \Omega_0\Psi(\mathbf{r}, t) + \Psi(\mathbf{r}, t-\Delta t) \\
\Omega_0 &= -\frac{2i\Delta t}{\hbar}(e\Phi + ic\hbar\boldsymbol{\alpha}\,\mathrm{Grad} - \beta mc^2) \\
\Psi(\mathbf{r}, t) &= \sum_{n=0}^{\infty} a_n\psi_n(\mathbf{r})\psi_n(t) \\
\psi_n(t+\Delta t) &= \omega_{0n}\psi_n(t) + \psi_n(t-\Delta t) \\
\Omega_0\psi_n(\mathbf{r}) &= \omega_{0n}\psi_n(\mathbf{r}) \\
\psi_{1n}(t) &= b_{1n}e^{iEt/\hbar}, \quad \psi_{2n} = b_{2n}e^{-iEt/\hbar} \\
\omega_{0n} &= 2i\sin\frac{E\Delta t}{\hbar}
\end{aligned} \tag{4}$$

We substitute $\Phi = -Ze/r$, leave out the transition to polar coordinates and the separation of the variables, and obtain two equations with the variable r:

$$\begin{aligned}
\left(\frac{Z\alpha}{r} + i\frac{\omega_{0n}}{2c\Delta t} + \frac{mc}{\hbar}\right)\psi_{1n}(r) + \frac{\Delta_s\psi_{3n}(r)}{\Delta r} - \frac{1}{r}\psi_{3n}(r) &= 0 \\
\left(\frac{Z\alpha}{r} + i\frac{\omega_{0n}}{2c\Delta t} - \frac{mc}{\hbar}\right)\psi_{3n}(r) - \frac{\Delta_s\psi_{1n}(r)}{\Delta r} - \frac{l+2}{r}\psi_{1n}(r) &= 0 \\
l_1 = l_2 = l+1,\ l_3 = l_4 = l &
\end{aligned} \tag{5}$$

Writing the difference operators explicitly yields:

$$\begin{aligned}
R\psi_{3n}(R+1) - 2l\psi_{3n}(R) - R\psi_{3n}(R-1) + (G_1R + 2Z\alpha)\psi_{1n}(R) &= 0 \\
R\psi_{1n}(R+1) + 2(l+2)\psi_{1n}(R) - R\psi_{1n}(R-1) - (G_3R - 2Z\alpha)\psi_{3n}(R) &= 0 \\
G_1 = 2\Delta r\left(i\frac{\omega_{0n}}{2c\Delta t} + \frac{mc}{\hbar}\right), \quad G_3 = 2\Delta r\left(i\frac{\omega_{0n}}{2c\Delta t} - \frac{mc}{\hbar}\right) &
\end{aligned} \tag{6}$$

Application of the Laplace transform brings:

$$\begin{aligned}(s^2-1)sw_3'+(s^2+2ls+1)w_3+G_1w_1's^2-2Z\alpha w_1s=0\\(s^2-1)sw_1'+\left[s^2-2(l+2)s+1\right]w_1-G_3w_3's^2+2Z\alpha w_3s=0\end{aligned}\qquad(7)$$

The series expansions of Eq.(11.1-20) in an arbitrary point s_0 instead of the specific point s_3 yields the following recursion formulas:

$$\begin{aligned}&\alpha_{\nu,\nu-2}a_{\nu-2}+\alpha_{\nu,\nu-1}a_{\nu-1}+\alpha_{\nu,\nu}a_\nu+\alpha_{\nu,\nu+1}a_{\nu+1}\\&\qquad+\alpha'_{\nu,\nu-1}d_{\nu-1}+\alpha'_{\nu,\nu}d_\nu+\alpha'_{\nu,\nu+1}d_{\nu+1}=0\\&\delta_{\nu,\nu-2}d_{\nu-2}+\delta_{\nu,\nu-1}d_{\nu-1}+\delta_{\nu,\nu}d_\nu+\delta_{\nu,\nu+1}d_{\nu+1}\\&\qquad+\delta'_{\nu,\nu-1}a_{\nu-1}+\delta'_{\nu,\nu}a_\nu+\delta'_{\nu,\nu+1}a_{\nu+1}=0\end{aligned}$$

$$\begin{aligned}\alpha_{\nu,\nu-2}&=p+\nu-1 & \alpha'_{\nu,\nu+1}&=-G_3s_0^2(p+\nu+1)\\\alpha_{\nu,\nu-1}&=3s_0(p+\nu-1) & \delta_{\nu,\nu-1}&=2(s_0+l)\\&\quad+2(s_0-l-2) & \delta_{\nu,\nu}&=(3s_0^2-1)(p+\nu)+s_0^2\\\alpha_{\nu,\nu}&=(3s_0^2-1)(p+\nu)+s_0^2 & &\quad+2ls_0+1\\&\quad-2(l+2)s_0+1 & \delta'_{\nu,\nu-1}&=G_1(p+\nu-1)-2Z\alpha\\\alpha_{\nu,\nu+1}&=s_0(s_0^2-1)(p+\nu+1) & \delta'_{\nu,\nu}&=2s_0[G_1(p+\nu)-Z\alpha]\\\alpha'_{\nu,\nu-1}&=-G_3(p+\nu-1)+2Z\alpha & \delta'_{\nu,\nu+1}&=G_1s_0^2(p+\nu+1)\\\alpha'_{\nu,\nu}&=-2s_0[G_3(p+\nu)-Z\alpha]\end{aligned}\qquad(8)$$

One obtains for $\nu=-1$:

$$\begin{aligned}G_1s_0a_0+(s_0^2-1)d_0&=0\\(s_0^2-1)a_0-G_3s_0d_0&=0\\a_0&=\frac{d_0(s_0^2-1)}{G_1s_0}\end{aligned}\qquad(9)$$

The coefficient s_0 must have one of the following four values to obtain a vanishing determinant of the coefficients:

$$\begin{aligned}s_2=-s_4&=\left[1-\frac{1}{2}G_1G_3+i\sqrt{G_1G_3}\left(1-\frac{1}{4}G_1G_3\right)^{1/2}\right]^{1/2}\\s_3=-s_5&=\left[1-\frac{1}{2}G_1G_3-i\sqrt{G_1G_3}\left(1-\frac{1}{4}G_1G_3\right)^{1/2}\right]^{1/2}\end{aligned}\qquad(10)$$

The determinant of the coefficients of the variables a_ν and d_ν in Eq.(8) vanishes for $\nu \geq 0$ due to Eq.(9). Hence, the rank of the extended matrix must also be zero. This condition yields for $\nu = 0$ the value of p,

$$p = iZ\alpha \frac{G_1 + G_3}{2\sqrt{G_1 G_3}} \approx Z\alpha \frac{E}{mc^2}\left[1 - \left(\frac{E}{mc^2}\right)^2\right]^{-1/2} \tag{11}$$

and for $\nu > 0$ a recursion formula, which one may use instead of Eq.(8) for the computation of a_ν and d_ν:

$$\beta_{\nu,\nu} d_\nu + \beta'_{\nu,\nu} a_\nu + \beta_{\nu,\nu-1} d_{\nu-1} - \beta'_{\nu,\nu} a_{\nu-1} + \beta_{\nu,\nu-2} d_{\nu-2} + \beta'_{\nu,\nu-2} a_{\nu-2} = 0$$

$$\begin{aligned}
\beta_{\nu,\nu} &= \alpha_{\nu,\nu+1}\alpha'_{\nu,\nu} - \alpha'_{\nu,\nu+1}\delta_{\nu,\nu} \\
\beta'_{\nu,\nu} &= \alpha_{\nu,\nu+1}\alpha_{\nu,\nu} - \alpha'_{\nu,\nu+1}\delta'_{\nu,\nu} \\
\beta_{\nu,\nu-1} &= \alpha_{\nu,\nu+1}\alpha'_{\nu,\nu-1} - \alpha'_{\nu,\nu+1}\delta_{\nu,\nu-1} \\
\beta'_{\nu,\nu-1} &= \alpha_{\nu,\nu+1}\alpha_{\nu,\nu-1} - \alpha'_{\nu,\nu+1}\delta_{\nu,\nu-1} \\
\beta_{\nu,\nu-2} &= -\alpha'_{\nu,\nu+1}\alpha_{\nu,\nu-2} \\
\beta'_{\nu,\nu-2} &= \alpha_{\nu,\nu+1}\alpha_{\nu,\nu-2}
\end{aligned} \tag{12}$$

There are again no polynomial solutions and we look for an asymptotically terminating series. Let the recursion formulas of Eq.(8) be written as a system of linear equations. The determinant of the coefficients must vanish if a_μ and d_μ are to be zero. This determinant may be written in first approximation in Δr as a product of subdeterminants with two rows. One obtains the condition for asymptotic termination:

$$\begin{vmatrix} \delta_{\mu-1,\mu-1} & \delta'_{\mu-1,\mu-1} \\ \alpha'_{\mu-1,\mu-1} & \alpha_{\mu-1,\mu-1} \end{vmatrix} = 0 \tag{13}$$

Equation (13) yields for $s_2 \approx s_3 \approx 1$,

$$\begin{aligned}
p_1 &= -n + l + 1 - \left[(l+1)^2 - Z^2\alpha^2\right]^{1/2} \\
p_2 &= -n + l + 1 + \left[(l+1)^2 - Z^2\alpha^2\right]^{1/2}
\end{aligned} \tag{14}$$

and for $s_4 \approx s_5 \approx= -1$:

$$\begin{aligned}
p_3 &= -n + l - 1 - \left[(l+1)^2 - Z^2\alpha^2\right]^{1/2} \\
p_4 &= -n + l - 1 + \left[(l+1)^2 - Z^2\alpha^2\right]^{1/2}
\end{aligned} \tag{15}$$

The value of p_1 and p_2 in Eq.(14) is the same as that of p_1 and p_2 in Eq.(11.1-25), but p_3 and p_4 are different. From p_1 and p_2 one obtains the same energy eigenvalues with the help of Eq.(11) as from the differential equations.

If one substitutes $l_1 = l_2 = l - 1$, $l_3 = l_4 = l$ in Eq.(5), one obtains:

$$\begin{aligned} p_1 &= -n + l + 1 - (l^2 - Z^2\alpha^2)^{1/2} \\ p_2 &= -n + l + 1 + (l^2 - Z^2\alpha^2)^{1/2} \\ p_3 &= -n + l - 1 - (l^2 - Z^2\alpha^2)^{1/2} \\ p_4 &= -n + l - 1 + (l^2 - Z^2\alpha^2)^{1/2} \end{aligned} \tag{16}$$

The asymptotically convergent eigenfunctions are represented by a factorial series of the form of Eq.(10.2-9), if q_ν is replaced by a_ν or d_ν. The convergent series is obtained by substituting r/ρ for R and recomputing the coefficients of the series according to Eq.(10.4-11). The eigenfunctions for the points s_3 and s_4 have the following form at large distances:

$$\begin{aligned} \psi_n(r) &\approx e^{-\kappa r}\left(\frac{r}{\rho}\right)^{-p-1}, \quad s_0 = s_3 \\ \psi_n(r) &\approx e^{i\pi r/\Delta r}e^{-\kappa r}\left(\frac{r}{\rho}\right)^{-p-1}, \quad s_0 = s_5 \end{aligned} \tag{17}$$

The eigenfunctions in the point $s_5 \approx -1$ do not become independent of Δr for small values of Δr due to the arbitrary factor $\exp(i\pi r/\Delta r)$. This arbitrary factor is always encountered if the singular point, around which the contour integral is taken, is not of the form $1 + O(\Delta r)$. The eigenfunctions in the point s_3 with p equal to p_1 or p_2 are also only at great distances equal to those of the iterated equation. One may see this by solving the difference equations Eq.(6) for $\psi_{3n}(R-1)$ and $\psi_{1n}(R-1)$. The singular point furthest to the right is $R = -1$, while it was $R = 0$ for the iterated equation according to Eq.(11.1-32), and the abscissas of convergence are thus different.

When discussing the dyadic calculus in Section 6.3 we found that the second derivative $D^2 f(m)/Dm^2$ of Eq.(6.3-21) is obtained by applying twice the difference operator of first order $Df(m)/Dm$ of Eq.(6.3-20), while this was not so for the difference operators $\Delta^2 f(\eta)/\Delta\eta^2$ of Eq.(6.2-12) and $\Delta f(\eta)/\Delta\eta$ of Eq.(6.2-6). Hence, if we substitute in Eq.(1)

$$E^2 = -\hbar^2 \frac{D^2}{Dm^2}, \quad p_{x_j}^2 = -\hbar^2 \frac{D^2}{Ds^2} \tag{18}$$

where m is the time variable and s the spatial variable according to Eq.(6.3-22), we may extract the square root in closer analogy to differential calculus:

$$E = i\hbar \frac{D}{Dm}, \quad p_{x_j} = -i\hbar \frac{D}{Ds} \tag{19}$$

Since the mathematical methods developed so far apply to the difference operators $\Delta_s/\Delta x$ and $\Delta^2/(\Delta x)^2$ on the limit ring $2^N \to \infty$, one must first develop methods of comparable generality for the dyadic operator D/Dm before one can substitute Eq.(19) into Eq.(1), and find eigenvalues and eigensolutions for, e.g. a Coulomb potential. We will not do so here, but end with the observation that the linearized Dirac equations appear to be a good example to test the mettle of dyadic calculus.

12 Mathematical Supplements

12.1 Transmission Capacity

The concept of infinity is often a root for an incorrect application to the physical world of a correct result of abstract mathematics. An example is the transmission capacity of communication channels[1]

$$C = \Delta f \log(1 + P_S/P_N) \tag{1}$$

where Δf is the nominal frequency bandwidth and P_S/P_N the average signal-to-noise power ratio. For the derivation of this formula one assumes infinite transmitted information which is usually measured in bit(s). In this case the type of error made by the incorrect application of a correct but abstract mathematical result is so well known in mathematics that it received its own name *Klein's hotel*[2]. The reason why infinities lead frequently to a wrong application is that "infinite numbers" sometimes are like very large numbers, but sometimes they are very different. In the case of the geometric series, the difference between the sum of infinitely many terms and a finite number of terms converges to zero for a large number of terms, and we may thus use the sum of an infinite number of terms as an approximation. There is no convergence in the case of Klein's hotel, which is a hotel with denumerably many rooms, and one arrives at hilarious results for a real hotel if one ignores this fact. A real hotel cannot accommodate *excess guests* once all rooms have been filled. If Klein's hotel is completely filled with *accommodated guests*, one may still add denumerably many excess guests by shifting the accommodated guests to the rooms with even numbers. This leaves the denumerably many rooms with odd numbers free for the excess guests. The process can be repeated a finite number of times. Hence, one has *error-free reservation* for an accommodated-to-excess (guest) ratio that is finite and larger than zero if the number of rooms is denumerably infinite. However, for a large but finite number of rooms, the accommodated-to-excess ratio

---

[1] Shannon (1948, 1949).
[2] Felix Klein, 1849–1925, German mathematician.

must always be infinite for error-free reservation; it does *not* converge to the finite ratio obtained for an infinite number of rooms. Similarly, one has *error-free transmission* for a *signal-to-noise* (energy) ratio that is finite and larger than zero, if the number of bit(s) is denumerably infinite. However, for a large but finite number of bit(s), the signal-to-noise ratio must always be infinite for error-free transmission; it does *not* converge to the finite ratio obtained for an infinite number of bit(s). One may avoid the infinite guests and energy by introducing a finite guest flow and energy flow or power. Furthermore, one may equip the hotel with a revolving door making Δf revolutions per second to imitate a filter having a frequency bandwidth of Δf cycles per second, and make other improvements to increase the analogy between Klein's hotel and the abstract communication channel. The important point here is that mathematical correctness is only a necessary but not a sufficient condition in an experimental science.

12.2 Distributive law for Dyadic Multiplication

Consider three binary numbers with two digits 1 and 0 each:

$$\xi = \xi_1\xi_0, \ \eta = \eta_1\eta_0, \ \zeta = \zeta_1\zeta_0$$

The product $\xi \otimes (\eta \oplus \zeta)$ can be obtained by first performing the addition modulo 2 and then the dyadic multiplication according to Box 4.3-1:

$$\begin{array}{lllll}
 & \eta_1 & \eta_0 & & \\
 & \oplus\,\zeta_1 & \zeta_0 & & \\
\hline
 & (\eta_1 \oplus \zeta_1) & (\eta_0 \oplus \zeta_0) & \otimes\xi_1\xi_0 & \\
\hline
 & \xi_1(\eta_1 \oplus \zeta_1) & \xi_1(\eta_0 \oplus \zeta_0) & & \\
\oplus & & \xi_0(\eta_1 \oplus \zeta_1) & \xi_0(\eta_0 \oplus \zeta_0) & \\
\hline
 & \xi_1(\eta_1 \oplus \zeta_1) & \xi_1(\eta_0 \oplus \zeta_0) \oplus \xi_0(\eta_1 \oplus \zeta_1) & \xi_0(\eta_0 \oplus \zeta_0) & (1)
\end{array}$$

A second way to obtain the product $\xi \otimes (\eta \oplus \zeta)$ is to produce first the two products $\xi \otimes \eta$ and $\xi \otimes \zeta$, and then add them modulo 2:

$$\begin{array}{llll}
 & \eta_1 & \eta_0 & \otimes\xi_1\xi_0 \\
\hline
 & \xi_1\eta_1 & \xi_1\eta_0 & \\
\oplus & & \xi_0\eta_1 & \xi_0\eta_0 \\
\hline
 & \xi_1\eta_1 & \xi_1\eta_0 \oplus \xi_0\eta_1 & \xi_0\eta_0
\end{array}
\qquad
\begin{array}{llll}
 & \zeta_1 & \zeta_0 & \otimes\xi_1\xi_0 \\
\hline
 & \xi_1\zeta_1 & \xi_1\zeta_0 & \\
\oplus & & \xi_0\zeta_1 & \xi_0\zeta_0 \\
\hline
 & \xi_1\zeta_1 & \xi_1\zeta_0 \oplus \xi_0\zeta_1 & \xi_0\zeta_0
\end{array}
\qquad (2)$$

TABLE 12.2-1
PROOF OF THE DISTRIBUTIVE LAW FOR DYADIC MULTIPLICATION BY LISTING THE EIGHT POSSIBLE CASES OF $\xi_i(\eta_i \oplus \zeta_i) = \xi_i\eta_i \oplus \xi_i\zeta_i$, $i = 0, 1, 2, \ldots$

ξ_i	η_i	ζ_1	$\xi_i(\eta_i \oplus \zeta_i)$	$\xi_i\eta_i \oplus \xi_i\zeta_i$
0	0	0	0	0
0	0	1	0	0
0	1	0	0	0
0	1	1	0	0
1	0	0	0	0
1	0	1	1	1
1	1	0	1	1
1	1	1	0	0

The modulo 2 sum of these two products yields $\xi \otimes \eta \oplus \xi \otimes \zeta$:

$$\begin{array}{lll} \xi_1\eta_1 & \xi_1\eta_0 \oplus \xi_0\eta_1 & \xi_0\eta_0 \\ \oplus\, \xi_1\zeta_1 & \xi_1\zeta_0 \oplus \xi_0\zeta_1 & \xi_0\zeta_0 \\ \hline \xi_1\eta_1 \oplus \xi_1\zeta_1 & (\xi_1\eta_0 \oplus \xi_0\eta_1) \oplus (\xi_1\zeta_0 \oplus \xi_0\zeta_1) & \xi_0\eta_0 \oplus \xi_0\zeta_0 \end{array} \qquad (3)$$

Two conditions must be satisfied for the two results of Eqs.(1) and (3) to be equal:

a) The associative and the cummutative law must be satisfied for modulo 2 addition: $a \oplus (b \oplus c) = (a \oplus b) \oplus c = (a \oplus c) \oplus b$. This is obviously so since a sum like $1 \oplus 0 \oplus 0 \oplus 1 \oplus 1 \oplus \cdots$ will yield 1 if the number of digits 1 is odd, and zero if the number of digits 1 is even, regardless of the sequence of summation of the digits.

b) The distributive law $\xi_i(\eta_i \oplus \zeta_i) = \xi_i \otimes (\eta_i \oplus \zeta_i) = \xi_i\eta_i \oplus \xi_i\zeta_i$ must hold. Note that there is no difference between ordinary and dyadic multiplication if only singe digits are multiplied with each other. Since there are only eight possible values of $\xi_i(\eta_i \oplus \zeta_i)$ one can prove the distributive law by listing these eight cases. Table 12.2-1 shows such a listing.

The distributive law makes the first and last terms in the results of Eqs.(1) and (3) equal. The middle term requires in addition the associative and commutative law of dyadic addition:

$$\begin{aligned} \xi_1(\eta_0 \oplus \zeta_0) \oplus \xi_0(\eta_1 \oplus \zeta_1) &= \xi_1\eta_0 \oplus \xi_1\zeta_0 \oplus \xi_0\eta_1 \oplus \xi_0\zeta_1 \\ &= \xi_1\eta_0 \oplus \xi_0\zeta_1 \oplus \xi_1\eta_0 \oplus \xi_0\zeta_1 \end{aligned} \qquad (4)$$

Let now ξ, η, ζ have n rather than 2 digits. By adding zeroes to the left of the most significant digit and to the right of the least significant digit, one can always assure that each one of these three numbers has n digits. The product $\xi \otimes (\eta \oplus \zeta)$ then yields terms as follows:

$$
\begin{aligned}
&\xi_n(\eta_n \oplus \zeta_n)\\
&\xi_n(\eta_{n-1} \oplus \zeta_{n-1}) \oplus \xi_{n-1}(\eta_n \oplus \zeta_n)\\
&\xi_n(\eta_{n-2} \oplus \zeta_{n-2}) \oplus \xi_{n-1}(\eta_{n-1} \oplus \zeta_{n-1}) \oplus \xi_{n-2}(\eta_n \oplus \zeta_n)\\
&\qquad\vdots\\
&\xi_n(\eta_0 \oplus \zeta_0) \oplus \xi_{n-1}(\eta_1 \oplus \zeta_1) \oplus \cdots \oplus \xi_1(\eta_{n-1} \oplus \zeta_{n-1}) \oplus \xi_0(\eta_n \oplus \zeta_n)\\
&\qquad \xi_{n-1}(\eta_0 \oplus \zeta_0) \oplus \cdots \oplus \xi_1(\eta_{n-2} \oplus \zeta_{n-2}) \oplus \xi_0(\eta_{n-1} \oplus \zeta_{n-1})\\
&\qquad\vdots\\
&\qquad \xi_2(\eta_0 \oplus \zeta_0) \oplus \xi_1(\eta_1 \oplus \zeta_1) \oplus \xi_0(\eta_2 \oplus \zeta_2)\\
&\qquad \xi_1(\eta_0 \oplus \zeta_0) \oplus \xi_0(\eta_1 \oplus \zeta_1)\\
&\qquad \xi_0(\eta_0 \oplus \zeta_0)
\end{aligned}
\qquad (5)
$$

On the other hand, the sum $\xi \otimes \eta \oplus \xi \otimes \zeta$ yields the terms as follows:

$$
\begin{aligned}
&\xi_n\eta_n \oplus \xi_n\zeta_n\\
&(\xi_n\eta_{n-1} \oplus \xi_{n-1}\eta_n) \oplus (\xi_n\zeta_{n-1} \oplus \xi_{n-1}\zeta_n)\\
&(\xi_n\eta_{n-2} \oplus \xi_{n-1}\eta_{n-1} \oplus \xi_{n-2}\eta_n) \oplus (\xi_n\zeta_{n-2} \oplus \xi_{n-1}\zeta_{n-1} \oplus \xi_{n-2}\zeta_n)\\
&\qquad\vdots\\
&(\xi_n\eta_0 \oplus \xi_{n-1}\eta_1 \oplus \cdots \oplus \xi_1\eta_{n-1} \oplus \xi_0\eta_n) \oplus (\xi_n\zeta_0 \oplus \xi_{n-1}\zeta_1 \oplus \cdots \oplus \xi_1\zeta_{n-1} \oplus \xi_0\zeta_n)\\
&\qquad (\xi_{n-1}\eta_0 \oplus \cdots \oplus \xi_1\eta_{n-2} \oplus \xi_0\eta_{n-1}) \oplus (\xi_{n-1}\zeta_0 \oplus \cdots \oplus \xi_1\zeta_{n-2} \oplus \xi_0\zeta_{n-1})\\
&\qquad\vdots\\
&\qquad (\xi_2\eta_0 \oplus \xi_1\eta_1 \oplus \xi_0\eta_2) \oplus (\xi_2\zeta_0 \oplus \xi_1\zeta_1 \oplus \xi_0\zeta_2)\\
&\qquad (\xi_1\eta_0 \oplus \xi_0\eta_1) \oplus (\xi_1\zeta_0 \oplus \xi_0\zeta_1)\\
&\qquad \xi_0\eta_0 \oplus \xi_0\zeta_0
\end{aligned}
\qquad (6)
$$

A comparison of the terms of Eqs.(5) and (6) shows that nothing essentially new occurs compared with the case $n = 2$ of Eqs.(2) and (3). The associative and commutative laws of modulo 2 addition and the distributive law of dyadic multiplication assure that the respective terms are equal.

12.3 Method for Dyadic Division

A dyadic division $a \odot b = c$ can be written as dyadic multiplication $a = c \otimes b$. Let a, b, and c have the following digits:

$$a = 11100.1, \quad b = 1010, \quad c = c_1 c_2 c_3 \ldots$$

The following relations are obtained from $c \otimes b = a$:

$$\begin{array}{cccccccc}
c_1 & c_2 & c_3 \ldots & & \otimes 1010 & & & \\
\hline
c_1 & c_2 & c_3 & c_4 & c_5 & c_6 & \ldots & \\
 & 0 & 0 & 0 & 0 & 0 & \ldots & \\
 & & c_1 & c_2 & c_3 & c_4 & \ldots & \\
\oplus & & & 0 & 0 & 0 & \ldots & \\
\hline
c_1 & c_2 & (c_3 \oplus c_1) & (c_4 \oplus c_2) & (c_5 \oplus c_3) & (c_6 \oplus c_4) & \ldots & = a = 11100.1
\end{array}$$

$$c_1 = 1, \; c_2 = 1, \; c_3 \oplus c_1 = 1, \; c_4 \oplus c_2 = 0, \; c_5 \oplus c_3 = 0, \; c_6 \oplus c_4 = 1$$
$$c_{n+1} \oplus c_{2n-1} = c_{2n+2} \oplus c_{2n} = 0 \text{ for } n \geq 3$$

One readily obtains from this sequence of equations:

$$c_1 = 1, \; c_2 = 1, \; c_3 = 0, \; c_4 = 1, \; c_5 = 0, \; c_6 = 0,$$
$$c_{2n+1} = c_{2n+2} = 0 \text{ for } n \geq 3$$
$$c' = 1101000\ldots$$

The digits of c are determined, but the position of the binary point is still unknown. Consider the following products, where b is a number with n digits to the left of the binary point:

$$\begin{array}{lr}
1 \;\; b_{n-2} \ldots b_1 \;\; b_0 \bullet b_{-1} \;\; b_{-2} \ldots \otimes & 1 \;\bullet c_{-1} \;\; c_{-2} \;\ldots \\
 & 1 \;\; c_0 \bullet c_{-1} \;\; c_{-2} \;\ldots \\
 & 1 \;\; c_1 \;\; c_0 \bullet c_{-1} \;\; c_{-2} \;\ldots
\end{array}$$

A short reflection shows that the first product must yield a number with n digits to the left of the binary point, the second one with $n+1$ digits, the third one with $n+2$ digits, etc. Let generally a number c have m digits and a number b have n digits to the left of the binary point. The dyadic product $c \otimes b = a$ then yields a number a with $n+m-1$ digits to the left of the binary point. Vice versa, if a has $l = n+m-1$ digits and b

has n digits to the left of the binary point, then c must have $m = l - n + 1$ digits to the left of the binary point.

In our example holds $l = 5$ and $n = 4$, and it follows $m = 5 - 4 + 1 = 2$. Hence, c has the value $c = 11.01$. A comparison with the dyadic product of Box 4.3-1 shows that this is indeed the correct value for c.

As another example consider the multiplication of $b = 0.101$ and $c = 0.11$:

$$\begin{array}{r} 0.\,101 \otimes 0.\,11 \\ \hline 101 \\ \oplus \quad 101 \\ \hline 0.\,01111 = a \end{array}$$

Here, $m = 0$ and $n = 0$ yield $n + m - 1 = -1$ digits to the left of the binary point. Hence, $n + m - 1 = 1, 0, -1, -2, \ldots$ means numbers of the form $1.c_{-1}c_{-2}\ldots, 0.c_{-1}c_{-2}\ldots, 0.0c_{-2}c_{-3}\ldots, 0.00c_{-3}c_{-4}\ldots$.

12.4 Right and Left Difference Quotient

The Schrödinger equation, Eq.(8.2-1), for a free particle in one dimension yields the following difference equation if the right difference quotient is used:

$$\Psi(x, t + \Delta t) = i\frac{\hbar \Delta t}{2m(\Delta x)^2}\left[\Psi(x + \Delta x, t) - 2\Psi(x, t) + \Psi(x - \Delta x, t)\right] - \Psi(x, t) \tag{1}$$

No term $\Psi(x, t - \Delta t)$ occurs. The eigenfunctions $\varphi(x) = \exp(i\kappa x)$ yield the eigenvalues $\omega_{1n} = 1$ and ω_{0n}:

$$\omega_{0n} = -2i\frac{\hbar \Delta t}{m(\Delta x)^2}\sin^2\frac{\kappa \Delta x}{2} - 1 \tag{2}$$

The characteristic equation is of first order:

$$Q_n - \omega_{0n} = 0$$

The magnitude of Q_n,

$$|Q_n| = \left(1 + \frac{4\hbar^2(\Delta t)^2}{m^2(\Delta x)^4}\sin^4\frac{\kappa \Delta x}{2}\right)^{1/2} \tag{3}$$

shows that stationary solutions can exist only for particular values of $\kappa \Delta x$.

The left difference quotient yields the equation:

$$i\frac{\hbar\Delta t}{2m(\Delta x)^2}[\Psi(x+\Delta x,t+\Delta t)-2\Psi(x,t+\Delta t)+\Psi(x-\Delta x,t+\Delta t)] \\ -\Psi(x,t+\Delta t)=-\Psi(x,t) \quad (4)$$

The eigenvalues ω_{1n},

$$\omega_{1n}=-2i\frac{\hbar\Delta t}{m(\Delta x)^2}\sin^2\frac{\kappa\Delta x}{2}-1 \quad (5)$$

and $\omega_{0n}=-1$ yield the following magnitude of the solution of the characteristic equation:

$$|Q_n|=\left(1+\frac{4\hbar^2(\Delta t)^2}{m(\Delta x)^2}\sin^2\frac{\kappa\Delta x}{2}\right)^{-1/2} \quad (6)$$

Stationary solutions for arbitrary values of $\kappa\Delta x$ cannot exist, since the magnitude of Q_n is too small. The right difference quotient yields exponentially increasing and the left difference quotient yields exponentially decreasing solutions for small values of $\kappa\Delta x$.

12.5 Independence Relation in Three-Dimensional Cartesian Coordinates

The operators $\Omega_0(\mathbf{r})=\Omega_0(x,y,z)$ of the Schrödinger equation, Eq.(8.2-4), applied to the function $\varphi_n(\mathbf{r})=\varphi_n(x,y,z)$ yields for $V(\mathbf{r},t)=0$ the following equation:

$$\begin{aligned}\Omega_0(x,y,z)\varphi_n(x,y,z)=&\\ =i\frac{\hbar\Delta t}{m}\Big\{&\frac{1}{(\Delta x)^2}[\varphi_n(x+\Delta x,y,z)-2\varphi_n(x,y,z)+\varphi_n(x-\Delta x,y,z)]\\ &+\frac{1}{(\Delta y)^2}[\varphi_n(x,y+\Delta y,z)-2\varphi_n(x,y,z)+\varphi_n(x,y-\Delta y,z)]\\ &+\frac{1}{(\Delta z)^2}[\varphi_n(x,y,z+\Delta z)-2\varphi_n(x,y,z)+\varphi_n(x,y,z-\Delta z)]\Big\}\\ &=\omega_{0n}\varphi_n(x,y,z) \quad (1)\end{aligned}$$

The eigenfunctions

$$\varphi_n(x,y,z)=e^{i(\kappa_x x+\kappa_y y+\kappa_z z)} \quad (2)$$

yield the eigenvalues:

$$\omega_{0n} = -2iB = -4i\frac{\hbar\Delta t}{m(\Delta x)^2} \times \left[\sin^2\frac{\kappa_x\Delta x}{2} + \left(\frac{\Delta x}{\Delta y}\right)^2 \sin^2\frac{\kappa_y\Delta y}{2} + \left(\frac{\Delta x}{\Delta z}\right)^2 \sin^2\frac{\kappa_z\Delta z}{2}\right] \quad (3)$$

The independence relation requires the condition:

$$B^2 \leq 1 + |O(\Delta t)^2| \quad (4)$$

One obtains a condition for the ratio $m(\Delta r)^2/(\Delta t)$ for arbitrary values of κ_x, κ_y, and κ_z:

$$\frac{m(\Delta r)^2}{\Delta t} \geq 2\hbar\left[\left(\frac{\Delta r}{\Delta x}\right)^2 + \left(\frac{\Delta r}{\Delta y}\right)^2 + \left(\frac{\Delta r}{\Delta z}\right)^2\right]^{1/2} [1 - |O(\Delta t)^2|]$$
$$(\Delta r)^2 = (\Delta x)^2 + (\Delta y)^2 + (\Delta z)^2 \quad (5)$$

Let κ_x, κ_y, and κ_z be bounded and let Δx, Δy, and Δz be sufficiently small. In this case one obtains a bound on Δt instead of the condition of Eq.(5):

$$\Delta t \leq \frac{2m}{\hbar\kappa^2}\left[\left(\frac{\kappa_x}{\kappa}\right)^2 + \left(\frac{\kappa_y}{\kappa}\right)^2 + \left(\frac{\kappa_z}{\kappa}\right)^2 +\right]^{-1}$$
$$\kappa^2 = \kappa_x^2 + \kappa_y^2 + \kappa_z^2 \quad (6)$$

12.6 Polynomials as Solutions of Difference Equations of Second Order

Consider a difference equation of second order:

$$P_1(X)v(X+1) + P_0(X)v(X) + P_{-1}(X)v(X-1) = 0 \quad (1)$$

Let $P_1(X)$ and $P_{-1}(X)$ be polynomials of the same degree, and let $P_0(X)$ be a polynomial of the same or lower degree:

$$P_1(X) = d_{10}X^j + d_{11}X^{j-1} + \cdots + d_{1j}$$
$$P_0(X) = d_{00}X^j + d_{01}X^{j-1} + \cdots + d_{0j}$$
$$P_{-1}(X) = d_{-10}X^j + d_{-11}X^{j-1} + \cdots + d_{-1j}$$
$$d_{10},\ d_{-10} \neq 0 \quad (2)$$

One may always rewrite $P_1(X)$, $P_0(X)$, and $P_{-1}(X)$ in the following form:

$$P_1(X) = c_{10}(X+1)(X+2)\dots(X+j) + c_{11}(X+1)\dots(X+j-1) + \dots + c_{1j}$$

$$P_0(X) = c_{00}X(X+1)\dots(X+j-1) + c_{01}X\dots(X+j-2) + \dots + c_{0j}$$

$$P_{-1}(X) = c_{-10}(X-1)X\dots(X+j-2) + c_{-11}(X-1)\dots(X+j-3) + \dots + c_{-1j} \quad (3)$$

$$c_{10} = d_{10},\ c_{00} = d_{00},\ c_{-10} = d_{-10}$$

$$c_{11} = d_{11} - c_{10}\sum_{\nu=1}^{j}\nu = d_{11} - \frac{1}{2}j(j+1)d_{10}$$

$$c_{01} = d_{00} - \frac{1}{2}(j-1)j\,d_{00}$$

$$c_{-11} = d_{-11} - \frac{1}{2}j(j-3)d_{-10}$$

$$c_{12} = d_{12} - c_{11}\sum_{\nu=1}^{j-1}\nu - c_{10}\sum_{\nu=1}^{j-1}\sum_{\mu=\nu+1}^{j}\nu\mu$$

$$= d_{12} - \frac{1}{2}(j-1)j\,d_{11} - \left(\sum_{\nu=1}^{j-1}\sum_{\mu=\nu+1}^{j}\nu\mu - \frac{1}{4}j^2(j^2-1)\right)d_{10}$$

$$c_{02} = d_{02} - \frac{1}{2}(j-2)(j-1)d_{00} - \left(\sum_{\nu=1}^{j-2}\sum_{\mu=\nu+1}^{j-1}\nu\mu - \frac{1}{4}j(j-1)^2(j-2)\right)d_{00}$$

$$c_{-12} = d_{-12} - \frac{1}{2}(j-1)(j-4)d_{-11}$$

$$- \left(\sum_{\nu=1}^{j-3}\sum_{\mu=\nu+1}^{j-2}\nu\mu - \frac{1}{2}(j-1)(j-2) - \frac{1}{4}j(j-1)(j-3)(j-4)\right)d_{-10}$$

$$\sum_{\nu=1}^{j-3}\sum_{\mu=\nu+1}^{j-2}\nu\mu = \frac{1}{2}\sum_{\nu=1}^{j-3}\nu(\nu+j-1)(j-\nu-2) = \eta$$

$$\sum_{\nu=1}^{j-2}\sum_{\mu=\nu+1}^{j-1}\nu\mu = \eta + \frac{1}{2}(j-1)^2(j-2)$$

$$\sum_{\nu=1}^{j-1}\sum_{\mu=\nu+1}^{j}\nu\mu = \eta + \frac{1}{2}(j-1)^2(j-2) + \frac{1}{2}j^2(j-1) \quad (4)$$

The integral transformation

$$v(X) = \frac{1}{2i\pi} \int_{\ell} s^{X-1} y(X)\, dX$$

transforms the difference equation (1) into a differential equation, if the line of integration ℓ is chosen so that the terms

$$s^X y\Big|_{\ell_1}^{\ell_2}, \quad s^{X+1} y'\Big|_{\ell_1}^{\ell_2}, \quad \text{etc.}$$

vanish:

$$\begin{aligned}(c_{10}s^2 + c_{00}s + c_{-10})s^j y^{(j)} - (c_{11}s^2 + c_{01}s + c_{-11})s^{j-1} y^{(j-1)} \\ + (c_{12}s^2 + c_{02}s + c_{-12})s^{j-2} y^{(j-2)} - \cdots + (-1)^j (c_{1j}s^2 + c_{0j}s + c_{-1j})y = 0\end{aligned} \tag{5}$$

The characteristic equation

$$(c_{10}s^2 + c_{00}s + c_{-10})s^j = 0 \tag{6}$$

has the simple roots:

$$\begin{aligned} s_2 &= \frac{1}{2}\frac{1}{d_{10}}\left(-d_{00} + (d_{00}^2 - 4d_{-10}d_{10})^{1/2}\right) \\ s_3 &= \frac{1}{2}\frac{1}{d_{10}}\left(-d_{00} - (d_{00}^2 - 4d_{-10}d_{10})^{1/2}\right) \\ s_2 s_3 &= \frac{d_{-10}}{d_{10}} \end{aligned} \tag{7}$$

In addition, there is the j–fold root $s_1 = 0$.

The substitution $z = s - s_2$ in the differential equation, Eq.(5), yields:

$$\begin{aligned} c_{10}y^{(j)} - \frac{c_{11}(z+s_2)^2 + c_{01}(z+s_2) + c_{-11}}{z(z+s_2)(z+s_2-s_3)} y^{(j-1)} \\ + \frac{c_{12}(z+s_2)^2 + c_{02}(z+s_2) + c_{-12}}{z^2(z+s_2)(z+s_2-s_3)} y^{(j-2)} - \cdots = 0 \end{aligned} \tag{8}$$

Equation (8) is solved by means of a power series:

$$y = \sum_{\nu=0}^{\infty} q_\nu z^{p+\nu} \tag{9}$$

The following determining equation for p is obtained if s_2 is not equal to s_1 or s_3:

$$p(p-1)\cdots(p-j+2)\left((p-j+1)c_{10} - \frac{c_{11}s_2^2 + c_{01}s_2 + c_{-11}}{s_2(s_2-s_3)}\right) = 0 \tag{10}$$

Equation (10) has the trivial roots $p = 0, 1, \ldots, j-2$ and the non-trivial root:

$$p = \frac{1}{d_{10}s_2(s_2-s_3)}(d_{11}s_2^2 + d_{01}s_2 + d_{-11}) - 1 \tag{11}$$

Let us interrupt this course of calculations and let us consider a series expansion in negative powers of X:

$$\begin{aligned} P_1'(X) &= d_{10} + d_{11}X^{-1} + \cdots + d_{1j}X^{-j} \\ P_0'(X) &= d_{00} + d_{01}X^{-1} + \cdots + d_{0j}X^{-j} \\ P_{-1}'(X) &= d_{-10} + d_{-11}X^{-1} + \cdots + d_{-1j}X^{-j} \end{aligned} \tag{12}$$

Multiplication by X^j yields the polynomials of Eq.(2). The initial power p is independent of the degree of approximation of the series expansion of Eq.(12), since the terms containing j do not show up in Eq.(11). An interchange of s_2 and s_3 leaves the form of Eq.(11) unchanged. Hence, the same comment also applies to the solution in the point s_3, if s_3 is not equal to s_1 or s_2.

Let us return to the characteristic equation Eq.(6). In the case of a double root,

$$s_2 = s_3 = -\frac{d_{00}}{2d_{10}}$$

one may solve the difference equation Eq.(1) only, if s_2 is also a root of the equation:

$$c_{11}s^2 + c_{01}s + c_{-11} = 0 \tag{13}$$

The roots of Eq.(13) shall be denoted s_4 and s_5, and it shall hold $s_2 = s_3 = s_4$. One obtains:

$$s_4 = -\frac{d_{00}}{2d_{10}}, \quad s_5 = -\frac{2d_{10}}{d_{00}} \frac{2d_{-11} - j(j-3)d_{-10}}{2d_{11} - j(j+1)d_{10}} \tag{14}$$

The determining equation for p has now the following form:

$$\left[(p-j+2)(p-j+1)c_{10} - (p-j+1)c_{11}\left(1 - \frac{s_5}{s_2}\right)\right.$$
$$\left. s_2^{-2}(c_{12}s_2^2 + c_{02}s_2 + c_{-12})\right] p \cdots (p-j-3) = 0 \tag{15}$$

The non-trivial values of p are the roots of the equation:

$$p^2 + \left(3 - \frac{d_{11}}{d_{10}} + \frac{4d_{10}d_{-11}}{d_{00}^2}\right) p$$
$$+ 2\left(1 + \frac{d_{01}}{d_{00}} - \frac{d_{11}}{d_{10}} - \frac{d_{02}}{d_{00}}\right) + \frac{d_{12}}{d_{10}} + \frac{4d_{10}d_{-12}}{d_{00}^2} = 0 \tag{16}$$

Again, p is independent of j.

The recursion formula for the coefficients q_ν of the power series Eq.(9) are obtained by the substitution of Eq.(9) into Eq.(5):

$$\alpha_{0,\nu}q_0 + \cdots + \alpha_{\nu,\nu}q_\nu + \alpha_{\nu+1,\nu}q_{\nu+1} = 0 \tag{17}$$

One obtains after lengthy calculations with the help of Eqs.(4), (7), and (11):

$$\alpha_{\nu+1,\nu} = d_{10}s_2(s_2 - s_3)(\nu+1)(p+\nu+1)s_2^{j-1}(p+\nu)\cdots(p+\nu-j+3)$$
$$\alpha_{\nu,\nu} = \left[d_{12}s_2^2 + d_{02}s_2 + d_{-12} - (p+\nu+1)(d_{11}s_2^2 - d_{-11})\right.$$
$$+ (p+\nu+1)(p+\nu+2)d_{10}s_2^2 - (p+1)(d_{10}s_2^2 - d_{-10})$$
$$\left. + j(d_{10}s_2^2 - d_{-10})\nu\left(\nu + p - \frac{1}{2}j + \frac{1}{2}\right)\right] s_2^{j-2}(p+\nu)\cdots(p+\nu-j+3) \tag{18}$$

The recursion formula Eq.(17) may be written as a system of linear equations. The coefficient $\alpha_{\nu+1,\nu}$ vanishes for $p = -(\nu+1)$ and a system of p equations with $p-1$ variables is obtained. The necessary and sufficient condition for the existence of such a system is that the determinant of the coefficients vanishes. This condition is simplified if s_2 or $s_2 - s_3$ are small; one obtains in first approximation:

$$\prod_{\nu=0}^{-p-1} \alpha_{\nu,\nu} = 0$$

The condition $s_2 - s_3 \ll 1$ yields from this product:

$$\begin{aligned} d_{12}s_2^2 + d_{02}s_2 + d_{-12} - (p+\nu+1)(d_{11}s_2^2 - d_{-11}) \\ + (p+\nu+1)(p+\nu+2)d_{10}s_2^2 = 0 \\ \nu = 0,\ 1, \ldots,\ -(p+1) \end{aligned} \tag{19}$$

12.7 Difference Equation of the Discrete Spherical Harmonics

The following rules for difference operators may be derived from the definitions of Eqs.(8.2-2) and (8.2-3):

$$\begin{gathered} \frac{\Delta_{\mathrm{r}}}{\Delta x}\left(\frac{\Delta_{\mathrm{l}} y(x)}{\Delta x}\right) = \frac{\Delta_{\mathrm{l}}}{\Delta x}\left(\frac{\Delta_{\mathrm{r}} y(x)}{\Delta x}\right) = \frac{\Delta^2 y(x)}{(\Delta x)^2} \\ \frac{\Delta_{\mathrm{r}} y(x)}{\Delta x} - \frac{\Delta_{\mathrm{l}} y(x)}{\Delta x} = \Delta x \frac{\Delta^2 y(x)}{(\Delta x)^2} \\ \frac{\Delta_{\mathrm{r}} y(x)}{\Delta x} + \frac{\Delta_{\mathrm{l}} y(x)}{\Delta x} = 2\frac{\Delta_{\mathrm{s}} y(x)}{\Delta x} \\ \frac{\Delta_{\mathrm{s}} x}{\Delta x} = \frac{\Delta_{\mathrm{r}} x}{\Delta x} = \frac{\Delta_{\mathrm{l}} x}{\Delta x} = 1 \\ \frac{\Delta_{\mathrm{d}} y(x)}{\Delta x} = \frac{\Delta_{\mathrm{d}} y(x)}{\Delta z}\frac{\Delta_{\mathrm{d}} z}{\Delta x}, \quad d = r,\ l,\ s, \quad z = z(x) \end{gathered}$$

$$y(x+\Delta x) = y(x) - \Delta x \frac{\Delta_{\mathrm{r}} y(x)}{\Delta x}, \quad y(x-\Delta x) = y(x) - \Delta x \frac{\Delta_{\mathrm{l}} y(x)}{\Delta x}$$

$$\begin{aligned} \frac{\Delta_{\mathrm{s}} y(x)z(x)}{\Delta x} &= y(x)\frac{\Delta_{\mathrm{s}} z(x)}{\Delta x} + z(x)\frac{\Delta_{\mathrm{s}} y(x)}{\Delta x} + \frac{\Delta_{\mathrm{r}} y(x)\Delta_{\mathrm{r}} z(x) - \Delta_{\mathrm{l}} y(x)\Delta_{\mathrm{l}} z(x)}{2\Delta x} \\ \frac{\Delta_{\mathrm{l}} y(x)z(x)}{\Delta x} &= y(x)\frac{\Delta_{\mathrm{l}} z(x)}{\Delta x} + z(x)\frac{\Delta_{\mathrm{l}} y(x)}{\Delta x} - \frac{\Delta_{\mathrm{l}} y(x)\Delta_{\mathrm{l}} z(x)}{\Delta x} \\ \frac{\Delta_{\mathrm{r}} y(x)z(x)}{\Delta x} &= y(x)\frac{\Delta_{\mathrm{r}} z(x)}{\Delta x} + z(x)\frac{\Delta_{\mathrm{r}} y(x)}{\Delta x} + \frac{\Delta_{\mathrm{r}} y(x)\Delta_{\mathrm{r}} z(x)}{\Delta x} \end{aligned} \tag{1}$$

Let us now investigate the solution of the difference equation of the finite spherical harmonics, Eq.(9.1-6). We substitute

$$\cos\vartheta = x, \quad \psi_n(\vartheta) = \psi_n(x) \tag{2}$$

and transform the difference quotients $\Delta_s/\Delta\vartheta$ and $\Delta^2/(\Delta\vartheta)^2$ into difference quotients $\Delta_s/\Delta x$ and $\Delta^2/(\Delta x)^2$:

$$\begin{aligned}
\frac{\Delta_s\psi_n(\vartheta)}{\Delta\vartheta} &= \frac{\Delta_s\psi_n(x)}{\Delta x}\frac{\Delta_s x}{\Delta\vartheta} = -\sin\vartheta\frac{\sin\Delta\vartheta}{\Delta\vartheta}\frac{\Delta_s\psi_n(x)}{\Delta x} \\
\frac{\Delta^2\psi_n(\vartheta)}{(\Delta\vartheta)^2} &= \frac{1}{2}\left[\frac{\Delta_l}{\Delta\vartheta}\left(\frac{\Delta_r\psi_n(\vartheta)}{\Delta\vartheta}\right) + \frac{\Delta_r}{\Delta\vartheta}\left(\frac{\Delta_l\psi_n(\vartheta)}{\Delta\vartheta}\right)\right] \\
&= \left[\sin^2\vartheta\left(\frac{\sin\Delta\vartheta}{\Delta\vartheta}\right)^2 + \cos^2\vartheta\left(\frac{1-\cos\Delta\vartheta}{\Delta\vartheta}\right)^2\right]\frac{\Delta^2\psi_n(x)}{(\Delta x)^2} \\
&\quad - 2\cos\vartheta\frac{1-\cos\vartheta}{(\Delta\vartheta)^2}\frac{\Delta_s\psi_n(x)}{\Delta x}
\end{aligned} \tag{3}$$

Equation (9.1-6) assumes the following form:

$$\begin{aligned}
&\left\{\left[(1-x^2)\left(\frac{\sin\Delta\vartheta}{\Delta\vartheta}\right)^2 + x^2\left(\frac{1-\cos\Delta\vartheta}{\Delta\vartheta}\right)^2\right]\frac{\Delta^2}{(\Delta x)^2}\right. \\
&\quad \left. - x\left[2\frac{1-\cos\Delta\vartheta}{(\Delta\vartheta)^2} + \frac{\sin\Delta\vartheta}{\Delta\vartheta}\right]\frac{\Delta_s}{\Delta x} - \lambda + \frac{\delta}{1-x^2}\right\}\psi_n(x) = 0
\end{aligned} \tag{4}$$

From the relations

$$\frac{\Delta_s x}{\Delta\vartheta} = -\sin\vartheta\frac{\sin\Delta\vartheta}{\Delta\vartheta}, \quad \Delta x = \Delta_s x = -\sin\vartheta\sin\Delta\vartheta \tag{5}$$

one obtains $\Delta\vartheta$ expressed by Δx:

$$\begin{aligned}
\Delta\vartheta &= -\sin^{-1}\frac{\Delta x}{(1-x^2)^{1/2}} \\
&= -\left(\frac{\Delta x}{(1-x^2)^{1/2}} + \frac{1}{2}\frac{(\Delta x)^3}{3(1-x^2)^{3/2}} + \frac{1\cdot 3}{2\cdot 4}\frac{(\Delta x)^5}{5(1-x^2)^{5/2}} + \cdots\right) \\
\frac{\sin\Delta\vartheta}{\Delta\vartheta} &= 1 - \frac{1}{2}\frac{(\Delta x)^2}{3(1-x^2)} + \cdots \\
\frac{1-\cos\Delta\vartheta}{\Delta\vartheta} &= -\frac{1}{2}\frac{\Delta x}{(1-x^2)^{1/2}}\left(1 + \frac{1}{24}\frac{(\Delta x)^2}{1-x^2} + \cdots\right) \\
&\frac{(\Delta x)^2}{1-x^2} \le 1
\end{aligned} \tag{6}$$

Only even powers of $(1-x^2)^{1/2}$ appear in the series expansions of the terms $(\sin\Delta\vartheta)/\Delta\vartheta$, $(\sin\Delta\vartheta)^2/(\Delta\vartheta)^2$, $(1-\cos\Delta\vartheta)/\Delta\vartheta$, and $(1-\cos\Delta\vartheta)^2/(\Delta\vartheta)^2$.

Equation (4) assumes the following form if the explicit difference quotients are substituted for the symbolic ones:

$$\begin{aligned}
&\left\{\left(\frac{x}{\Delta x}\right)^2\left[\left(\frac{\sin\Delta\vartheta}{\Delta\vartheta}\right)^2-\left(\frac{1-\cos\Delta\vartheta}{\Delta\vartheta}\right)^2\right]\right.\\
&\quad\left.+\frac{x}{\Delta x}\left(\frac{\sin\Delta\theta}{2\Delta\vartheta}+\frac{1-\cos\Delta\vartheta}{(\Delta\vartheta)^2}\right)-\frac{1}{(\Delta x)^2}\left(\frac{\sin\Delta\vartheta}{\Delta\vartheta}\right)^2\right\}\psi_n(x+\Delta x)\\
&\quad+\left\{2\left(\frac{x}{\Delta x}\right)^2\left[\left(\frac{1-\cos\Delta\vartheta}{\Delta\vartheta}\right)^2-\left(\frac{\sin\Delta\vartheta}{\Delta\vartheta}\right)^2\right]\right.\\
&\qquad\qquad\left.+\frac{2}{(\Delta x)^2}\left(\frac{\sin\Delta\vartheta}{\Delta\vartheta}\right)^2-\lambda+\frac{\delta}{1-x^2}\right\}\psi_n(x)\\
&\quad+\left\{\left(\frac{x}{\Delta x}\right)^2\left[\left(\frac{\sin\Delta\vartheta}{\Delta\vartheta}\right)^2-\left(\frac{1-\cos\Delta\vartheta}{\Delta\vartheta}\right)^2\right]\right.\\
&\qquad\qquad-\frac{x}{\Delta x}\left(\frac{\sin\Delta\vartheta}{2\Delta\vartheta}+\frac{1-\cos\Delta\vartheta}{(\Delta\vartheta)^2}\right)\\
&\qquad\qquad\left.-\frac{1}{(\Delta x)^2}\left(\frac{\sin\Delta\vartheta}{\Delta\vartheta}\right)^2\right\}\psi_n(x-\Delta x)=0 \qquad (7)
\end{aligned}$$

This difference equation has transcendental coefficients. They can be replaced by polynomials of arbitrarily high degree by means of the series expansions of Eq.(6) multiplied by $(1-x^2)^k$.

$$\begin{aligned}
&\left[X^j+X^{j-1}+\left(\frac{7}{12}-\frac{j}{2(\Delta x)^2}\right)X^{j-2}+\cdots\right]\psi_n(X+1)\\
&\qquad+\left[-2X^j+\left(\frac{j}{(\Delta x)^2}+\lambda-\frac{7}{6}\right)X^{j-2}+\cdots\right]\psi_n(X)\\
&\quad+\left[X^j-X^{j-1}+\left(\frac{7}{12}-\frac{j}{2(\Delta x)^2}\right)X^{j-2}+\cdots\right]\psi_n(X-1)=0\\
&\qquad X=x/\Delta X,\ \psi_n(X)=\psi_n(x),\ j=2k+2 \qquad (8)
\end{aligned}$$

Equations (8) and (12.6-1) become equal if the following values are substituted for the coefficients d_{ij}:

$$\begin{aligned} &d_{10} = 1 \qquad d_{11} = 1 \qquad d_{12} = \frac{7}{12} - \frac{j}{2(\Delta x)^2} \\ &d_{00} = -2 \qquad d_{01} = 0 \qquad d_{02} = -\frac{7}{6} + \lambda + \frac{j}{2(\Delta x)^2} \\ &d_{-10} = 1 \qquad d_{-11} = -1 \quad d_{-12} = \frac{7}{12} - \frac{j}{2(\Delta x)^2} \end{aligned} \tag{9}$$

One obtains from Eqs.(12.6-7), (12.6-11), and (12.6-16):

$$s_2 = s_3 = s_4 = 1, \quad s_5 = \frac{2 + j(j-3)}{j(j+1) - 2}, \quad \lim_{j\to\infty} s_5 = 1, \quad p(p+1) = -\lambda \tag{10}$$

The solution of Eq.(8) in the point $s_2 = s_3 = 1$ may be written as factorial series:

$$\psi_n(X) = \frac{\Gamma(X)}{\Gamma(X+1+p)} \sum_{\nu=0}^{\infty} (-1)^\nu C_\nu \frac{(p+1)\cdots(p+\nu)}{(X+p+1)\cdots(X+p+\nu)} \tag{11}$$

The differential equation Eq.(12.6-5) belonging to the difference equation (8) has the singular points $s = 1$, $s = 0$, and $s = \infty$. A power series in the point $s = 1$ converges inside a circle that passes through $s = 0$. The factorial series Eq.(11) converges thus in the half plane $X > X_0$.

Let ϑ run from 0 to π. The variable x runs then from 1 to -1, and X runs from $1/\Delta x$ to $-1/\Delta x$. The difference Δx is arbitrarily small. Hence, only factorial series with an abscissa of convergence $X_0 < -1/\Delta x$ are of interest as solutions. Let us assume Eq.(11) converges for $X > X_0 < -1/\Delta x$. The gamma function $\Gamma(X)$ has poles for negative integer values of X, which must be compensated either by zeros in the factorial series or by poles of $\Gamma(X+p+1)$. The gamma functions $\Gamma(X+p+1)$ and $\Gamma(X)$ have common poles if p is an integer:

$$p = -(l+1), \quad p = +l, \quad l = 0,\ 1,\ 2, \ldots \tag{12}$$

Only non-positive integer values of p avoid poles for $X = 0, -1/\Delta x, \ldots$. The series of Eq.(11) becomes a polynomial if p is a negative integer, and the abscissa of convergence X_0 becomes $-\infty$.

$$\psi_n(X) = \psi_n(\vartheta) = \sum_{\nu=0}^{l} C_\nu \binom{X-1}{l-\nu}, \qquad X = -\frac{\operatorname{ctn}\vartheta}{\sin \Delta\vartheta} \tag{13}$$

Equations (12) and (10) yield the eigenvalues:

$$\lambda = -l(l+1) \tag{14}$$

This result is surprising, since it does not depend on $\Delta\vartheta$, and it shows a noteworthy difference between Cartesian and spherical coordinates. The differences Δx, Δy, and Δz of a three-dimensional Cartesian coordinate system appear perfectly equal in Eqs.(8.4-13), (12.5-5), and (12.5-6). In three-dimensional spherical coordinates, on the other hand, it is impossible for $\Delta\vartheta$ and $\Delta\varphi$ to appear in the independence relation, since the only connection between the equations for $\psi_n(\vartheta)$ and $\chi_n(\varphi)$ with that for $\varphi_n(r)$ in Eq.(9.1-5), (9.1-6), (9.1-7) is provided by the parameter λ. Furthermore, Eq.(14) is the reason why the energy eigenvalues of Eq.(9.3-5) depend on Δt and Δr but not on $\Delta\vartheta$ and $\Delta\varphi$.

The condition of Eq.(14) is necessary for the existence of polynomials $\psi_n(X)$. The investigation of the sufficient condition calls for a knowledge of the coefficients C_ν, which permits one to write a determinant like that of Eq.(9.2-23). Equation (12.6-18) is of no help since it does not hold for dual roots of the characteristic equation, Eq.(12.6-6). A recursion formula for the coefficients C_ν may be obtained by substituting Eq.(11) into Eq.(7), but the calculation is so tedious that no results of general interest have been obtained so far. In the simplest case, $p = -1$, $l = 0$, $\psi_n(X) = C_0$, one obtains the obviously satisfied sufficient condition $\nu = 0$. However, a comparison of Eqs.(7) and (9.1-8) makes one suspect that one would generally obtain a relation between $\Delta\vartheta$ and $\Delta\varphi$.

Let us assume that Eq.(11) does not terminate. The factorial series must then have zeros in the points $X = 0, -\Delta x, \ldots$ and the coefficients C_ν must be functions of the arbitrary differences Δx. Hence, the polynomial solutions are the only ones that have no poles and that become independent of Δx for small values of Δx.

The difference equation Eq.(9.1-6) is satisfied in first approximation by spherical harmonics. Let us substitute $\delta = -m^2 = 0$ and replace the difference quotients by means of a Taylor expansion by differential quotients:

$$\frac{\Delta_s}{\Delta\vartheta} = \sum_{\nu=0}^{\infty} \frac{(\Delta\vartheta)^{2\nu}}{(2\nu+1)!} \frac{d^{2\nu+1}}{d\vartheta^{\nu+1}}$$

$$\frac{\Delta^2}{(\Delta\vartheta)^2} = 2\sum_{\nu=0}^{\infty} \frac{(\Delta\vartheta)^{2\nu}}{(2\nu+1)!} \frac{d^{2\nu+2}}{d\vartheta^{2\nu+2}} \tag{15}$$

One obtains:

$$\left[\frac{d^2}{d\vartheta^2} + 2\sum_{\nu=0}^{\infty}\frac{(\Delta\vartheta)^{2\nu}}{(2\nu+2)!}\frac{d^{2\nu+2}}{d\vartheta^{2\nu+2}} + \frac{\cos\vartheta}{\sin\vartheta}\left(\frac{d}{d\vartheta} + \sum_{\nu=0}^{\infty}\frac{(\Delta\vartheta)^{2\nu}}{(2\nu+1)!}\frac{d^{2\nu+1}}{d\vartheta^{2\nu+1}}\right) - \lambda\right]\psi_n(\vartheta) = 0 \quad (16)$$

The terms not containing $\Delta\vartheta$ are satisfied by spherical harmonics $P_l(\cos\vartheta)$ for $\lambda = -l(l+1)$:

$$P_l(\cos\vartheta) = 2\frac{1\cdot 3\cdot 5\cdots(2l-1)}{2^l\cdot l!}\left(\cos l\vartheta + \frac{1}{1}\frac{l}{2l+1}\cos(l-2)\vartheta + \frac{1\cdot 3}{1\cdot 2}\frac{l(l-1)}{(2l-1)(2l-3)}\cos(l-4)\vartheta + \cdots\right) \quad (17)$$

Differentiation yields:

$$(\Delta\vartheta)^2\frac{d^{2\nu+2}}{d\theta^{2\nu+2}}P_l(\cos\vartheta) = i^{2\nu}(l\Delta\vartheta)^2 2\frac{1\cdot 3\cdot 5\cdots(2l-1)}{2^{l-2}\cdot l!}\left[\cos l\vartheta + \frac{1}{1}\frac{l}{2l-1}\left(\frac{l-2}{l}\right)^{2\nu}\cos(l-2)\vartheta + \cdots\right] \quad (18)$$

The terms in Eq.(16) which contain $\Delta\vartheta$ may be made arbitrarily small by choosing $\Delta\vartheta$ sufficiently small. Hence, $\psi_n(\vartheta)$ has the form $P_l(\cos\vartheta) + O(\Delta\vartheta)$. Taking the derivative of order m of Eq.(16) with respect to $\cos\vartheta$ causes the terms not containing $\Delta\vartheta$ to be satisfied by the associated spherical harmonics $P_l^m(\cos\vartheta)$. The finite number of differentiations does not affect the principle of the vanishing of the terms multiplied by $\Delta\vartheta$. Hence, the difference equation Eq.(9.1-6) has eigenfunctions of the form $P_l^m(\cos\vartheta) + O(\Delta\vartheta) + O(\Delta\varphi)$.

12.8 Convergence of the Solution of the Klein-Gordon Equation with Coulomb Field

The series of Eq.(10.2-6) yields a solution of the difference equation Eq.(10.2-1), if the line of integration in Eq.(9.2-3) is chosen so that the terms of Eqs.(9.2-4) and (9.2-7) vanish. The line of integration begins and ends in the point $s = 0$. Hence, we investigate $\varphi_n(R)$ in the neighborhood of $s = 0$. The series

$$w(s) = \sum_{\nu=0}^{\infty} c_\nu s^{\beta+\nu} \tag{1}$$

substituted into the differential equation, Eq.(10.2-4), provides two initial powers $\beta_1 = 0$ and $\beta_2 = 1$. The solution using β_2 has the form

$$w_2(s) = s \sum_{\nu=0}^{\infty} c_\nu s^\nu \tag{2}$$

while the solution using β_1 contains a logarithmic term:

$$w_1(s) = \sum_{\nu=0}^{\infty} (c'_\nu + c_\nu \ln s) s^\nu \tag{3}$$

The general solution in the neighborhood of $s = 0$ is a linear combination of $w_1(s)$ and $w_2(s)$:

$$w(s) = C_1 w_1(s) + C_2 w_2(s) \tag{4}$$

The products $s^R w(s)$ in Eq.(9.2-4) and $s^{R+1} w'(s)$ in Eq.(9.2-7) vanish for $s = 0$, if R is larger than zero. Hence, the series for $\varphi_n(R)$ in Eqs.(10.2-9), (10.4-14), and (10.4-16) are formal solutions of the difference equation Eq.(10.2-1) for $R > 0$.

The exponent $\beta_1 = 0$ yields the smaller one of the two initial powers in the point $s = 0$. This implies, that the order of the series expansion of $w(\zeta^{1/P})$ in Eq.(10.4-5) in the singular point $\zeta = 0$ equals zero (Nörlund, 1914, 1929; Nielsen, 1906). Hence, the coefficients A_ν of the factorial series of Eq.(10.4-14) have the bounds

$$K\nu^{-1-\epsilon} < |A_\nu| < K\nu^{-1+\epsilon} \tag{5}$$

for large values of ν. The exponent ϵ is an arbitrarily small constant, and K is a sufficiently large but finite constant.

By means of the approximation

$$\left(1 + \frac{R}{P(p+\nu')}\right)\left(1 + \frac{R}{P(p+\nu'+1)}\right) \cdots \left(1 + \frac{R}{P(p+\nu)}\right) \approx \exp\left(\frac{R}{P} \sum_{j=\nu'}^{\nu} \frac{1}{p+j}\right)$$

$$\nu' + p > 0 \tag{6}$$

and the relation

$$\sum_{i=1}^{\nu} \frac{1}{i} - \ln \nu \approx C, \; C = \text{Euler's constant}, \; \nu \gg 1$$

one obtains the bound:

$$\frac{(p+\nu')\cdots(p+\nu)}{(R/P+p+\nu')\cdots(R/P+p+\nu)} < K'\nu^{-R/P} \tag{7}$$

Equations (5) and (7) yield:

$$K''\nu^{-1+\epsilon-R/P} > |A_\nu| \frac{(p+1)\cdots(p+\nu)}{(R/P+p+1)\cdots(R/P+p+\nu)} \frac{\Gamma(R/P)}{\Gamma(R/P+p+1)} \tag{8}$$

The sum

$$\sum_{\nu=1}^{\infty} \nu^{-1+\epsilon-R/P} \tag{9}$$

converges absolutely for $R/P > \epsilon > 0$.

The investigation of the point $s = \infty$ of the differential equation Eq.(10.2-4) by means of the series

$$w(s) = \sum_{\nu=0}^{\infty} c_\nu s^{\gamma-\nu}$$

yields the initial powers $\gamma_1 = 0$ and $\gamma_2 = -1$. This implies that the *second canonic solution* of the difference equation Eq.(10.2-1) in the points s_2 and s_3 converges for $r < \epsilon' < 0$ (Nörlund, 1924; Milne-Thompson, 1951). These solutions are obtained by replacing the line of integration ℓ in Figs.9.2-1a and b by a line L, which begins at infinity, runs around s_3 or s_2 and returns to infinity.

The exponents β and γ depend only on the coefficients of $\varphi_n(R+1)$ and $\varphi_n(R-1)$. Hence, these coefficients and thus the abscissa of convergence $R_0 = 0$ of the factorial series are independent of the potential Φ.

12.9 Orthogonality of the Eigenfunctions

We investigate the orthogonality of the eigenfunctions of the Schrödinger difference equation of Section 8.2. Let us substitute the operator Ω_0 of Eq.(8.2-4) into Eq.(8.1-2). The relation between ω_{0n} and E_n of Eq.(8.1-13) is used.

$$\left(-\frac{\hbar^2}{2m}\nabla^2+V(\mathbf{r})\right)\varphi_n(\mathbf{r})=i\frac{\hbar}{2\Delta t}\omega_{0n}\varphi_n(\mathbf{r})$$
$$=\frac{\hbar}{\Delta t}\sin\left(\frac{E_n\Delta t}{\hbar}\right)\varphi_n(\mathbf{r}) \tag{1}$$

Let the same equation be written for the conjugated complex eigenfunction $\varphi_m^*(\mathbf{r})$. Let Eq.(1) be multiplied by $\varphi_m^*(\mathbf{r})$, and the equation for $\varphi_m^*(\mathbf{r})$ by $\varphi_n(\mathbf{r})$. The difference of the two products is integrated over a volume v:

$$-\frac{\hbar^2}{2m}\int(\varphi_m^*\nabla^2\varphi_n-\varphi_n\nabla^2\varphi_m^*)\,dv$$
$$=\frac{\hbar}{\Delta t}\left(\sin\frac{E_n\Delta t}{\hbar}-\sin\frac{E_m\Delta t}{\hbar}\right)\int\varphi_m^*\varphi_n\,dv \tag{2}$$

The difference operator ∇ is written as a series of differential operators by means of a Taylor expansion:

$$\nabla^2=\frac{\Delta^2}{(\Delta x_1)^2}+\frac{\Delta^2}{(\Delta x_2)^2}+\frac{\Delta^2}{(\Delta x_3)^2}=\sum_{i=0}^{\infty}\frac{2}{(2i+2)!}(\Delta x_j)^{2i}\frac{\partial^{2i+2}}{\partial x_j^{2i+2}} \tag{3}$$

The sum is to be taken for $j=1, 2, 3$ whenever the index j occurs twice.

The integral over the volume in Eq.(2) can be transformed into a surface integral by means of the generalized Green's formula of Section 12.10:

$$-\frac{\hbar^2}{2m}\oint(\varphi_m^*\operatorname{grad}\varphi_n-\varphi_n\operatorname{grad}\varphi_m^*)\cdot d\mathbf{f}_N$$
$$-\frac{\hbar^2}{2m}\sum_{i=1}^{\infty}\frac{2}{(2i+2)!}\sum_{s=0}^{i}\oint\mathbf{D}_{i,s}\cdot d\mathbf{f}_N$$
$$=\frac{\hbar}{\Delta t}\left(\sin\frac{E_n\Delta t}{\hbar}-\sin\frac{E_m\Delta t}{\hbar}\right)\int\varphi_m^*\varphi_n\,dv \tag{4}$$

The differential $d\mathbf{f}_N$ denotes the vector of the area element df pointing to the outside of the enclosed volume. The vector $\mathbf{D}_{i,s}$,

$$\mathbf{D}_{i,s}=\mathbf{i}_jD_{i,s,j} \tag{5}$$

has the components

$$D_{i,s,j} = (\Delta x)^{2i} \left(\frac{\partial^s \varphi_m^*}{\partial x_j^s} \frac{\partial^{2i-s+1} \varphi_n}{\partial x_j^{2i-s+1}} - \frac{\partial^s \varphi_n}{\partial x_j^s} \frac{\partial^{2i-s+1} \varphi_m^*}{\partial x_j^{2i-s+1}} \right)$$
$$j = 1,\ 2,\ 3 \qquad (6)$$

The surface integrals $\oint \mathbf{D}_{i,s} \cdot d\mathbf{f}_{\mathrm{N}}$ must vanish to make the volume integral $\int \varphi_m^*(\mathbf{r})\varphi_n(\mathbf{r})\, dv$ zero. Let us exclude functions that do not become independent of Δx_j for small values of Δx_j, since such solutions have been excluded. The necessary and sufficient condition for the integral $\oint \mathbf{D}_{i,s} \cdot d\mathbf{f}_{\mathrm{N}}$ to vanish is that $(\Delta x_j)^{2i}\partial^{2i+1}\varphi_m^*/\partial x_j^{2i+1}$ and $(\Delta x_j)^{2i}\partial^{2i+1}\varphi_n/\partial x_j^{2i+1}$ can be made arbitrarily small on a large surface for any value of i by making Δx_j sufficiently small. Infinite values would otherwise occur in the integrals.

Equation (1) may be written by means of Eq.(3) in the following form:

$$\left\{ -\frac{\hbar^2}{2m} \left[\frac{\partial^2}{\partial x_j \partial x_j} + \sum_{i=1}^{\infty} \frac{2}{(2i+2)!} (\Delta x_j)^{2i} \frac{\partial^{2i+2}}{\partial x_j^{2i+2}} \right] + V(x_j) \right\} \varphi_n(x_j) =$$
$$= \frac{\hbar}{\Delta t} \sin\left(\frac{E_n \Delta t}{\hbar}\right) \varphi_n(x_j) \quad (7)$$

The terms $(\Delta x_j)^{2i}\partial^{2i+2}\varphi_n/\partial x_j^{2i+2}$ vanish with the terms $(\Delta x_j)^{2i}\partial^{2i+1}\varphi_n/\partial x_j^{2i+1}$ if the function $\varphi_n(x_j)$ and its derivatives are continuous. Hence, $\varphi_n(x_j)$ is an eigenfunction of the following differential equation on a sufficiently large surface:

$$\left[-\frac{\hbar^2}{2m} \frac{\partial^2}{\partial x_j \partial x_j} + V(x_j) \right] \varphi_n(x_j) = \frac{\hbar}{\Delta t} \sin\left(\frac{E_n \Delta t}{\hbar}\right) \varphi_n(x_j) \qquad (8)$$

The functions $\varphi_n(x_j)$ must be identical with the eigenfunctions $u_n(x_j)$ of the Schrödinger *differential* equation,

$$\left[-\frac{\hbar^2}{2m} \frac{\partial^2}{\partial x_j \partial x_j} + V(x_j) \right] u_n(x_j) = E_n u_n(x_j) \qquad (9)$$

on a large surface, except for terms of order $O(\Delta x_j, \Delta t)$. Let us multiply Eq.(7) by $u_m^*(x_j)$, and the differential equation for $\varphi_m^*(x_j)$ corresponding to Eq.(7) by $u_n(x_j)$. The difference of the two products is integrated over the volume v:

$$-\frac{\hbar^2}{2m}\int\left(u_m^*\frac{\partial^2}{\partial x_j\partial x_j}u_n - u_n\frac{\partial^2}{\partial x_j\partial x_j}u_m^*\right)dv =$$

$$\oint\left(u_m^*\operatorname{grad}u_n - u_n\operatorname{grad}u_m^*\right)\cdot d\mathbf{f}_{\mathrm{N}} =$$

$$= \oint\left(\varphi_m^*\operatorname{grad}\varphi_n - \varphi_n\operatorname{grad}\varphi_m^*\right)\cdot d\mathbf{f}_{\mathrm{N}} = (E_n - E_m)\int u_m^* u_n\,dv = 0 \quad (10)$$

It is known that the eigenfunctions u_n, u_m of the Schrödinger differential equation are orthogonal. Hence, one deduces from Eq.(10) that the first integral in Eq.(4) vanishes too. The asymptotic transition of the eigenfunctions φ_n, φ_m into the eigenfunctions u_m, u_n is thus the necessary and sufficient condition for the orthogonality of the eigenfunctions φ_n, φ_m if one is interested in terms of order $O(1)$ only.

12.10 Generalization of Green's Formula

Let us substitute into Gauss' integral theorem

$$\int \operatorname{div}\mathbf{A}_1\,dv = \oint \mathbf{A}_1\cdot d\mathbf{f}_{\mathrm{N}} \quad (1)$$

for $\mathbf{A}_1$ the expression:

$$\mathbf{A}_1 = \sum_{j=1}^{3}\mathbf{i}_j A(\Delta x_j)^2\frac{\partial^3 B}{\partial x_j^3} \quad (2)$$

We further define a vector $\mathbf{B}$:

$$\mathbf{B} = \sum_{j=1}^{3}\mathbf{i}_j B(\Delta x_j)^2\frac{\partial^3 A}{\partial x_j^3} \quad (3)$$

One obtains:

$$\int(\operatorname{div}\mathbf{A}_1 - \operatorname{div}\mathbf{B}_1)\,dv = \int(\Delta x_j)^2\left(A\frac{\partial^4 B}{\partial x_j^4} - B\frac{\partial^4 A}{\partial x_j^4}\right)dv + \int\operatorname{div}\mathbf{D}_{11}\,dv \quad (4)$$

The vector $\mathbf{D}_{11}$ is defined as follows:

$$\mathbf{D}_{11} = \sum_{j=1}^{3}\mathbf{i}_j(\Delta x_j)^2\left(\frac{\partial A}{\partial x_j}\frac{\partial^2 B}{\partial x_j^2} - \frac{\partial B}{\partial x_j}\frac{\partial^2 A}{\partial x_j^2}\right) \quad (5)$$

The sum has to be taken for $j = 1, 2, 3$ in Eq.(4) when the index j occurs twice. Application of Gauss' integral theorem to Eq.(4) yields:

$$\int (\Delta x)^2 \left(A \frac{\partial^4 B}{\partial x_j^4} - B \frac{\partial^4 A}{\partial x_j^4} \right) dv = \oint (\mathbf{A}_1 - \mathbf{B}_1) \cdot d\mathbf{f}_\mathrm{N} - \oint \mathbf{D}_{11} \cdot d\mathbf{f}_\mathrm{N} \quad (6)$$

Two more vectors $\mathbf{A}_2$ and $\mathbf{B}_2$ are defined:

$$\mathbf{A}_2 = \sum_{j=1}^{3} \mathbf{i}_j A (\Delta x_j)^4 \frac{\partial^5 B}{\partial x_j^5}$$

$$\mathbf{B}_2 = \sum_{j=1}^{3} \mathbf{i}_j B (\Delta x_j)^4 \frac{\partial^5 A}{\partial x_j^5} \quad (7)$$

One obtains from the relation

$$\int (\operatorname{div} \mathbf{A}_2 - \operatorname{div} \mathbf{B}_2) dv = \oint (\mathbf{A}_2 - \mathbf{B}_2) \cdot d\mathbf{f}_\mathrm{N} \quad (8)$$

the formula:

$$\int (\Delta x_j)^4 \left(A \frac{\partial^6 B}{\partial x_j^6} - B \frac{\partial^6 A}{\partial x_j^6} \right) dv = \oint (\mathbf{A}_2 - \mathbf{B}_2) \cdot d\mathbf{f}_\mathrm{N} - \oint \mathbf{D}_{21} \cdot d\mathbf{f}_\mathrm{N} + \oint \mathbf{D}_{22} \cdot d\mathbf{f}_\mathrm{N}$$

$$\mathbf{D}_{21} = \sum_{j=1}^{3} \mathbf{i}_j (\Delta x_j)^4 \left(\frac{\partial A}{\partial x_j} \frac{\partial^4 B}{\partial x_j^4} - \frac{\partial B}{\partial x_j} \frac{\partial^4 A}{\partial x_j^4} \right)$$

$$\mathbf{D}_{22} = \sum_{j=1}^{3} \mathbf{i}_j (\Delta x_j)^4 \left(\frac{\partial^2 A}{\partial x_j^2} \frac{\partial^3 B}{\partial x_j^3} - \frac{\partial^2 B}{\partial x_j^2} \frac{\partial^3 A}{\partial x_j^3} \right) \quad (9)$$

Generally holds:

$$\int (\Delta x_j)^{2n-2} \left(A \frac{\partial^{2n} B}{\partial x_j^{2n}} - B \frac{\partial^{2n} A}{\partial x_j^{2n}} \right) dv = \oint (\mathbf{A}_{n-1} - \mathbf{B}_{n-1}) \cdot d\mathbf{f}_\mathrm{N} + \sum_{s=1}^{n-1} (-1)^s \oint \mathbf{D}_{n-1,s} \cdot d\mathbf{f}_\mathrm{N}$$

$$\mathbf{A}_{n-1} = \sum_{j=1}^{3} \mathbf{i}_j (\Delta x_j)^{2n-2} A \frac{\partial^{2n-1} B}{\partial x_j^{2n-1}}$$

$$\mathbf{B}_{n-1} = \sum_{j=1}^{3} \mathbf{i}_j (\Delta x_j)^{2n-2} B \frac{\partial^{2n-1} A}{\partial x_j^{2n-1}}$$

$$\mathbf{D}_{n-1,s} = \sum_{j=1}^{3} \mathbf{i}_j (\Delta x_j)^{2n-2} \left(\frac{\partial^s A}{\partial x_j^s} \frac{\partial^{2n-s-1} B}{\partial x_j^{2n-s-1}} - \frac{\partial^s B}{\partial x_j^s} \frac{\partial^{2n-s-1} A}{\partial x_j^{2n-s-1}} \right) \tag{10}$$

Use of the notation $\partial^0 A/\partial x^0 = A$ permits rewriting of the first line of Eq.(10) into the following short form:

$$\int (\Delta x_j)^{2n-2} \left(A \frac{\partial^{2n} B}{\partial x_j^{2n}} - B \frac{\partial^{2n} A}{\partial x_j^{2n}} \right) dv = \sum_{s=0}^{n-1} (-1)^s \oint \mathbf{D}_{n-1,s} \cdot d\mathbf{f}_{\mathrm{N}} \tag{11}$$

12.11 Eigenfunctions of the Dyadic Difference Operator

We introduced in Section 6.3 the dyadic difference operator $Df(m)/Dm$, Eq.(6.3-1), but then changed to the operator $gf(m)/gm$ of Eq.(6.3-3) introduced by Gibbs. One of the main features of $gf(m)/gm$ is the reproduction of the Walsh functions according to Eq.(6.3-6):

$$\frac{g \operatorname{Wal}(k,m)}{gm} = k \operatorname{Wal}(k,m) \tag{1}$$

Let us see whether an equivalent relation

$$\frac{Df(m)}{Dm} = kf(m) \tag{2}$$

exists for the operator $Df(m)/Dm$. The variable m shall have the four values 00, 01, 10, 11, and a function $f(m)$ the four values a, b, c, d as shown in Table 12.11-1. The derivative

$$\frac{Df(m)}{Dm} = \frac{1}{2}\{[f(m) - f(m \oplus 01)] + [f(m) - f(m \oplus 10)]\} \tag{3}$$

is also shown in Table 12.11-1. The relation of Eq.(2) requires that the following four conditions are satisfied:

TABLE 12.11-1

DYADIC DERIVATIVE $DF(m)/Dm$ OF A FUNCTION $f(m)$.

m	00	01	10	11
$f(m)$	a	b	c	d
$f(m \oplus 01)$	b	a	d	c
$f(m \oplus 10)$	c	d	a	b
$Df(m)/Dm$	$a-(b+c)/2$	$b-(a+d)/2$	$c-(a+d)/2$	$d-(b+c)/2$

TABLE 12.11-2

FUNCTIONS SATISFYING THE RELATION $Df(k,m)/Dm = kf(k,m)$.

m	00	01	10	11
$f(0,m)$	a	a	a	a
$f(1,m)$	a	b	$-b$	$-a$
$f(2,m)$	a	$-a$	$-a$	a

$$
\begin{aligned}
a - \frac{1}{2}(b+c) &= ka \\
b - \frac{1}{2}(a+d) &= kb \\
c - \frac{1}{2}(a+d) &= kc \\
d - \frac{1}{2}(b+c) &= kd
\end{aligned} \tag{4}
$$

Nontrivial solutions of this system of equations exist for $k = 0, 1, 2$. The resulting functions $f(0,m)$, $f(1,m)$, $f(2,m)$ are listed in Table 12.11-2. For any other values of k one obtains the trivial solution $a = b = c = d = 0$. Normalization of the three functions yields the Walsh functions $\mathrm{Wal_p}(0,m)$, $\mathrm{Wal_p}(1,m)$, and $\mathrm{Wal_p}(3,m)$ according to Fig.6.3-3—allowing for the change from eight to four values of m—but $\mathrm{Wal_p}(2,m)$ is missing. Equation (2) thus yields an incomplete set of functions while Eq.(1) yields a complete set.

If one advances from the four values of m in Table 12.11-1 to the eight values $m = 000, 001, \ldots, 111$, one obtains a system of eight equations instead of the four equations in Eq.(4). Only four values $k = 0, 2/3, 4/3, 2$ are obtained. Hence, Eq.(2) is less desirable then Eq.(1) in this case too.

Since we used Eq.(1) in Section 6.3-1 to find the general solution of the dyadic wave equation, we do not need Eq.(2) for the dyadic difference

TABLE 12.11-3

DYADIC DERIVATIVE $D^2 f(m)/Dm^2$ OF A FUNCTION $f(m)$.

m	00	01	10	11
$f(m)$	a	b	c	d
$Df(m)/Dm$	$a-(b+c)/2$	$b-(a+d)/2$	$c-(a+d)/2$	$d-(b+c)/2$
$Df(m\oplus 01)/Dm$	$b-\dfrac{a+d}{2}$	$a-\dfrac{b+c}{2}$	$d-\dfrac{b+c}{2}$	$c-\dfrac{a+d}{2}$
$Df(m\oplus 10)/Dm$	$c-\dfrac{a+d}{2}$	$d-\dfrac{b+c}{2}$	$a-\dfrac{b+c}{2}$	$b-\dfrac{a+d}{2}$
$D^2 f(m)/Dm^2$	$\dfrac{3a}{2}-b-c+\dfrac{d}{2}$	$\dfrac{3b}{2}-a-d+\dfrac{c}{2}$	$\dfrac{3c}{2}-d-a+\dfrac{b}{2}$	$\dfrac{3d}{2}-c-b+\dfrac{a}{2}$

operator of first order but would be satisfied with a relation for the operator of second order,

$$\frac{D^2 f(m)}{Dm^2} = k^2 f(m) \tag{5}$$

in analogy to the relation $d^2(\cos kx)/dx^2 = -k^2 \cos kx$ of differential calculus. The second dyadic derivative $D^2 f(m)/Dm^2$, together with the first derivative $Df(m)/Dm$ taken from Table 12.11-1, is listed in Table 12.11-3. The following relations must be satisfied according to Eq.(5):

$$\begin{aligned}
&\frac{3}{2}a - b - c + \frac{1}{2}d = k^2 a \\
&\frac{3}{2}b - a - d + \frac{1}{2}c = k^2 b \\
&\frac{3}{2}c - d - a + \frac{1}{2}b = k^2 c \\
&\frac{3}{2}d - c - b + \frac{1}{2}a = k^2 d
\end{aligned} \tag{6}$$

Nontrivial solutions of this system of equations exist for $k^2 = 0, 1, 4$. The three functions obtained are the same as those in Table 12.11-2. Hence, neither Eq.(2) nor Eq.(5) has the desirable features of Eq.(1).

References and Bibliography

Ahmed, N., Schreiber, H., and Lopresti, P.V. (1973). On notation and definition of terms related to a class of complete orthogonal functions. *IEEE Trans. Electromagn. Compat.*, **EMC-15**, 75–80.

Apostle, H.G. (1969). "Aristotle's Physics, Translated with Commentaries and Glossary". Indiana University Press, Bloomington.

Aristotle (1930). "The Works of Aristotle, vol. II. Physica". R.R. Hardie and R.K. Gaye transl., Clarendon Press, Oxford.

Ash, R.B. (1965). "Information Theory". Wiley, New York.

Atkinson, D., and Halpern, M.B. (1967). Non-usual topologies on space-time and high-energy scattering. *J. Mathematical Physics*, **8**, 373–387.

Bekenstein, J.D. (1981a). Energy cost of information transfer. *Phys. Rev. Lett.*, **46**, 623.

Bekenstein, J.D. (1981b). Universal upper bound on the entropy-to-energy ratio for bounded systems. *Phys. Rev.*, **D23**, 287–298.

Bolyai, J. (1832). "The science of absolute space" (original in Latin). See appendix of Bonola (1955).

Bonola, R. (1955). "Non-Euclidean Geometry". Dover, New York. (Originally published in Italian in 1906).

Borel, E.F.E.J. (1922). "L'Espace et le Temp". Alcan, Paris. "Space and Time". Blackie, London and Glasgow, 1926.

Bradley, D.J. (1977). Methods of generation, *in* "Ultrashort Light Pulses, Picosecond Techniques and Applications". S.L. Shapiro ed., Springer-Verlag, New York, pp. 17–81.

Braginsky, V.B., Vorontsov, Y.I., and Thorne, K.S. (1980). Quantum nondemolition measurements. *Science*, **209**, No. 4456, 547–557.

Brillouin, L. (1956). "Science and Information Theory". Academic Press, New York.

Brown, L.A. (1949). "The Story of Maps". Bonanza Books, New York.

Butzer, P.L., and Wagner, H.J. (1973). A calculus for Walsh functions defined on R_+. *Proc. 1973 Walsh Functions Symp.*, (A.E.Showalter and R.W.Zeek eds.) 75–81. Nat. Tech. Inf. Service. AD 763 000, Springfield, Virginia.

Cole, E.A.B. (1970). Transition from continuous to a discrete space-time scheme. *Nuovo Cimento*, **66A**, 645–655.

Cole, E.A.B. (1971). Cellular space-time and quantum field theory. *Nuovo Cimento*, **1A**, 120–132.

Cole, E.A.B. (1972a). The observer-dependence of cellular space-time structure. *Intern. J. Theor. Phys.*, **5**, 3–14.

Cole, E.A.B. (1972b). The classification of displacements and rotations in a cellular space-time. *Intern. J. Theor. Phys.*, **5**, 437–446.

Cole, E.A.B. (1973a). Perception and operation in the definition of observables. *Int. J. Theor. Phys.*, **8**, 155-170.

Cole, E.A.B. (1973b). Transformations between cellular space-time structures and the principle of covariance. *Nuovo Cimento*, **18A**, 445–458.

Courant, R., Friedrichs, K., and Lewy, H. (1928). Über die partiellen Differenzengleichungen der mathematischen Physik. *Math. Ann.*, **100**, 32–74.

Das, A. (1960). Cellular space-time and quantum field theory. *Nuovo Cimento*, **18**, 482–504.

Das, A. (1966a). Complex space-time and classical field theory. *J. Math. Phys.*, **7**, 45–51.

Das, A. (1966b). The quantized complex space-time and quantum theory of free fields. *J. Math. Phys.*, **7**, 52–60.

Das, A. (1966c). Complex space-time and geometrization of electromagnetism. *J. Math. Phys.*, **7**, 61–63.

Davis, P.C. (1974). "The Physics of Time Asymmetry". University of California Press, Berkeley.

Dhar, D. (1977). Lattices of effectively nonintegral dimensionality. *J. Mathematical Physics*, **18**, 577–585.

Dhar, D. (1980). On the connectivity index for lattices of nonintegral dimensionality. Pramāna (India) , **15**, 545–549.

Dixon, W.G. (1978). "Special Relativity". Cambridge University Press, London.

Dawes, E.L., and Marburger, J.H. (1969). Computer studies in self-focusing. *Phys. Rev.*, **179**, 862–868.

Douglas, J. Jr. (1956). On the relation between stability and convergence in the numerical solution of linear parabolic and hyperbolic equations. *J. Soc. Ind. Appl. Math.*, **4**, 20–37.

Einstein, A. (1915). Erklärung der Perihelbewegung des Merkur aus der allgemeinen Relativitätstheorie. *Sitzungsberichte preuss. Akad. Wiss. Berlin*, **47**, 831–839.

Einstein, A. (1916). Die Grundlagen der allgemeinen Relativitätstheorie. *Ann. Phys.*, **49**, 769–822.

Einstein, A. (1955). "The Meaning of Relativity". Princeton Univ. Press, Princeton, New Jersey.

Einstein, A. and Infeld, L. (1938). "The Evolution of Physics". Simon and Schuster, New York.

Engel, F. (1899). "N.I. Lobachevskii: Zwei geometrische Abhandlungen aus dem Russischen übersetzt mit Anmerkungen und einer Biographie des Verfassers". Teubner, Leipzig.

Euclid (1916). "Euclidis Opera Omnia", J.L. Heiberg and Menge, H. eds. Teubner, Leipzig.

Euclid (1956). "The Thirteen Books of Euclid's Elements" (T.L. Heat, translator). Dover, New York.

Feinstein, A. (1958). "Foundations of Information Theory". McGraw-Hill, New York.

Fey, P. (1963). "Informationstheorie". Akademie Verlag, Berlin.

Fine, N.J. (1949). On the Walsh functions. *Trans. Amer. Math. Soc.*, **65**, 372–414.

Fine, N.J. (1950). The generalized Walsh functions. *Trans. Amer. Math. Soc.*, **69**, 66–77.

Flint, H.T. (1948). The quantization of space and time. *Phys. Rev.*, **74**, 209–210.

Fock, V. (1959). "The Theory of Space-Time and Gravitation". Pergamon Press, London.

Fraser, J.T. (1978). "Time as Conflict—A Scientific and Humanistic Study". Birkhäuser, Basel und Stuttgart.

Fraser, J.T., Haber, F.C., and Müller, G.H. eds. (1972). "The Study of Time I". Springer-Verlag, Berlin and New York.

Fraser, J.T. and Lawrence, N. eds. (1975). "The Study of Time II". Springer-Verlag, Berlin and New York.

Fraser, J.T., Lawrence, N., and Park, D. eds. (1978). "The study of Time III". Springer-Verlag, Berlin and New York.

Gauss, K.F. (1827). Disquisitiones generales circa superficies curvas. German edition: Allgemeine Flächentheorie, Akademische Verlagsgesellschaft, Leipzig 1921.

Gauss, K.F. (1919). "Carl Friedrich Gauss Werke". Gesellschaft der Wissenschaften, Göttingen.

Gauss, K.F. (1969). "Briefwechsel zwischen C.F. Gauss and H.C. Schumacher, Nachträge". Vandenhoeck und Ruprecht, Göttingen.

Gelfond, A.O. (1958). "Differenzenrechnung". VEB Deutscher Verlag der Wissenschaften, Berlin (Russian original 1952).

Gibbs, J.E. (1979). A theory of time. *NPL Report DES 20*, National Physical Laboratory, Teddington, Middlesex, England.

Gibbs, J.E. and Ireland, B. (1974). Walsh functions and differentiation, *in* "Applications of Walsh Functions and Sequency Theory" (H. Schreiber and G.F. Sandy eds.), 147–176. Institute of Electrical and Electronic Engineers, New York.

Gouy, M. (1886). Sur le mouvement lumineux. *J. de Physique Théorique et Appliquée* (currently *J. Physique et Radium*), 2e ser. **5**, 354–362.

Hamming, R.W. (1950). Error detecting and error correcting codes. *Bell System Tech. J.*, **29**, 147–160.

Hamming, R.W. (1980). "Coding and Information Theory". Prentice Hall, Englewood Cliffs, New Jersey.

Harmuth, H.F. (1957). On the Solution of the Schrödinger and the Klein-Gordon equation by digital computers. *J. Math. Phys.*, **36**, 269–278.

Harmuth, H.F. (1972). "Transmission of Information by Orthogonal Functions". Springer Verlag, New York and Berlin.

Harmuth, H.F. (1977). "Sequency Theory, Foundations and Applications". Academic Press, New York.

Harmuth, H.F. (1980). From the flat Earth to the topology of space-time. *Adv. in Electronics and Electron Physics*, **50**, 261–350.

Harmuth, H.F. (1981). "Nonsinusoidal Waves for Radar and Radio Communication". Academic Press, New York.

Harris, D.C. and Wade, W.R. (1971). Sets of divergence on the group 2^ω. *Trans. Amer. Math. Soc.*, **240**, 385–392.

Hartley, R.V.L. (1928). Transmission of information. *Bell System Tech. J.*, **7**, 535–563.

Hasebe, K. (1972). Quantum space-time. *Prog. Theor. Phys.*, **48**, 1742–1750.

Hawking, S.W. and Ellis, G.F.R. (1973). "The Large Scale Structure of Space-Time". Cambridge University Press, London.

Heller, M. (1975). Global time problem in relativistic cosmology. *Annales de la Société Scientifique de Bruxelles*, **89**, IV, 522–532.

Heller, M. (1975). Some remarks about the concept of cosmic time. *Acta Cosmologica*, **3**, 41–44.

Hellund, E.J. and Tanaka, K. (1954). Quantized space-time. *Phys. Rev.*, **94**, 192–195.

Hilbert, D. (1921). "The Foundations of Geometry". Open Court, Chicago.

Hilbert, D. (1930). "Grundlagen der Geometrie". Teubner, Leipzig.

Hill, E.L. (1955). Relativistic theory of discrete momentum space and discrete space-time. *Phys. Rev.*, **100**, 1780–1783.

Hume, D. (1888). "A Treatise of Human Nature". L.A. Selby-Brigge ed. Clarendon Press. Oxford.

Huygens, C. (1690). "Treatise on Light". English edition Dover, New York 1962.

Jauch, J.M. (1968). "Foundations of Quantum Mechanics". Addison-Wesley, Reading, Massachusetts.

Kadishevsky, V.G. (1978). Towards a more profound theory of electromagnetic interactions. *Fermi Natl. Accelerator Lab.*, Preprint FERMILAB-Pub-78170-THY.

Kant, I. (1922). "Critique of Pure Reason", F.M. Müller, translator. McMillan, New York.

Kant, I. (1956). "Kritik der reinen Vernunft". Felix Meiner, Hamburg.

Kartaschoff, P. (1978). "Frequency and Time". Academic Press, New York.

Kelley, P.L. (1965). Self-focusing of optical beams. *Phys. Rev. Lett.*, **26**, 1005–1008.
Kepler, J. (1596). "Mysterium Cosmographicum" (Complete title: A Forerunner to Cosmographical Treatises, Containing the Cosmic Mystery of the Admirable Proportions Between the Heavenly Orbits and the True and Proper Reasons for Their Numbers, Magnitudes, and Periodic Motions). Tübingen. German transl. by M. Caspar, "Das Weltgeheimnis". Oldenbourg, Munich and Berlin, 1936.
Kepler, J. (1937). "Johannes Keplers gesammelte Werke", W. von Dyck and M. Caspar eds. C.H. Beck, Munich.
Koestler, A. (1968). "The Sleepwalkers". McMillan, New York.
Landau, L. and Peierls, R. (1931). Erweiterung des Unbestimmtheitsprinzips für die relativistische Quantentheorie. *Z. Phys.*, **69**, 56–69.
Letokhov, S.V. (1969). Generation of ultrashort light pulses in lasers with nonlinear absorber. *Sov. Phys.*, JETP **28**, 562–568.
Lobachevskii, N.I. (1840). "Geometrische Untersuchungen zur Theorie der Parallellinien". G. Finke, Berlin. Reprinted Mayer und Müller, Berlin 1887. English translation in the appendix of Bonola (1955).
Lobachevskii, N.I. (1856). "Pangéométrie; ou, précis de géométrie fondée sur une théorie générale et rigoureuse des parallèls". Kasan University (Reprinted by Herman, Paris 1905).
Lübke, A. (1958). "Die Uhr". VDI-Verlag, Düsseldorf.
Lucas, J.R. (1973). "A Treatise on Time and Space". Methuen, London.
Mach, E. (1907). "The Science of Mechanics". Open Court, Chicago. First German edition in 1883.
March, A. (1948). "Natur und Erkenntnis". Springer-Verlag, Vienna and New York.
March, A. (1951). "Quantum Mechanics of Particles and Wave Fields". Wiley, New York.
Maxwell, J.C. (1873). "A Treatise on Electricity and Magnetism". Reprint of the third edition by Dover, New York 1954.
Milne-Thomson, L.M. (1951). "The Calculus of Finite Differences". Macmillan, London.
Misner, C.W., Thorne, K.S., and Wheeler, C.A.. (1973). "Gravitation". W.H. Friedman, San Francisco.
Moller, C. (1972). "The Theory of Relativity". Oxford University Press, London.
Mundicici, D. (1981). Irreversibility, uncertainty, relativity, and computer limitations. *Nuovo Cimento*, **61B2**, 297–305.
Nerlich, G. (1976). "The Shape of Space". Cambridge University Press, London.
Newton, I. (1971). "Mathematical Principles", A. Motte transl., F. Cajori ed. University of California Press, Berkeley.
Nielsen, N. (1906). "Handbuch der Theorie der Gammafunktion". Teubner, Leipzig.
Nörlund, N. (1910). Fractions continues et différences réciproques. *Acta Math.*, **34**, 1–108.
Nörlund, N. (1914). Sur les séries de facultés. *Acta Math.*, **37**, 327–387.
Nörlund, N. (1915). Sur les équations linéaires aux différences finies à coefficients rationels. *Acta Math.*, **40**, 191–249.
Nörlund, N. (1924). "Vorlesungen über Differenzenrechnung". Springer-Verlag, Berlin and New York.
Nörlund, N. (1929). "Equations Linéaires aux Différences Finies". Gauthier-Villars, Paris.
Paley, R.E.A.C. (1932). A remarcable series of orthogonal functions. *Proc. London Math. Soc.*, **34**, 241–264, 265–279.
Park, D. (1980). "The Image of Eternity—Roots of Time in the Physical World". University of Massachusetts Press, Amherst.
Pauli, W. (1933). Die verallgemeinerten Prinzipien der Wellenmechanik, *in* "Handbuch der Physik", 2nd ed., Geiger and Scheel eds., vol. 24, part 1. Reprinted in "Handbuch der Physik", S. Flügge ed., vol. V, part 1 (1958) 1–168, particularly p. 10. Springer-Verlag, Berlin and New York.

Penney, R. (1965). On the dimensionality of the real world. *J. Mathematical Physics*, **6**, 1607–1611.

Pichler, F. (1967). Das System der sal- und cal-Funktionen als Erweiterung des Systems der Walsh-Funktionen und die Theorie der sal- und cal-Fouriertransformation. PhD Thesis, Philosophische Fakultät, Universität Innsbruck, Austria.

Ptolemy, C. (1952). "The Almagest", by Ptolemy, R.C. Taliaferro transl.; "On the Revolutions of the Heavenly Spheres", by N. Copernicus, C.G. Wallis transl.; "The Harmonies of the World", by J. Kepler, C.G. Wallis transl. Encyclopaedia Britannica, Chicago.

Raine, D. and Heller, M. (1981). "The Science of Space-Time". Panchard, Tucson, Arizona.

Rayleigh, J.W. (1889). Wave theory of light, *in* "Encyclopaedia Britannica", 9th ed., vol. 24, p. 421. Stoddard, Philadelphia.

Rayleigh, J.W. (1894). "The Theory of Sound". Reprinted by Dover, New York 1945.

Reichenbach, H. (1957). "The Philosophy of Space-Time". Dover, New York, (German original: Philosophie der Raum-Zeit-Lehre, 1928).

Reisbeck, G. (1964). "Information Theory, an Introduction for Scientists and Engineers". MIT Press, Cambridge, Massachusetts.

Reza, F.M. (1961). "An Introduction to Information Theory". McGraw-Hill, New York.

Richtmyer, R.D. (1957). "Difference Methods for Initial Value Problems". Wiley Interscience, New York.

Riemann, B. (1854). Über die Hypothesen, welche der Geometrie zu Grunde liegen, *in* "Gesammelte Mathematische Werke", H. Weber ed., 272–287. Teubner, Leipzig, 1892.

Riemann, B. (1858). Ein Beitrag zur Elektrodynamik, *in* "Gesammelte Mathematische Werke", H. Weber ed., 288–293. Teubner, Leipzig 1892.

Riemann, B. (1861). Commentatio mathematica, qua respondere tentatur questioni ab IIIma Academia Parisiensi propositae ..., *in* "Gesammelte Mathematische Werke" H. Weber ed., 391–423, Teubner, Leipzig 1892.

Riemann, B. (1866). Gravitation und Licht, *in* "Gesammelte Mathematische Werke", H. Weber ed., 532–538. Teubner, Leipzig 1892.

Robinson, G. de B. (1959). "The Foundations of Geometry". University of Toronto Press, Toronto.

Rosen, D.L., Doukas, A.G., Budansky, Y., Katz, A., and Alfano, R.R. (1981). A subpicosecond tunable ring dye laser and its applications to time resolving fluorescence spectroscopy. *IEEE J. Quantum Electronics*, **QE-17**, 2264–2266.

Schiff, L.I. (1968). "Quantum Mechanics". McGraw-Hill, New York.

Schild, A. (1949). Discrete space-time and integral Lorentz transformations. *Canadian J. Math.*, **1**, 29–47.

Schlick, M. (1920). "Space and Time in Contemporary Physics". Oxford University Press, London.

Schrödinger, E. (1930). Über die kräftefreie Bewegung in der relativistischen Quantenmechanik. *Sitzungsberichte Akademie Wiss. Berlin*, 418–428.

Schrödinger, E. (1950). "Space-Time Structure". University Press, Cambridge, England.

Schrödinger, E. (1956). Causality and wave mechanics, *in* "The World of Mathematics", vol. 2, 1056–1068. Simon and Schuster, New York.

Schuster, A. (1904). "The Theorie of Optics". E.A. Arnold, London.

Selfridge, G. (1955). Generalized Walsh transform. *Pacific J. Math.*, **5**, 451–480.

Shannon, C.E. (1948). A mathematical theory of communication, *Bell System Tech. J.*, **27**, 379–423, 623–656.

Shannon, C.E. (1949). Communication in the presence of noise. *Proc. IRE*, **37**, 10–21.

Singh, J. (1966). "Great Ideas in Information Theory, Language, and Cybernetics". Dover, New York.

Snyder, H.S. (1947a). Quantized space-time. *Phys. Rev.*, **71**, 38–41.

Snyder, H.S. (1947b). The electromagnetic field in quantized space-time. *Phys. Rev.*, **72**, 68–71.

Sommerfeld, A. (1939). "Atombau und Spektrallinien", vol. 2, Mathematische Zusätze und Ergänzungen. Vieweg, Braunschweig.

Stäckel, P.G. (1895). "Die Theorie der Parallellinien von Euclid bis auf Gauss". Teubner, Leipzig.

Stäckel, P.G. (1899). Franz Adolf Taurinus. *Abh. Geschichte Math.*, **9**, 397–427.

Stäckel, P.G. (1901). Die Entdeckung der nichteuklidischen Geometrie durch Johann Bolyai. *Math. und Naturwissenschaftliche Berichte aus Ungarn*, vol. 17.

Stäckel, P.G. and Engel, F. (1897). Gauss, die beiden Bolyai und die nichteuklidische Geometrie. *Math. Ann.*, **49**, 149–167.

Synge, J.L. (1960). "Relativity: The General Theory". North Holland, Amsterdam.

Synge, J.L. (1965). "Relativity: The Special Theory". North Holland, Amsterdam.

Synge, J.L. and Schild, A. (1969). "Tensor Calculus". Dover, New York.

Thompson, J.E.S. (1959). "Grandeza y Decadencia de los Mayas". Fondo de Cultura Económica, Mexico and Buenos Aires.

van der Waerden, B.L. (1966). "Algebra". Springer-Verlag, Berlin and New York.

Vilenkin, N.J. (1947). On a class of complete orthogonal systems. *Izv. AN SSSR*, ser. math. **11**, 363–400. English tranl. by R.P. Boas, *Trans. Amer. Math. Soc.*, Ser. 2, **28**, 1963, 1–35.

Wade. W.R. (1969). A uniqueness theorem for Haar and Walsh series. *Trans. Amer. Math. Soc.*, **141**, 187–194.

Wade, W.R. (1971). Summing closed U-sets for Walsh series. *Proc. Amer. Math. Soc.*, **29**, 123–125.

Wade, W.R. (1975). Uniqueness and α-capacity on the group 2^{ω}. *Trans. Amer. Math. Soc.*, **208**, 309–315.

Wallenberg, G. and Guldberg, A. (1911). "Theorie der linearen Differenzengleichungen". Teubner, Leipzig.

Wallis, W.D., Street, A.P., and Wallis, J.S. (1972). "Combinatorics: Room Squares, Sum-Free Sets, Hadamard Matrices" (Lecture Notes in Mathematics 292). Springer-Verlag, Berlin and New York.

Walsh, J.L. (1923). A closed set of normal orthogonal functions. *Amer. J. Math.*, **45**, 5–24.

Watari, C. (1965). On decomposition of Walsh-Fourier series. *Tohôku Math. J.*, 2nd ser. **17**, 76–86.

Weinberg, S. (1972). "Gravitation and Cosmology". Wiley, New York.

Welch, L.C. (1976). Quantum mechanics in a discrete space-time. *Nuovo Cimento*, **31B**, 279–288.

Weyl, H. (1921). "Raum, Zeit, Materie", 4th ed. English by H.L. Brose, "Space, Time, Matter". Dover, New York 1952.

Weyl, H. (1921). Das Raumproblem. *Jahresberichte der Deutschen Mathematikervereinigung*, **30**, 92–93.

Weyl, H. (1968). "Gesammelte Abhandlungen". Springer-Verlag, Berlin and New York.

Whitrow, G.J. (1980). "The Natural Philosophy of Time". Oxford University Press, London.

Wieland, W. (1961). "Die aristotelische Physik". Vandenhoeck und Rupprecht, Göttingen.

Yukawa, H. (1966). Atomistics and the divisibility of space and time. *Suppl. Prog. Theor. Phys.*, **37-38**, 512–523.

Zinner, E. (1951). "Astronomie—Geschichte ihrer Probleme". Karl Alber, Freiburg.

Zwart, P.J. (1976). "About Time". North Holland, Amsterdam.

Index